BIOMASS AS ENERGY SOURCE: RESOURCES, SYSTEMS AND APPLICATIONS

Sustainable Energy Developments

Series Editor

Jochen Bundschuh
University of Southern Queensland (USQ), Toowoomba, Australia
Royal Institute of Technology (KTH), Stockholm, Sweden

ISSN: 2164-0645

Volume 3

Biomass as Energy Source:
Resources, Systems and Applications

Editor:

Erik Dahlquist

School of Sustainable Development of Society and Technology,
Malardalen University, Högskoleplan Vasteras, Sweden

CRC Press
Taylor & Francis Group
Boca Raton London New York Leiden

CRC Press is an imprint of the
Taylor & Francis Group, an **informa** business

A BALKEMA BOOK

CRC Press
Taylor & Francis Group
6000 Broken Sound Parkway NW, Suite 300
Boca Raton, FL 33487-2742

First issued in paperback 2017

Published by: CRC Press/Balkema
                P.O. Box 11320, 2301 EH, Leiden, The Netherlands
                e-mail: Pub.NL@taylorandfrancis.com
                www.crcpress.com – www.taylorandfrancis.com

Library of Congress Cataloging-in-Publication Data

Applied for

ISBN 13: 978-0-415-62087-1 (hbk)
ISBN 13: 978-1-138-07322-7 (pbk)

About the book series

Renewable energy sources and sustainable policies, including the promotion of energy efficiency and energy conservation, offer substantial long-term benefits to industrialized, developing and transitional countries. They provide access to clean and domestically available energy and lead to a decreased dependence on fossil fuel imports, and a reduction in greenhouse gas emissions.

Replacing fossil fuels with renewable resources affords a solution to the increased scarcity and price of fossil fuels. Additionally it helps to reduce anthropogenic emission of greenhouse gases and their impacts on climate change. In the energy sector, fossil fuels can be replaced by renewable energy sources. In the chemistry sector, petroleum chemistry can be replaced by sustainable or green chemistry. In agriculture, sustainable methods can be used that enable soils to act as carbon dioxide sinks. In the construction sector, sustainable building practice and green construction can be used, replacing for example steel-enforced concrete by textile-reinforced concrete. Research and development and capital investments in all these sectors will not only contribute to climate protection but will also stimulate economic growth and create millions of new jobs.

This book series will serve as a multi-disciplinary resource. It links the use of renewable energy and renewable raw materials, such as sustainably grown plants, with the needs of human society. The series addresses the rapidly growing worldwide interest in sustainable solutions. These solutions foster development and economic growth while providing a secure supply of energy. They make society less dependent on petroleum by substituting alternative compounds for fossil-fuel-based goods. All these contribute to minimize our impacts on climate change. The series covers all fields of renewable energy sources and materials. It addresses possible applications not only from a technical point of view, but also from economic, financial, social and political viewpoints. Legislative and regulatory aspects, key issues for implementing sustainable measures, are of particular interest.

This book series aims to become a state-of-the-art resource for a broad group of readers including a diversity of stakeholders and professionals. Readers will include members of governmental and non-governmental organizations, international funding agencies, universities, public energy institutions, the renewable industry sector, the green chemistry sector, organic farmers and farming industry, public health and other relevant institutions, and the broader public. It is designed to increase awareness and understanding of renewable energy sources and the use of sustainable materials. It aims also to accelerate their development and deployment worldwide, bringing their use into the mainstream over the next few decades while systematically replacing fossil and nuclear fuels.

The objective of this book series is to focus on practical solutions in the implementation of sustainable energy and climate protection projects. Not moving forward with these efforts could have serious social and economic impacts. This book series will help to consolidate international findings on sustainable solutions. It includes books authored and edited by world-renowned scientists and engineers and by leading authorities in in economics and politics. It will provide a valuable reference work to help surmount our existing global challenges.

Jochen Bundschuh
(Series Editor)

Editorial board

This book is dedicated to my wife Christina and my children Katarina, Emma and Gustaf

–Erik Dahlquist

Table of contents

Contributors

Thorsten Ahrens   Ostfalia University of Applied Sciences, Wolfenbuettel, Germany,
                         th.ahrens@ostfalia.de, th.ahrens@fh-wolfenbuettel.de

Martin Andresen   Borregaard LignoTech R&D, Sarpsborg, Norway,
                         martin.andresen@borregaard.com

Robert Aulin   Bestwood AB and Swedish University of Agricultural Sciences (SLU),
                         Unit of Biomass Technology and Chemistry, Umeå, Sweden,
                         robert.aulin@bestwood.se

Stefan Backa   Borregaard ChemCell R&D, Sarpsborg, Norway,
                         stefan.backa@borregaard.com

Jochen Bundschuh   University of Southern Queensland (USQ), Toowoomba, Australia
                         Royal Institute of Technology (KTH), Stockholm, Sweden,
                         jochen.bundschuh@usq.edu.au

Erik Dahlquist   School of Sustainable Development of Society and Technology,
                         Mälardalen University, Västerås, Sweden, erik.dahlquist@mdh.se

Du Feng-Guang   Henan Tianguan Enterprise Group Co, Ltd, Nanyang city, Henan,
                         China, dfg@tianguan.com.cn, dufengguang@163.com

Elias Hakalehto   Institute of Biomedicine, University of Eastern Finland, Kuopio, Finland,
                         elias.hakalehto@gmail.com

Lauri Heitto   Environmental Research of Savo-Karjala Oy, Kuopio, Finland,
                         lauri.heitto@skvsy.fi

Tarmo Humppi   Defence Forces Technical Research Centre, Lakiala, Finland,
                         tarmo.humppi@mil.fi

Ari Jääskeläinen   Savonia University of Applied Sciences, Kuopio,
                         Finland, ari.jaaskelainen@savonia.fi

Torbjörn A. Lestander   Swedish University of Agricultural Sciences, Unit of Biomass
                         Technology and Chemistry, Umeå, Sweden, torbjorn.lestander@slu.se

Emily Nelson   Bio Science and Technology Branch, NASA Glenn Research Center,
                         Cleveland, OH, USA, emily.s.nelson@nasa.gov

Muhammad Raza Naqvi   Energy Processes, Royal Institute of Technology (KTH), Stockholm,
                         Sweden & University of Gujrat, Pakistan, rnaqvi@kth.se

Trond Rojahn   Borregaard LignoTech, Borregaard Industries Ltd., Sarpsborg,
                         Norway, trond.rojahn@borregaard.com

Semida Silveira   Division of Energy and Climate Studies, Department of Energy
                         Technology, Royal Institute of Technology (KTH), Stockholm, Sweden.
                         Collaborating with CTBE – Brazilian Bioethanol Science and
                         Technology Laboratory and UFMG – Federal University of
                         Minas Gerais, Brazil, semida.silveira@energy.kth.se

Eva Thorin   School of Sustainable Development of Society and Technology,
                    Mälardalen University, Västerås, Sweden, eva.thorin@mdh.se

Feng Wensheng   Henan Tianguan Enterprise Group Co., Ltd, Nanyang city, Henan,
                    fwg@tianguan.com.cn

Jinyue Yan   Energy Processes, Royal Institute of Technology (KTH), Stockholm,
                    Sweden & institute missing, Malardalen University, Vasteras, Sweden,
                    jinyue.yan@ket.kth.se

Foreword

We are humans, we are biomass. We have always eaten plants that stored solar energy collected in tasty structures, giving our muscles strength and energy to walk, run and think. We then learned to cultivate more and more food on less land which has led to a surplus of arable land in Europe and elsewhere and continuously growing forests worldwide except in the poorest countries. Many people say that biomass volumes are limited. I would say we produce what we ask for and demand. Someone asked me "How many potatoes are there in the world". Just to illustrate that – we have to demand potatoes if any farmer is to grow it. It is the same with biomass for energy. We have to demand it if it is to be professionally and rationally produced, which is also the key possibility to make bioenergy production competitive with fossil energy sources.

When Sweden introduced carbon dioxide tax in the heating sector in 1991, it was a lot cheaper to heat a house with heating oil than with wood chips. The fuel cost per kWh was lower with oil than with wood chips. Today the wood chip price delivered to the Combined Heat and Power plant is only 40 percent of the crude oil price per unit energy at the world market and Sweden saves tremendous amounts of money in our trade balance compared to if we had continued to use oil. At the same time renewables account for more than 50 percent of our energy use in 2013 and bioenergy is the largest energy source, supplying one third of Sweden's energy use while at the same time the total stock of Swedish forests increase.

This book gives a fantastic overview of some of the many opportunities there are to develop energy production from biomass. Bioenergy production is often the most profitable solution, using waste streams or bi-product flows from forest or food industry to produce energy. When reading the book you also understand that it is not possible for politicians to pick the best solutions as we do not know what solution will be the most competitive. Therefore we always ask for general incentives and to let the polluter pay. If companies that emit fossil carbon dioxide have to pay for their emissions they will try to avoid the payment and find better solutions. Sometimes the best solution is to take the bike instead of the car, or to insulate the house or some other energy efficiency investment. Sometimes and actually very often the best option is to use biomass. You consume biomass to provide warmth and energy to your body. We can feed sustainable biomass into our CHP to warm our homes or charge the mobile or put biofuel in our car to bring us where we want to go. I wish you enjoyable reading and new biomass to energy perspectives.

Gustav Melin

President Svebio, Swedish Bioenergy Association and
President AEBIOM, European Biomass Association

Editor's Foreword

The purpose of this book is to give an overview of the area "Biomass resources". It is being published in conjunction with the book "Biomass conversion" in the same book series. Biomass is the energy source that has been used since before humans became "*Homo sapiens*" and must be considered as the most important energy source we have. All food is biomass. Biomass gives us heat and the possibility to cook food and nowadays it also provides us electric power through conversion in thermal power plants. It is also construction material for buildings and other structures. When the era of fossil fuels is over it will once again be the dominating material for production of many chemicals, textiles, etc. In short – it is the most important material in our lives. It is thus important to gain insight into what resources we actually have at our disposal today, and what possibilities we can foresee for the future. Will the biomass last? Will it be enough for all our needs? Is the distribution over the world balanced?

This book covers many aspects of biomass and hopefully will give the reader a good overview of this resource, and hopefully also stimulate the reader to find out more about it. We hope the book can be of interest to both professionals who know a lot about the subject, but want to broaden their perspectives even further, as well as students at different levels who want to get a good overview of the field as part of courses or during different project courses. It should also be of interest to persons working on legal aspects and policy questions and to politicians who want to dig deeper into the field to make better decisions. It can also be of interest to anyone with a general interest in nature.

The chapters about biorefineries will also give an introduction to future industry perspectives as the plants described are in many ways determining the routes that others will have to take when fossil fuels become scarce. This might happen much sooner than perhaps is sometimes perceived when we listen to some politicians in many countries claiming that there will never be any shortage of fossil fuel resources. When the alternatives become commonplace it might very well be more economical to use these instead of sources like oil tar and other complex products. It should be noticed that these types of resources being developed right now are not only very costly, but also give rise to environmental problems that may provoke strong opposition from the peoples affected.

The book uses SI units as the standard. Still, SI units can have different forms as well. It is common to use MJ (million Joules) for energy in SI units, but as kWh, (kilo watt hours), MWh (mega watt hours), TWh (terra watt hours) and toe (tonne oil equivalents) are used by e.g. the UN Worldbank for energy these units have been used as well. One toe. is approximately 10 MWh. For electric power usually MW has been used. China is using t.c.e (tonne coal equivalent) for energy as well, and in a few places throughout the book this unit has been used relating to Chinese energy data. For surface area ha (10,000 m^2) and km^2 (100 ha) have been used concerning calculations related to production of different crops, yield etc. Both m^3 (1000 liters) and liters have been used for volume. Both kg and tonnes have been used as well, where we refer to metric tonne (=1000 kg). Sometimes other units like Pg are also used, which is 10^{15} g and TJ, which is 10^{12} J.

In closing, I would like to wish you all a very interesting read and hope you will enjoy the few hours ahead.

Erik Dahlquist
January 2013

About the editor

Erik Dahlquist with a TPV module for combined heat and electric power in small biomass fired boilers

Erik Dahlquist is currently Professor in Energy Technology at Malardalen University (MDU) in Västerås, Sweden. His focus is on biomass utilization and process efficiency improvements. He started working at ASEA Research in 1975 as engineer in analytical chemistry related to nuclear power, trouble shooting of electrical equipment and manufacturing processes. In 1982 he started with energy technology within the pulp and paper industry and participated in the development of year-round fuel production from peat. In 1984 ASEA started a company ASEA Oil and Gas with a focus on off-shore production systems. One area was waste water treatment and separation of oil and water. He then became technical project manager for development of a Cross Flow Membrane filter. This led to the formation of ABB Membrane Filtration. The filter is now a commercial product at Finnish Metso Oy under the name Optifilter. As part of this development work he started as an industrial PhD student at KTH and received his doctorate in 1991. In 1989 he became project leader for ABB's Black Liquor Gasification project, which resulted in a number of patents. From 1992 to 1995 he was department manager for Combustion and Process Industry Technology at ABB Corporate Research. He was also at that time member of the board of directors for ABB Corporate Research in Sweden. From January 1996 to 2002 he was General Manager for the Product Responsible Unit "Pulp Applications" worldwide within ABB Automation Systems. The product area was Advanced Control, Diagnostics, Optimization, Process Simulation and Special Sensors within the pulp and paper industry. During 1997–2000 he was part time adjunct professor at KTH and from 2000 to 2002 part time professor at MDU. He has been responsible for research in Environmental, Energy and Resource Optimization at MDU since 2000. During 2001–2007 he was first deputy dean and later dean of the faculty of Natural Science and Technology. He has been a member of the board of the Swedish Thermal Engineering Research

Institute division for Process Control systems since 1999. He received the ABB Corporate Research Award 1989. He has been a member of the board of SIMS (Scandinavian Simulation and Modeling Society) since 2003 and deputy member of the board of Eurosim since 2009. He has been a member of the editorial board of the Journal of Applied Energy, Elsevier since 2007. He is also a member of the Swedish Royal Academy of Engineering (IVA) since 2011. He has 21 (different) patents and approximately 170 scientific publications in refereed journals or conference proceedings with refereed contributions. He has published seven books, either as editor or author.

Acknowledgements

I would like to thank all contributing authors to this book. Without you this book would not have been written! Many thanks also to the Series Editor Jochen Bundschuh for checking and editing the final version of the manuscript. I would also like to thank the Swedish Energy Agency, and especially Sven Risberg, for strongly supporting our biomass research. I would also like to thank our partners at Malarenergi, Eskilstuna Energy and Environment, Vafab Miljö, ABB, SHEAB and ENA Energy for a lot of very important input on both biomass conversion and how to optimize systems.

Erik Dahlquist
January 2013

Introduction

Erik Dahlquist

Biomass was once upon a time the totally predominant energy resource for humankind, aside from the direct sunshine. During the process of industrialization other energy resources like coal initially and later oil became very important. In Sweden a big project to harvest peat in a year-round production process started around 1912, but after the First World War the project was stopped, as the peat could not compete with coal. During the Second World War a similar project was started up again. Besides this, a major emphasis was placed on using wood as the basic resource for production of any kind of chemical to replace oil. After the war still the use of oil became too competitive to peat and wood for this purpose, so the projects were shut down again. Oil became the dominant primary energy resource. In Sweden nuclear power was instead seen as the replacement for oil in the future, and during the 1970s and 1980s 11 nuclear reactors were started up. Actually they did replace primarily oil, and the use decreased from 2/3 of the primary energy demand to only 1/3, roughly. Then the Three Mile Island incident occurred in the US, and the politicians said – no more nuclear power. After a referendum in 1980 it was said that we should not build new reactors after the existing ones had been "phased out" for technical reasons. The direction should be towards renewable energy instead. In 1992 after many discussions the politicians agreed on a carbon tax on fossil carbon. It meant approximately 1.6 €cent/kWh in extra tax. This really spurred on the introduction of bioenergy. Before this bioenergy use was 55 TWh/year, primarily in pulp and paper industry as spent black liquors. In 2012 the total use was approximately 140 TWh/year, which should be compared to 400 TWh/year total uses and approximately 130 TWh/year oil use. As such we have bioenergy as the single most important primary energy in an industrial country.

This historical outlook is quite interesting. In reality we should be able to take a similar course also in most other countries in the world where we have agriculture and forestry. The biomass resources are not used as efficiently as they could be today, and a lot of it is just decomposing in the fields or in the forests, or even being combusted just to get rid of it. For those using biomass as the primary source for cooking food the efficiency is often not more than 10%, compared to 117% in many co-generation plants with exhaust condensation in Sweden.

In this book and in a parallel book in this series on conversion of biomass we want to give an overview of the existing and potential biomass resources, as well as the methods available today or in the near future for conversion of biomass to all our needs. This means looking back to the 1940s when the war made oil unavailable, and wood was used for everything. Today the reason is not the war-blockage but the worries about both global warming and the trend towards more limited oil resources available easily, and thereby increased prices. During the 1980s and 1990s the oil price was normally around 20 $/barrel while during the 2000s it increased to around 100–120 $/barrel. This of course means that alternatives become more interesting, and biomass is one of few resources that can be easily stored. This is why we believe it is worth writing these two books in the overall series of renewable energy. The combination of biomass, solar power and wind power, with the complement of hydro power, can easily replace all fossil fuel energy within the next fifty years, if only the right incentives are implemented, like the carbon tax in Sweden in 1992. So the problem is the political will, nothing else. If it had been economically disastrous to do Sweden would have been a very poor country in 2012, but this is not the case. This shows that if a political will and brave politicians take the right actions you can also develop the countries in a positive way generally.

I am very happy that the interest for bioenergy is vast all over the world. It is my pleasure to note that chapter authors come from all parts of the world, including the Americas, Africa,

China, India and Europe. It also shows that development and demonstration of new technologies is nowadays a global activity, and not something only for some chosen few.

The books are intended for interested engineers, people working for authorities, researchers, students at high schools and universities as well as for everyone interested in our future renewable energy systems. With that I wish you all pleasant reading. Enjoy the ride!

Part I
Biomass resources

CHAPTER 1

Introduction and context: global biomass resources – types of biomass, quantities and accessibility. Biomass from agriculture, forestry, energy crops and organic wastes

Erik Dahlquist & Jochen Bundschuh

This chapter gives an overview of the available biomass resources globally today, as well as the potential for enhancements of the production, and also shows how agricultural and forestry waste can be utilized as resources more efficiently. Environmental issues and concerns are covered as well.

Globally the official figures for the use of biomass as an energy resource show that it is approximately 13% of all primary energy used. This is according to the World Bank figures. If we go back 30 years, the official figure was almost 0%. The reason for the difference is that in 1980 only material that was traded was registered and accounted for. As it was obvious that bioenergy was very important in many countries although not being traded the new calculation was made. The figure of 13% has principally been maintained for 20 years, although the total energy usage has increased a lot, around 30–40%.

The real production of biomass still is significantly larger as a lot of it is just decomposing in the fields or in the forests. Some estimates gives a biomass production around 300,000–600,000 TWh/y, which is 2–4 times the total energy utilization today.

1.1 HARD FACTS

If we look at what figures there are available as "hard facts", the World Bank is collecting data from most countries of the world with respect to agricultural area and forestry area, as well as the productivity with respect to different crops in different countries. The World Bank is the statistical data organization for the United Nations (UN) and gather not only this type of data, but also data about the population, income etc.

It is of some relevance to group the countries into different economies as the average income also gives a general figure of what resources are available for fertilizers, irrigation and mostly also about how well structured and organized the country is. A country with high income normally has higher productivity than a middle income country, which has higher productivity than most low income countries.

In 2010, the UN estimated the world population to be 6860 million inhabitants globally. Of these 50.3% were living in urban areas. This shows that now more people live in urban areas than rural. The situation differs between low income countries, middle income and high income countries. In low income countries only 27.9% were living in urban areas, in middle income 48% with 77.3% in high income countries.

World Bank Development Indicators 2010 (World Bank, 2011) has calculated the average income for each of the income groups. For low income countries this is 507 US$/capita, for middle income 3980 US$/capita and for high income countries 38293 US$/capita.

There are 213 countries in the UN statistics of which 58 are in Europe and Central Asia (EurCA), 35 in East Asia (EastAsia), 8 in South Asia (SoAsia), 48 in Africa in the sub-Sahara region (SubSah), 21 in Middle East and North Africa (MEasNAf) and 3 in North America. In Table 1.1, we can see how the regions are distributed with respect to the income for separate countries, that is, how many countries in each region belong to each income group.

Table 1.1.   Low income, middle income and high income countries distributed on regions.

	EurCA	EastAsia	LatAm	SoAsia	SubSah	MEasNAf	NAm	Total
Low income	2	3	1	3	26	0	0	35
Middle income	21	21	29	5	21	13	0	110
High income	35	11	10	0	1	8	3	68
Global total	58	35	40	8	48	21	3	213

Table 1.2.   Statistics from World Bank development indicators relating to different average income per capita in different categories of countries (World Bank, 2011).

	Cereal yield		Electricity production from fossil fuel		Electricity consumption	Energy use	Fossil fuel	GDP
	1970 [kg/ha]	2009 [kg/ha]	1970 [%]	2008 [%]	2008 [kWh/cap]	2008* [kgoe/cap]	2008 [% of total]	2010 [US$/cap]
Low income	1296	1952	17.9	31.4	222	350	30.1	507
Middle income	1515	3202	37.2	73	1606	1255	81.3	3980
High income	2766	5448	71.7	62.9	9515	5127	82.7	38 293
World wide	1829	3513	61.6	67.2	2874	1834	81.1	9197

*oe = oil equivalents, US$/cap = US$/capita

As can be seen Europe and central Asia, have most countries in the middle and high income groups, while sub-Saharan Africa has most of the countries in the low income economy group. As can be seen, almost all low income countries are in the sub-Saharan Africa region, 26 of a total of 35. In North America, on the other hand all three countries are in the high income country group.

In Table 1.2, some other key data are collected. First we have the cereal yield 1970 compared to 2009 for low income, middle income and high income countries on average. What can be seen is that the cereal yield has almost doubled in middle and high income countries, but the increase in yield is much lower in low income countries. We also can note that the yield is 2.5 times higher in high income countries compared to low income countries, although the climatic difference is not significant. The reason is more related to organizational issues than climatic issues. This gives some comfort with respect to the potential to increase the yields in low income countries also. With respect to energy, the amount of electricity produced by fossil fuels has doubled in low income and high income countries, while it has decreased in high income countries. This is primarily due to more nuclear power in high income countries. The utilization of energy per capita still is very different between the income groups. Low income countries used 222 kWh$_{el}$ per capita, middle income 1606 and high income 9515 kWh$_{el}$/capita in 2008. For total energy use, the corresponding figures were 350 kg oe/capita for low income economies, 1255 kg oe/capita for middle and 5127 kg oe/capita for high income economies (oe = oil equivalents). The ratio between high income and low income economies are as seen much lower for total energy compared to electricity, which shows that electricity is more strongly correlated to wealth than total energy use is.

After this general overview of energy statistics, it can be interesting to look more into the available resources with respect to land areas in the countries. In Table 1.3, we have the overview using the data from World Bank development indicators (World Bank, 2011).

Here we can see that the percentage of the land areas with respect to agriculture, arable land and forestry is approximately the same for low income, middle income and high income countries.

It then can be interesting to note also from Table 1.4 that the population in the low income countries is approximately 800 million, in middle income 5000 million and in high income approximately 1100 million inhabitants. If we compare these figures to the total land area available

Table 1.3. Distribution of agriculture, arable and forestland on economic categories (World Bank, 2011).

	Agricultural land [1000 km^2] 2008	Agricultural land [% of total] 2008	Arable land [% of total] 2008	Land area cereal prod [1000 ha] 2008	Cereal production [1000 tonne] 2009	Forest area [1000 km^2] 2010	Forest area as [% of total] 2010
Low income	5679	37.8	9.3	88349	172000	4155	27.6
Middle income	30498	37.8	10.9	473000	1510000	26420	32.8
High income	12597	37.3	10.9	147000	803000	9629	28.8
World wide	48774	37.7	10.7	708000	2490000	40204	31.1

Table 1.4. Life expectancy and population distribution between different categories of countries.

	Life expectancy at birth 2009	Population 0–14 y [% of all] 2009	Population 15–64 y [% of all] 2009	Population >65 y [% of all] 2009	Population growth per year [%] 2010	Total population [million] 2010
Low income	57.5	39.4	57.1	3.5	2.16	817
Middle income	68.7	27.3	66.3	6.4	1.11	4920
High income	79.8	17.5	67.1	15.4	0.6	1120
World wide	69.2	27.2	65.3	7.5	1.15	6860

for growing crops and trees this is 15 million km^2 for low income, 80 for middle and 34 for high income countries. This gives 0.019 km^2 per capita for low income, 0.016 for middle income and 0.031 for high income countries on average. There are obviously differences, but not that big, between the different economies. Those having the lowest area per capita are actually the middle income countries, and not the low income.

From Table 1.4 we can also see some of the demography of the populations. Here we can note that in high income economies, we have more elderly people and fewer younger than in the middle income economies, and this is actuated even further comparing to the low income economies. Still the trend is that when the economy improves, fewer children are born and the demography changes, as seen in the table. If we predict the population development from this historical experience, we can see that the population will continue to grow especially in Africa south of the Sahara, and estimates are made that when we have reached 9 billion inhabitants the population will start to decrease again. In a country like Germany we already have seen the trend towards reduced population and predictions are made that Germany will decrease by some 10–20% during this century. Short term this may cause problems but long term it should be good, as the stress on the resources decrease. In sub-Saharan Africa on the other hand the stress on the environment will increase a lot. The positive fact is still that the productivity there is low and the potential to increase therefore high, if the political and organizational aspects can be improved.

In Table 1.5 we can see the figures for fertilizer consumption in kg/ha in the different economies as well as the cereal yield in kg/ha. It is obvious that the yield is significantly lower in low income countries, but if we compare the amount of fertilizers used, it is still not that big a difference. Actually, the middle income economies use more fertilizers per hectare than the high income countries, although the production is lower. This shows that it is not only the amount that matters, but also how it is administered. With high distribution, the risk for leakage into waters increases as well as causing secondary problems in estuaries downstream the farmland. As synthetic fertilizers also consume a lot of energy to produce it is good to administer fertilizers in an adequate way.

One example of this has been at the farm Nibble outside Vasteras in Sweden. By administering 1/3 of the liquid fertilizer going in one direction at the farmland and then back along the same

Table 1.5.   Agriculture and food production as a function of economic categories.

	Fertilizer consumption % of production 2008	Fertilizer consumption [kg/ha] 2008	Cereal yield [kg/ha]		Agriculture tractors 2002
			1970	2009	
Low income	207	16.5	1296	1952	121409
Middle income	104	138.6	1515	3202	9268485
High income	73.2	109.3	2766	5448	14210537
World wide	95	119.3	1829	3513	25486834

track distributing the residual 2/3 the soil could take up all nitrogen with very few emission to the air, as well as to the small river downstream. As this was administered when the cereal (wheat) was approximately 10 cm high, all could be utilized and thus the production could be sustained but with half the dosage of ammonia/nitrate per hectare! This of course saved energy, cost and the environment! (Odlare, 2010). Larsson *et al.* (2009) have presented the concept ERA, Environmental Recycling Agriculture. The focus for this concept is to recycle nutrients like N and P from animals in such an amount that what is produced in a country should be distributed evenly. Not more than one animal unit (= corresponding to manure from one cow) should be fed per ha to avoid eutrophication. The primary target was the Baltic Sea and the countries around this. To get a balance we should reduce the consumption of meat by some 60–70%, as the production of crops will be reduced, but by doing this we will still get the food needed.

The most important crops from a human food perspective are cereals. It should be noticed that generally we get at least the same amount of biomass from waste from the food crops as we get grains used for food. As we increase the yield for these crops, we simultaneously also increase the production of hydrocarbons that can be used for energy purpose. Some specific important crops will be covered in the following section.

1.2   CROPS USED PRIMARILY FOR FOOD

There are some crops which are more important than others for the human population of earth. In this section, a number of these are presented more in detail.

1.2.1   *Soybean*

Soybean is a very popular crop for production of very good protein. The beans contain 40% with a very good distribution between amino acids making it good as a replacement for animal protein. In 2009 the world production was 222 million tonne (FAO statistical database 2009). If we would distribute to all the population in the world it would be 13 kg per capita or almost 100 g protein per day. This could replace all animal and fish protein we eat today. Still, only a little more than 10% is used as direct human food (www.soyatech.com/soy_facts.htm, 2010). The rest is used to feed 18.6 billion chicken and hen, 1.4 billion cows and bulls and 940 million pigs worldwide annually.

1.2.2   *Rice*

Rice is a crop for tropical and subtropical countries. The global average production is 3.9 tonnes per ha, but the yields may be much higher where there is intensive irrigation like in Australia with 9.5 tonnes/ha per y and Egypt with 8.7 tonnes/ha per y. Some countries having "traditional" methods like the Republic of Congo on the other hand have production as low as 0.75 tonnes/ha per y! Thus 75% of the world production of rice is harvested on the 55% of the area that is irrigated.

The global production of rice was 678 million tonnes 2009. In China, with a total production of 197 million tonnes in 2009 the average yield was 6.6 tonne/ha/y. With this China is the biggest producer but also one of the countries with highest average yield. From a total production point of view India comes as number two with 131 million tonnes 2009, but with only 45% of the production per ha compared to China. Indonesia has productivity inbetween China and India and produced 64 million tonnes in the same year. If India could increase its productivity to the same level as in China, it would mean another 100 million tonnes per year. This could feed 400 million people as staple food! Other major producers of rice are Bangladesh producing 45, Vietnam 39, Thailand and Myanmar both 31, Philippines 16, Brazil 13, Japan 11 and Pakistan 10 million tonnes in 2007.

Rice is, aside of wheat, the most important food crop globally and is the most important calorie food for the populations in Asia, Latin America and Africa, or more than half of the world's population. About 95% of it is produced in developing countries. World production of rice has risen steadily from about 200 million tonnes of paddy rice in 1960 to over 678 million tonnes in 2009.

Rice straw gives approximately the same amount of biomass as grain. If we could reach 8 tonnes/ha per y of grain, 8 tonnes/ha per y of straw would also be produced (NationMaster, 2012).

1.2.3   *Wheat (Triticum spp.)*

The world production of wheat is 690 million tonnes per year. The biggest producer is China with 112–114.5 million tonnes produced 2008 (Fao.org "Major Food And Agricultural Commodities And Producers – Countries By Commodity") http://faostat.fao.org/site/567/DesktopDefault.aspx? Page ID=567#ancor, downloaded 2008) and 114.5 million tonnes 2009. India with 79 million tonnes in 2008 and 80.6 million tonnes in 2009 is number two from a production perspective, USA third with 68 million tonne/y 2008 (59.4 million tonnes 2009) and Russia number four with 64 million tonne/y 2008 (56.5 million tonnes 2009). From the figures, we can see that the production varies from one year to the other. Wheat is considered to be the most important crop from a food perspective globally. The straw production will be around 700 million tonnes/y, corresponding to approximately 3500 TWh/y if used as a fuel. Straw is generally not used very efficiently from an energy point of view, and here we have a huge potential.

Of the top ten wheat producers in 2008, next after the four mentioned above are France with 39, Canada with 29, Germany and Ukraine both with 26, Australia and Pakistan both with 21 million tonnes produced in 2008.

1.2.4   *Corn (Zea mays) and cassava*

Corn or maize is also one of the major crops from a food production perspective. The major producer of corn is the US with a production of 333 million tonnes/year (2009). At number two comes China with 163 million tonnes/year. Brazil produces 51 and Mexico 20 million tonnes per year. Thereafter we have Indonesia with 17.6, India with 17.3, France with 15.3, Argentina with 13.1, South Africa with 12 and Ukraine with 10.5 million tonne produced in 2009.

The total annual production was 817 million tonnes globally grown on 159 million ha in 2009. The biomass is approximately the same amount as the grains. Before corn is ripe, it produces a very powerful antibiotic substance, 2,4-dihydroxy-7-methoxy-1,4-benzoxazin-3-one (DIMBOA) which is accumulated in the crop. This is a natural defense against a wide range of pests, including insects, pathogenic fungi and bacteria. Due to its shallow root system, corn is sensitive to droughts, and should be grown on soils with enough nutrients. It may also be sensitive to strong winds (Wikipedia, 2011; Monsanto, 2011). In Gautam *et al.* (2011), different species of maize have been grown and compared. Different species gave 11–21 tonnes dry substance per hectare (DS/ha) for similar conditions, showing both the importance of genetic selection of seeds, as well as the future potential for large scale production. In these experiments, the goal was to produce crops for silage, and thus had a focus on total biomass production and not specifically grains for food usage.

Corn/maize is one of the crops that has been genetically modified and then planted the most. In the US and Canada actually 85% of the maize produced in 2009 was genetically modified.

Cassava is the third largest carbohydrate source for humans with 136 million tonnes 1985. More than 1/3 is grown in Africa, where it is very important as food crop. It is good as feedstock to ethanol production as well.

1.2.5   Barley, rye and oats

There are some other cereals of importance especially in the northern countries. Russia produces 15.5 million tonne barley per year, Canada 12.2, Ukraine 8.0, Turkey 7.0, Australia 7.0 and USA 6.0 million tonnes per year in 2009. Russia also produced 5 tonnes rye and 6.0 tonnes oats while Poland produced 3.2 million tonnes rye, Canada 3.7 million tonnes oats and the US 2.1 million tonnes oats in 2009.

1.2.6   Oil crops

18.1 million tonnes rapeseed was produced in EU27 2007. Most of it was produced in the temperate countries north of the Alps. In the southern countries, olive oil is more common.

The global rape seed oil production was 47 million tonnes in 2006. Major producers were China with 12.2, Canada with 9.1, India 6.0, Germany 5.3, France 4.1, UK 1.9 and Poland with 1.6 million tonnes. (http://www.soyatech.com/rapeseed_facts.htm).

Today, the world's top producers of soy are the United States, Brazil, Argentina, China and India. About 85% of the world's soybeans are processed, or "crushed," annually into soybean meal and oil. Approximately 9% of the soybean meal that is crushed is further processed into animal feed with the balance used to make soy flour and proteins. Of the oil fraction, 95% is consumed as edible oil; the rest is used for industrial products such as fatty acids, soaps and biodiesel. Soy is one of the few plants that provide a complete protein as it contains all eight amino acids essential for human health. World soybean production has increased by over 500% in the last 40 years.

In 2007/08 the production of different vegetable oils was: palm oil 41.3 million tonnes, soybean oil 41.3, rapeseed 18.2, sunflower 9.9, peanut 4.8, cottonnseed 5.0, palm kernel 4.9, coconut 3.5 and olive oil 2.8 million tonnes (Oilseeds, 2009).

1.2.7   Sugar cane

Sugar cane is an important crop for production of sugar, but also for ethanol in countries such as Brazil. Sugarcane is the world's largest crop in 2010 with respect to production. FAO (Crop production, 2010a) estimates about 23.8 million hectares were cultivated with sugar cane in more than 90 countries, with a worldwide harvest of 1.69 billion tonnes. Brazil was the largest producer of sugar cane in the world. The next five major producers, in decreasing amounts of production, were India, China, Thailand, Pakistan and Mexico.

The world demand for sugar is the primary driver of sugarcane agriculture. Cane accounts for 80% of sugar produced; most of the rest is produced from sugar beets.

1.3   ENERGY CROPS

There are principally three different types of pathways for the photosynthesis. The most common is called the C3 system. This is most common in colder and temperate climates, and produces 3-carbon organic acid (3-phosphoglyceric acid). The second main system is more common in warm climates. It is called the C4 system, as the first product is 4-carbon organic acids (malate and aspartate). The C4 system also utilizes much more CO_2: 70–100 mg CO_2/dm^2/h with light saturation around 4.2–5.9 J/cm^2/min (total radiation of visible light). For C3 plants, the corresponding value is 15–30 mg CO_2/dm^2/h with light saturation already at 0.8–2.4 J/cm^2/min. This

Figure 1.1.   Pineapple at the Azores – a CAM crop.

means that increased CO_2 concentration will be beneficial for C4 plants, but may be inhibiting C3 plants. The third enzyme system for photosynthesis is called the CAM system, or crasulacean acid metabolism. CAM crops produce 4-carbon organic acids like the C4 system, but they also can capture light during the day and later fix the CO_2 during the night. Two crops are of agricultural interest from this group – pineapple (Fig. 1.1) and sisal (a kind of agave that gives strong fibers).

Well-known C4 crops are sorghum, sugarcane, maize, miscanthus and cord grass. Some more are seen in the Table 1.6. Most species still have the C3 system.

There are many different potential energy crops. Some interesting crops are as follows.

1.3.1   *Switch grass*

Switch grass (Fig. 1.3) is seen as a potential important energy crop. Yield has been 5.2–11.1 tonnes/ha/y in test fields in the US. The net energy yield (NEY) has been 60 GJ/ha/y. Switch grass has been used as a feedstock for bio-ethanol production and then has gained 5–40 times more renewable energy as ethanol than fossil energy consumed for the production. This means a reduction of green- house gases (GHG) by 94% compared to gasoline when used as a fuel. It can also be pelletized or bricquetized as seen in Figure 1.2.

1.3.2   *Giant Kings Grass*

Giant Kings Grass has been able to produce up to 100 tonnes DS/ha per y at a farm in China by Viaspace Company (Kukkonen, 2011). It can grow with high yields also on marginal land if there is enough water through rain or irrigation. There may be 2–3 harvests per year and the need for fertilizers is relatively low. With crops available today the crop is sensitive to frost, but it might be that frost resistant genes could be transferred to it, to make it feasible also at higher altitudes sometime in the future.

Table 1.6.   Different potential energy crops and their characteristics.

Name	In Latin	Yield [t/ha DM]	Climate zone
Aleman grass	*Echinochloa polystachya*	20–100	C4, TR-TE, floodbank
Alfalfa	*Medicago sativa*	10.0–20.0	All climates
Annual rye grass	*Lolium multiflorum*	7.1–18.5	ME, CE
Bamboo	*Gramineae Bambusoideae*	3.0–14	TE-TR, >750 mm/y
Banana	*Musa paradisiaca*	15.0–50	TR/ST, >60%RH
Barley	*Hordeum vulgare*	8.0–16	TE, 1:1 grain/straw
Black locust	*Robinia pseudoacacia*	5.0–10	TE, >1000 mm/y
Broom (Ginestra)	*Spartium junceum*	9.0 –10	MEmarginal areas
Brown beetle grass	*Leptochloa fusca*	17.0–20	C4, TR-TE, wetswamp
Cardoon	*Cynara cardunculus*	20.0–30	ME, >450 mm/y
Cassava	*Manihot esculenta*	5.0–10(80)	TR/ST, >1000 mm/y
Coconut palm	*Cocos nucifera*	4.0–9	TR
Common reed	*Phragmites australis*	5.0–43	TR-TE salttolerant
Cordgrass	*Spartina spp.*	8.0–20	C4, TE, PE
Cotton	*Gossypium spp.*	4.0	TE, 3 m, deep roots
Elephant grass	*Pennisetum purpureum*	20.0–85	TR/ST
Eucalyptus	*Eucalyptus spp.*	25.0–50	TR-TE, adaptive
Fodderbeat	*Beta vulgaris rapacea*	20.0–50	TE
Giant knotweed	*Polygonum sachalinensis*	20.0–30	EU/AS 1/2 undergrd
Giant reed	*Arundo donax*	20–40	ME, PE grass,
Groundnut	*Arachis hypogaea*	3.0–6	EU/AS/Boliv 1/2 nut
Hemp	*Cannabis sativa*	8.0–21	TE >700 mm/y
Jatropha	*Jatropha curcas*	0.5–2	AR, wide temprange
Jerusalem artichoke	*Helianthus tuberosus*	10.0–16	PE, TE, inulin 7 t/ha
Jojoba	*Simmondsia chinensis*	3.0–5	S/NAm, Drout tolnt
Kenaf	*Hibiscus cannabinus*	20–30	AN, AF, >500 mm/y
Leucaena	*Leucaena leucocephala*	20–35	TE-TR, >600 mm/y
Lupin	*Lupinus spp.*	0.5–4	ME, SAm, lowwatdem
Maize	*Zea mays*	10.0–20	TE, 1:1 grain/straw
Medow foxtail	*Alopecurus pratensis*	7.0–13	CE, PE
Microalgae	*Oleaginous spp.*	1.0–20	Manyt types, oil
Miscanthus	*Miscanthus spp.*	10.0–40	C4, Eu, SEAsia, PE
Neem tree	*Azadirachta indica*	2.0–6	AF, SEAS, IN >450 /y
Oil plam	*Elaeis guineensis*	6.0–7	TR, >1500 mm/y
Olive tree	*Olea europaea*	0.4	ME&middle east
Perennial rye grass	*Lolium perenne*	7.0–13	TE-TR highland
Pigeonpea	*Cajanus cajan*	9.0–10	TR/ST, 2 tgr, 8 t/hastalk
Poplar	*Populus spp.*	12.0–15	TE, tree
Potato	*Solanum tuberosum*	5.0–15	TE
Rape	*Brassica napus*	10.0–15	TE, seed3t, str 10 t/ha
Reed canary grass	*Phalaris arundinacea*	5.0–19	PE, ST/TE, wet
Rice	*Oryza spp.*	2.0–20	TE/ST, 1:1grain, stalk
Root chickory	*Cichorium intybus*	11.0–17	BI, TE, root inulin
Rosin weed	*Silphium perfoliatum*	8.0–16	PE, NAm, tolerant
Safflower	*Carthamus tinctorius*	0.5–5	ME, IN, <1.5 m, AN
Safou	*Dacryodes edulis*	10.0–15	AF/TR tree, fruit
Salicornia	*Salicornia bigelovii*	1.0–10	AR, Salttol, 1–3 tseed/ha
Sorghum	*Sorghum bicolor*	14.0–40	C4, TE-TR, AN, 4 m
Sweet sorghum	*Sorghum bicolor*	12.0–45	C4, EU, USA, TR-TE
Sorrel	*Rumex acetosa*	10.0–40	PE, 2 m, TE, Tolerant
Soybean	*Glycine max*	2.0–3	AN, TE-TR, 40 prot, 20 oil
Sugarbeet	*Beta vulgaris altissima*	8.0–30	TE

(Continued)

Table 1.6. Continued.

Name	In Latin	Yield [t/ha DM]	Climate zone
Sugarcane	*Saccharum officinarum*	16.0–73	C4, PE, ST/TR, >1500/y
Sunflower	*Helianthus annuus*	4.0–12	ST/TE, 25seed, 75stalk
Sweet potato	*Ipomoea batatus*	5.0–15	TR/ST, PE, 750–1250/y
Switch grass	*Panicum virgatum*	5.0–16	C4, PE, TE, lnutrientdem
Tall fescue	*Festuca arundinacea*	11.0–13	PE, TE-TR, <1.8 m
Timothy	*Phleum pratense*	11.0–19	PE, TE, 1 m
Water hyacinth	*Eichornia crassipes*	30.0–90	PE, TR/STaquatic herb
Wheat	*Triticum aestivum*	10.0–20	TE, 1:1 grain/straw
Willow	*Salix* spp.	5.0–20	CO, Eu, Siberia, >600/y

ME = mediteranian, TR = tropical, ST = subtropical, TE = temperate, CO = continental, CE = central European, NO = northen climates, Am = America, Nam = North America, SAm = South America, EU = Europe, AR = arid areas, AS = Asia, IN = India, AF = Africa, PE = perennial, AN = annual, BI = biennial.

Figure 1.2. Briquettes formed from straw of switch grass and bark.

1.3.3 *Hybrid poplar*

Among trees, hybrid poplar is very interesting for the future. In Heilman and Stettler (1985), 50 clones were compared at sites in western Washington in the US. What could be seen was that there were major variations between different clones. The height of the clones varied between 8.5 and 11.8 m at 4 years. With respect to dry weight production, it varied between 5.2 and 23 tonnes/ha/y, with the average 12.5 tonnes/ha per y. Three Populustrichocarpa × P. deltoides hybrids were planted in a trial. The dry weight production was 15.6 to 27.8 tonnes/ha per y for an average of 23.6 tonnes/ha per y. This shows that also trees can reach very high productivity under good conditions, and when good clones have been selected. In Zalesny *et al.* (2009), the growth rate of different clones of hybrid poplar was evaluated. The results for different clones and sites are as follows: for each site, biomass ranges [tonne/ha,y] of the best six clones were: Westport: 2.3 to 3.9 (5 years), 8.0 to 10.1 (8 years), and 8.9 to 11.3 (10 years); Waseca: 10.4 to

Figure 1.3.   Rajj grass/switch grass cultivation in Vingåker, Sweden.

13.4 (7 years); Arlington: 5.1 to 7.1 (3 years), 14.8 to 20.9 (6 years), and 16.1 to 21.1 (8 years); and Ames: 4.3 to 5.3 (4 years), 11.1 to 20.9 (7 years), and 14.3 to 24.5 (9 years). The average of all these 10 clones at 10 sites was approximately 11 tonne DS/ha per y. We can not have only hybrid poplar everywhere, but a significant portion would be reasonable.

1.3.4   *Other proposed energy crops*

In the book *Energy Plant Species* by N. el Bassam many other energy crops are listed and discussed (Bassam, 1998). In Table 1.6 important species are listed from the book, complemented by other material from many different sources.

Some crops are producing a lot of oil and therefore can be interesting to use for production of fossil oil replacements. Typical seed yields and oil content of the seed are listed in Table 1.7.

1.3.5   *Quorn*

It can be interesting to note that it is not only plants that can give high productivity and good protein. During the 1960s, J. Arthur Rank started a project to develop fungus into good food. After several years of development work and test of soil from 3000 sites they found a fungus *Fusarium Veneatum* that turned out to be possible to grow on a large scale, tasted good and had a high productivity and gave good protein. A plant started in Billingham, and is still producing the fungus and actually also expanding from the 17,000 tonne produced in 2011. This product is called Quorn. The feedstock is primarily starch. To the starch solution nitrogen, phosphate, trace elements and air is added. A draw back in the production is that large amounts of RNA is produced, but by heat treatment at 70°C the content is reduced to level accepted to use the product as fodder for animals and food for humans as this meat replacement. Unfortunately, 30% of the product is lost during this treatment. Significant advantages with a production like this are that protein fodder with a quality like meat can be produced but without having to breed animals. CO_2 emissions are in the same range as for chicken meat, which is significantly below the levels for beef and pig meat (Snaprud, 2012). Marlow food, who produces and distributes refined products from Quorn (http://www.quorn.com/) is presenting how to use Quorn in different meals. It can be interesting to note that the growth rate of the fungus is compared to breeding cows with production of 50 adult cows in only one day!

Table 1.7. Oil content in seed of different crops.

Name	Latin name	Seed yield [t/ha]	Oil content [%]
Abyssinian kale	*Crambe abyssinica*	2–3.5	30–45
Bird rape	*Brassica rapa*	1–2.5	38–48
Indian mustard	*Brassica juncea*	1.5–3.3	30–40
Coconut palm	*Cocos nucifera*	4.2	36
Coriander	*Coriandrum sativum*	2.0–3.0	18–22
Gold of pleasure	*Camelina sativa*	2.3	33–42
Groundnut	*Arachis hypogea*	2	45–53
Hemp	*Cannabis sativa*	0.5–2.0	28–35
Jojoba	*Simmondsia chinensis*	2.1	48–56
Oil palm	*Elaeis guineensis*	30	26
Olive	*Olea* spp.	1.0–12	40
Rape	*Brassica napus*	2.0–3.5	40–50
Safflower	*Carthamus tinctorius*	1.8	18–50
Soybean	*Glycine max*	2.1	18–24
Spurge	*Euphorbia latyris*	1.5	48
Sunflower	*Helianthus annuus*	2.5–3.2	35–52
White mustard	*Sinapis alba*	1.5–2.5	22–42

1.4 ANIMALIAN BIOMASS AND ALGAE

1.4.1 *Animalian food*

When we think of biomass, we normally think of crops. Still, meat from different animals, fish and insects is very important especially to give us protein. Cow, pig, sheep and hen are the most common, but at least 1400 types of insects are also used as food, especially in tropical countries in Africa, Asia and South America (Nyström, 2012). To produce 1 kg of beef 10 kg fodder is needed, while 10 kg fodder may produce 9 kg of insects. In Thailand, grasshoppers are common as food. In Cambodia tarantel spider are fried and eaten. In China, many different types are eaten as seen in Figure 1.4. At the Wageningen University in Holland, a study over greenhouse gas emissions from cows *versus* insects was done recently. It concluded that the greenhouse gas emission is much less, because insects are cold blooded and thus do not waste energy on heating, which also avoids production of methane and other greenhouse gases.

FAO is presenting an overview of available livestock resources in the report World Livestock 2011, Livestock in food security (McLeod 2011). The authors have made predictions of how much livestock that will be needed per 1 billion people today and in the future. The figures are seen in Table 1.8.

What we can see from Table 1.8 is that pig and poultry are the most important apart from dairy today, but poultry meat is predicted increase the most. This has an advantage as less energy is needed to produce one kg hen compared to one kg cow.

For a young cow (250 kg) it may be as seen below. The figures are in MJ/day respectively g protein/day:

Unit	MJ	Protein [gSRP]
Total nutrients	49.5	416
4 kg hay	35.2	320
1.3 kg oats	14.4	107

Figure 1.4.   Insects at a food market in Beijing, China.

Table 1.8.   Projected consumption of livestock products per billion people based on 2002 population estimates.

Year	2010	2020	2030	2050	Growth 2010 to 2050
Human population (billions)	6.83	7.54	8.13	8.91	
Consumption (million tonnes per billion people)					
Bovine meat	9.85	10.25	10.93	11.93	121%
Ovine meat	1.94	2.08	2.28	2.64	136%
Pig meat	14.98	15.29	15.98	15.79	105%
Poultry meat	12.58	14.72	17.65	21.69	173%
Dairy	96.24	100.19	106.77	116.55	121%

Sources: FAO, 2006c; World Population Prospects, 2002. Some calculations made by the authors.

This can be compared to horses with the weight 500 kg:

Units	MJ	Protein [g]
Total nutrients	79	440
7 kg hay	56	280
2 kg oats	23	160

For a sheep without kids it may be as below:

Units	MJ	Protein [g]
Total nutrients	9.6	64
1 kg hay (DS)	9.6	65

Table 1.9. World catch of fish 2007 given in tonnes.

	Inland fisheries	Marine fisheries	Total
Fresh water fish	8695	23	8718
Diadromous fish	341	1444	1785
Marine fishes	82	65627	65709
Crustaceans	474	5367	5840
Mollusks	383	7182	7564
Other	61	388	449
Total	10035	80029	90064

This shows the needs for different animals with respect to total amount of nutrients, but also that bigger animals normally need higher value protein as well for good health and thereby growth rate. Aside of the "normal" animalian food also fish is important as food for humans as well as being used as raft food for animals and for fish farms.

In the FAO report Fish, Crustaceans, Mollusk, etc. – World capture production (FAO, 2010b) the catch of different type of fish is presented. The figures are seen in Table 1.9.

How high the production is totally is of course not directly seen by these figures. In some areas, the catch of fish is higher than the actual growth, while in other areas more is produced than catch. A lot of the captured fish is in reality used as fodder for fish farms, another large amount is actually thrown back, and often dies, as there are strict regulations about what size of fish that countries are allowed to catch. What we can say about fishing is that it is very "political" as it is an important but too scarce resource, where many countries are fighting to get as large share as possible.

1.4.2 *Algae*

Algae are a resource that is not used so much directly, but indirectly as food for many animals, which are later eaten by humans. These are primarily micro-algae and macro-algae. Micro-algae like green algae produce large amounts of fatty acids and fatty oil, which from a human perspective is interesting as a bio-fuel. This is discussed in Chapter 11 in this book about green fuels for aviation. Macro-algae are among others brown algae like kelp, which can grow by 0.5 meter per day under favorable conditions, and exist in large quantities in many regions. One example is along the Californian coast. Another example is at the Faeroe Islands, where it is estimated to grow some 15 TWh of kelp.

The growth rate of micro-algae measured as number of divisions per day can be seen for a number of species in Table 1.10.

Generally, we could say that there is one division every one or two hours for these algae. In this light intensity, the radiation is not a limiting factor, but for lower light intensities, we get lower growth, as well as in colder water.

The dry weight of each cell has been published for certain species. Table 1.10 is showing a bit of the range to expect. The density in cultures is normally 80–250 mg dry weight/L. If we have 100 mg/L and the division is once per hour this means a growth rate of 100 mg/L/h of dry solids, which would mean $100\,g/m^3$/per h or principally $2.4\,kg/m^3$/per day if growth all day around, which is principally the case in northern Scandinavia summer time. Production rates for algae are given in e.g. http://www.fao.org/docrep/003/w3732e/w3732e06.htm#b6-2.3.6.%20Algal%20production%20in%20outdoor%20ponds.

The total world production of algae for commercial use is around 10,000 tonne per year according to Beneman (2008). The total growth rate of algae globally is very difficult to calculate, but we are talking about at least hundreds or even thousands of TWh/y if we include both micro-algae and macro-algae. In Figure 1.5, we see examples of macro-algae.

Table 1.10.   Growth rate of micro-algae measured as number of divisions per day at 25°C
and a light intensity of 5000 respectively 2500 Lux.

Units per cell in Pg/cell*	5000	2500	Dry weight
Chlorella ellipsoidea	0.88	0.85	
Chlorella vulgaris (freshwater)	0.68	0.67	
Dunaniella tertiolecta	0.60	0.64	
Isochrysis aff. Galbana	0.78	0.81	8–30
Nannochlorois oculata	0.85	0.92	
Skeletonema costatum	0.6	0.67	52
Tetraselmus suecica	0.46	0.44	66–292

*1 Pg = 1000 Tg = 1000 million tonnes.

Figure 1.5.   Macro algae at Cape Good Hope, South Africa (photo E. Dahlquist).

1.5   REGIONAL OVERVIEWS

The following section some important regions will be addressed a bit more in detail. The overview
is not covering all countries and regions, but represents approximately half of the populations at
earth, and most type of climate zones.

1.5.1   *EU27 – an overall energy balance*

We have made some calculation of the agricultural and forestry production with respect to bioenergy for EU27 (EU with 27 member countries). The conservative figures (Dahlquist *et al.*, 2010)
say around 8500 TWh/y, which can be compared to the 16,000 TWh/y primary energy used today

Figure 1.6. Typical cereal agricultural land in northern Europe (Nibble Farm, Vasteras, Sweden. Photo E. Dahlquist).

according to official figures. With reasonable improvements in the growth rate due to selection of crops with higher yields, improved irrigation and addition of nutrients, the production should at least be possible to reach 12,500 TWh/y of biomass in the EU region. By implementing energy efficiency improvements in transportation, buildings and industry it should be possible to reduce the need for primary energy to around 12,500 TWh/y. Together with hydro power, solar power and wind power the balance between available renewable resources and consumption should then be possible to reach without the need for fossil fuels.

In Europe cereals like wheat is the most important crop. A typical view of farmland for cereals is seen in Figure 1.6. In this case, we see a field with spring wheat as well as autumn wheat.

1.5.2 *China – today and in year 2050*

The Chinese Academy of Engineering has made predictions for the energy utilization in China until the year 2050 (Du Xiangwan, 2008). The prediction is that renewable energies should deliver 0.88–1.71 billion tce 2050 (tce= tonne coal equivalents), reaching 17–34% share of the national total demand. Including hydropower, the renewables will give 1.32–2.15 billion tce providing 26–43% of the national total energy demand. Assuming the 1.7 billion tce total renewables in 2050, 26% will come from hydropower, 20% from biomass, 34% from solar power and 18% from wind power. Other renewable energies will contribute to 2% in this scenario where the total utilization is predicted to be 5 billion tce 2050. The production of bio pellets and briquettes will increase to 50 million tonnes by 2050. The increase of electric power from biogas-fueled power plants will be 20 GW by 2020 and 40 GW by 2030. Some of this will be through co-firing of biomass in coal-fired power plants. In addition, CHP will most probably increase to enhance the total efficiency for both coal and biomass fuels. Only 200 MW electricity was installed 2006 using biomass fuels, but the capacity is increasing fast.

Figure 1.7.   Forest in Southern China close to Guiling (photo E. Dahlquist).

There are already more than 22 million small-scale biogas plants producing 8.5 billion Nm³/y. Medium and large scale biogas projects will increase from 3671 year 2007 producing 2 billion Nm³/y biogas to 44 billion m³ by 2020 and 80 billion m³ by 2030 (the figure in 2006 being 10 billion m³ per year). This is according to professor Li Shi-Zhong at Tsinghua University (*pers. commun.* 2011). In addition, 39 million tonnes of bioethanol and 6 million tonnes of bio-diesel was produced 2007.

China has about 120 million hectares of marginal lands and 40 million hectares of degraded arable lands. Tuber crops have high biomass production yield (15–45 t/ha) and starch content (20–33%). Cassava is a good crop in southern China as it is less sensitive to diseases and insects, resistance to drought etc. Sweet potato can also be planted in poor soil. In Figure 1.7, we can see an example of forestland typical of Guiling and in Figure 1.8 typical farmland for the cereal production.

In Southern China we have some jungle forests (Fig. 1.9), especially at Hainan Island. Here also bamboo is growing, which is a good source for building construction and many other applications.

Henan province is the cereal production center of China and in Figure 1.10, we see a typical landscape with farmland and hills with some small forests in-between.

In Henan province and other agriculture areas buffalo are also seen frequently like in Figure 1.11. Here we also see some sugar cane in the background.

Jungle areas were mentioned already earlier and in Figure 1.12, we can see another typical jungle at southern Hainan Island. The jungle areas are dominated by trees of very many different types, which gives a wide resource of genetically diverse species, but it is not easy to grow in the same way as the monocultures seen in more tempered areas. When large areas are covered with eucalyptus, oil palms and acacia instead, this diversification is lost, and we do not really know how the forests will behave in the future.

Still, in Hainan and in many other areas in South and East Asia, forests are also kept as reserves to keep the diversity for future generations. In the same areas where we have the tropical jungles

Figure 1.8.   Agriculture in China (photo E. Dahlquist).

Figure 1.9.   Bamboo at Southern Hainan, China.

we also see a lot of rice being produced, but also fruits and vegetables of different kind. Here in Figure 1.13 rice and melons are being produced.

There are about 16 million ha of marginal lands available for planting starch tuber crops like cassava and sweet potato (Subramanian Narayana Moorthy).

Figure 1.10.   Agriculture landscape in Henan province, China (photo E. Dahlquist).

Figure 1.11.   Agriculture landscape in Henan province, China (photo E. Dahlquist).

High ethanol yield has been achieved: $4.7\,m^3$/ha for sweet sorghum stalk *vs*. $3.7\,m^3$/ha for corn, $3.8\,m^3$/ha for sugarcane, $1.5\,m^3$/ha for sweet sorghum grains and $4.8\,m^3$/ha for bagasse. Straw from agricultural crops results in 600–700 million tonnes in China annually. About 1.7 million tonnes of livestock and fowl manure is produced annually from the breeding industry in China.

The most important food crops in China are in the north wheat with a total of 114.5 million tonnes produced in 2009, and rice in the south with 197 million tonnes produced in the same year.

Figure 1.12.   Jungle landscape at Southern Hainan, China (photo E. Dahlquist).

Figure 1.13.   Rice and vegetable farming at Hainan Island, China (photo E. Dahlquist).

1.5.3   *India*

For India rice is the most important food crop (99.2 million tonne/y) followed by wheat at second place (80.6 million tonne/year 2009). The productivity of wheat varies a lot between different states, from 0.7 to 4.3 tonne/ha/per y. Coarse cereals give 39.5 million tonne/y and pulses 14.7

Figure 1.14.   Marshland in San Francisco, USA (photo E. Dahlquist).

million tonne/y. This gives 233.9 million tonne/y (2009) of all major crops altogether. The national directorate of wheat research in India states that 50% of the calories are at an average coming from wheat to the Indian population (Singh, 2009). The productivity with respect to wheat has increased from 0.9 tonnes/ha in 1965 to 3 tonnes/ha/per y today at an average. The increase has been due to selection of suitable clones for each type of soil and other conditions. This of course is very promising. The highest yields are in Punjab and Haryana with 4–4.3 tonnes/ha per y, while Karnataka has only 0.7 tonnes/ha per y. This shows that there is still a potential for improvements. Today the production is 67 kg/capita, while the demand is 73 kg/capita. With further improvements, Singh believes there will be a balance already within ten years. A potential threat still is rusts, leaf blight and insects as well as climatic issues.

1.5.4   *USA*

Huber and Dale (2009) made a review of US biomass potential especially for the purpose as fuels for vehicles. The authors presented the following figures: 428 million tonnes agricultural waste, 377 million tonnes energy crops, 368 million tonnes forest products, 87 million tonnes corn and other grains and 106 million tonnes other type of organic residues. Totally, this gives 1366 million tonnes/year of crop residues. They also discuss possible energy crops as a complement, where the following are considered to be of highest interest in the US: switch grass, sorghum, miscanthus and energy cane. The authors estimate that these residues and crops could produce 3.5 billion barrels of oil equivalents, which is roughly 50% of the 7.1 billion barrels of oil used today in the US. If we just assume that these crops have a higher heating value (*HHV*) of 5.4 MWh/tonne, it means 7376 TWh/y. In many areas in the western US we see relatively arid biotopes. In the marshland area close to San Francisco (Fig. 1.14), we can see examples of this as well as the typical mountain areas with scarce forest like in Figure 1.15 with Golden Gate Bridge in the foreground.

There are also more fruitful areas with forests in the west part of the US as illustrated in Figure 1.16 in Santa Barbara. Here many different species exist side by side.

Figure 1.15.   Californian landscape, Golden Gate Bridge, San Francisco, USA (photo E. Dahlquist).

Figure 1.16.   Forest in California, Santa Barbara, USA (photo E. Dahlquist).

We also can make another calculation to estimate the bioenergy potential in the US. From World Bank indicators (World Bank, 2011) we see that the average cereal yield is 7.2 tonnes/ha in the US. The agricultural area is 58,001,425 ha, giving 4.4×10^8 tonne/year, or with 5.4 MWh/tonne = 2270 TWh/y additional as cereals. This is 14% of the total agricultural land, 411,200,000 ha. If we assume the same amount of straw is produced, that is 2270 TWh/y, we get 4540 TWh/y from cereals including straw, and if we get the same amount of production on the rest of the land with energy crops, it would be 4540/0.14 = 32,430 TWh/y.

The most widely grown crop in the US is Corn with 332 million tonnes per year. From these 130 million tonnes or 40% is used for production of ethanol fuel. The reason for this is that the US government wanted to decrease the political risks with oil import and thus stimulated ethanol production by paying a guaranteed price per liter ethanol. There is no request on how the ethanol is produced, and thus oil and coal is often used to produce and distil the ethanol, giving poor ratio between the heating values of ethanol produced in relation to fossil fuels used for the production. This is the major reason for claims that ethanol is bad for the environment heard especially in European mass media.

The forestland area is 304,022,000 ha. If we assume an average of 3 tonne/ha/per y or 16 MWh/ha, we should produce some 4900 TWh/y from this as well. A total production then would be approximately 37,350 TWh/y in the US. In Zalesny *et al.* (2009), the growth rate of different clones of hybrid poplar was evaluated. The average of 10 clones at 10 sites was approximately 11 tonne DS/ha/per y. If we assume 30% of the US forestland area is planted with this it would mean 304,022,000 ha $\times$ 0.3 $\times$ 5.4 MWh/tonne $\times$ 11 tonne/ha/y = 5420 TWh/y and a total of 5420 + 4900 $\times$ 0.7 = 8850 TWh/y.

If we compare this to the total use of energy in the US, this is 4160 TWh/y electricity and 2,172,107 ktonne of oil equivalents/y or with 10 MWh/tonne o.e. 21,720 TWh/y totally (from which fossil fuels 84% today). These figures show that the available biomass resources should be enough to cover all energy needs if used efficiently. US government has stated that by 2030 bio-fuel will have replaced 30% of fossil fuels.

The conditions in Canada are even better than in the US, while Mexico has more limitations for biomass production due to drier climate.

1.5.5   *Brazil*

Brazil is generally a very "green" country with a lot of forests and farmland areas. Since the 1980s Brazil has been a leading country with respect to produce bio-ethanol for vehicles. First, there were many cars running on ethanol during the 1980s. Then oil became more favorable due to new political decisions. Since the beginning of the new millennium, ethanol has taken back its position and now Brazil is the major producer of ethanol globally.

There have been a lot of discussions about environmental issues around the ethanol production in among others Brazil. It has been claimed that rain forests have been cut down to plant sugar cane, where the sugar is fermented to ethanol, and the bagasse use for the distillation of the ethanol. In reality, this is not correct generally, as sugar cane normally is grown at land areas with a different climate, further to the south of the Amazonas. Instead soya beans are planted where rain forest are cut down. The soya beans powder then is exported to a large extent to western animal farms for production of the meat we eat.

According to Brito Cruz (2008), 15% of all energy used in Brazil is from sugar cane.1988, 50% of vehicle fuels was ethanol. 2004 it was 30%. Sugar cane gives 6 m^3/ha. The total arable land in Brazil was 354.8 million ha (year 2007). From this, 76.7 million ha were used for crops, 20.6 for soybean, 14.0 for corn, 7.8 for sugar cane, from which 3.4 million ha were used for ethanol production. This corresponds to 1% of the arable land area, but replaces 30% of the fossil fuel used for vehicles. 172.3 million ha are pastures and thus we have 105.8 million ha left for e.g. additional ethanol production. If we triple the production 11.3 million ha would be needed, or 7.9 million aside of what is already utilized. This would mean 7.5% of the available arable land not used

Figure 1.17.   Savannah in South Africa close to the Kruger Park (photo E. Dahlquist).

intensively today. It can also be interesting to note that the cost for sugar cane ethanol production is 0.25 US$/liter compared to 0.4–0.7 $/liter for fossil gasoline in Brazil (Brito Cruz, 2008).

1.5.6   *Africa south of the Sahara*

Africa has many different climatic conditions, from very dry areas in the north to tropical and subtropical south of Sahara. In the north, biomass has difficult conditions while south of Sahara the conditions are mainly quite good, especially in Western and Central Africa. Coming to the very south, we see other areas with less rain and thus difficult conditions for crops. This is especially true in Namibia, Botswana and parts of South Africa (see Figs. 1.17 and 1.18). The potential for using biomass for most of the needs is thus quite high in most countries south of Sahara. The problem in many countries is the instable political and cultural conditions causing problems in the development of both society and economy.

One interesting initiative to improve the productivity of African crops is the new species breaded at the African Rice Center under the name "New Rices for Africa" (NERICA). Here rice is selected to tolerate harsh growing conditions and low nutrient levels.

In South Africa, we also have areas with quite good conditions for crops, and there are many wine yards in the region close to Stellenbosch and Cape Town.

1.6   OTHER REGIONS

There are many areas with less good conditions for agriculture, but where animal breeding may be an alternative. The amount of arable land globally is distributed in the following way for the largest countries: USA 179,000 ha, India 169,700 ha, China 135,557 ha, Russia 126,820 ha, Brazil 65,200 ha, Australia 50,600 ha, Canada 45,700 ha, Indonesia 33,546 ha, Ukraine 33,496 ha, Nigeria 30,850 ha, Mexico 27,300 ha, Argentina 27,200 ha, Turkey 26,672 ha, Pakistan 21,960, Kazakhstan 21,671 ha, France 19,582 ha, Spain 18,217 ha, Thailand 18,000 ha, Sudan 16,433 ha, South Africa 15,712 ha, Poland 14,330 ha and Germany 12,020 ha.

Figure 1.18.   Dry forest in South Africa (photo E. Dahlquist).

Figure 1.19.   Typical landscape in South East Australia, close to Tumut (photo E. Dahlquist).

As can be seen the distribution of arable land is to some extent following where we have a lot of people, like India, China, Brazil and Indonesia, while some countries have huge surpluses in relation to their populations like USA, Canada and Russia, especially Canada. In Australia, we have a lot of this type of land in the southeastern part. A typical landscape of this is seen in Figure 1.19. Here we have some trees and a lot of grass. At some places, the rain is enough for more intense farming, while in other areas it is too little.

In areas like the western Asia and south east Europe the climate, also vary between dry and rainier. In northern Iran close to the Caspian Sea, we have relatively good conditions for farming and even some forestry like seen in Figure 1.20.

In South East Asia, on the other hand the climate is tropical and the forests very dense as seen in Figure 1.21 and Figure 1.22. The growth is very rapid and trees of different kind have developed a very sophisticated interaction to utilize the available resources in best possible way.

Figure 1.20.   Agriculture in Northern Iran representative the western Asia (photo E. Dahlquist).

Figure 1.21.   Tropical landscape in southern Thailand (photo E. Dahlquist).

Figure 1.22.   Jungle in southern Thailand (photo E. Dahlquist).

Table 1.11.   Distribution of cereal production and forestland areas as well as productivity per ha of cereals in different economic groups.

Economic groups	Land area cereal production [ha] 2008	Forest area [km²] 2010	Total land area [km²] 2009	Cereal prod [kg/ha/y] 2009
Low income	88349486	4154870	15043470	1952
Middle income	473000000	26420030	80675467	3202
High income	147000000	9629420	33842634	5448
Worldwide	708000000	40204320	130000000	3513

This exposé over different biotopes has given some hints on the diversity we have in our crops and forests naturally. It also shows that these different species can be important in the future for sustaining diversity in the gene pool for future developments of crops that are both productivity and resistant to diseases and harsh conditions. Some crops are totally dominating as food crops today, and some trees for forestry, but if a disease hits one important crop it is always good to have a number of alternatives available. This is just to remind ourselves about potential threats. Right now, a number of trees are for instance dying in northern Europe due to moulds. These are not so common but if pine, spruce, eucalyptus or acacia are hit in the same way, it can be very problematic.

1.7   GLOBAL PERSPECTIVE

It can be interesting to make a calculation of reasonable figures for the global biomass production as well. In Table 1.11, we see the land area used for cereal production resp total productive area for the different income group countries. In the last column, we also see the average cereal production for the different income group countries year 2009.

We should then also notice that the amount of straw is approximately the same as the grain that is harvested. Therefore, the total biomass production that can be utilized in high income countries then is approximately 11 tonne/ha per y. In forest areas we can assume that the official production of approximately 2 tonne DS/ha per y in reality is at least 3 tonne DS/ha per y. In arable land, we can assume slightly lower productivity and approximately 1.5–2 tonne DS/ha per y can be reasonable figures to use. We also assume that the heating value is 5.4 MWh/tonne DS, which is a reasonable value. If we take the official figures of land area for each group and multiply the cereal production in each group multiplied with 2 and summarize this with the total forest area times 3 t/ha per y and arable land multiplied with 2 t/ha per y, and multiply with 5.4 MWh/tonne, we obtain:

- agriculture including straw 187,000 TWh/y;
- forests 65,000 TWh/y;
- arable 19,000 TWh/y.

In total this adds up to 270,000 TWh/y as higher heating value of all the biomass produced on land. This can be compared to the total energy use of approximately 150,000 TWh/y globally. From this, we can conclude that there is enough energy for the global population if we use the energy in a good way. This means to first use wood for e.g. buildings, and then later use the waste wood for energy purposes.

It is also important to have a sustainable farming and forestry that utilizes all resources in an effective way. Nutrient for example can be distributed in open soil during autumn or in growing crops during spring/summer. The total effect will be that in the first case we will get a lot of the nutrients coming out as leachate into waters or emissions to the air. In the second way, most of the nutrients will go into the crops instead.

1.8 QUESTIONS FOR DISCUSSION

- How should we use the available biomass resources without causing long-term problems to sustain the production?
- Why is so little of the produced biomass actual utilized in an efficient way?
- Who has the rights to the biomass? Are market forces enough to develop the biomass market?
- Is it ethical to use biomass as fuel if the same biomass also can be used as food?
- Discuss the competition for biomass between industries like pulp and paper, chemical companies and power plants. Should some interest be more important than the other?

REFERENCES

Bassam el, N.: *Energy plant species – their use and impact on environment and development.* James&James,UK, 1998.
Beneman, J.: Opportunities and challenges in algae biofuels production. 2008, www.futureenergyevents.com/algae/ (accessed July 2012).
Brito Cruz, C.H. de: Bioethanol in Brazil. 2008, http://www.biofuels.apec.org/pdfs/apec_200810_brito-cruz.pdf (accessed July 2012).
Brunnby experimental farm. Experiment with hemp production. 2006, http://www.forsoken.se/Konferens/Svea/2007/07%20-%20Hampa.pdf (accessed July 2012).
Carlsson-Kanyama, A.: Collaborative housing and environmental efficiency. The case of food preparation and consumption. *Int. J. Sustain. Develop.* 7 (2004).
Cembureau. 2010, http://www.cembureau.be/about-cement/key-facts-figures, (accessed July 2012).

Converting agricultural waste biomass into resource – the Phillipine case. UNEP-DTIE-IETC in partner-ship with Development Academy of Philippines, 2009, http://gec.jp/gec/en/Activities/FY2009/ietc/wab/wab_day3–5.pdf (accessed July 2012).

Dahlquist, E.: How to become independent of fossil fuels in Sweden. Paper at conference in Shanghai arranged by SJTU and Cornell University Dec 6–8, 2008.

Dahlquist, E., Thorin, E. & Yan, J.: Alternative pathways to a fossil-fuel free energy system in the Mälardalen region of Sweden. *International Green Energy Conference II*, 25–29 June 2006, Oshawa, Ontario, Canada, 2006.

Dahlquist, E., Thorin, E., & Yan, J.: Alternative pathways to a fossil-fuel free energy system in the Mälardalen region of Sweden. *Int. J. Energy Res.* June 2007.

Dahlquist, E., Vassileva, I., Wallin, F., Roots, P. & Yan J.: Optimization of the energy system to achieve a national balance without fossil fuels. *Proceedings Vth International Conference on Green Energy*, 1–3 June 2010, Waterloo, Ontario, Canada, 2010.

Du Xiangwan: Development of renewable energy in China: significance & strategic objectives . Chinese Academy of Engineering, Sweden, June, 2008.

Dupriez, H. & De Leener, P.: *Agriculture in African rural communities–Crops and soil*. Macmillan Publishers, 1988.

EC: Eurostat: Statistical pocketbook 2010. EU Commission, Brussels, Belgium, 2010, http://ec.europa.eu/energy/publications/statistics/statistics_en.htm (accessed July 2012).

Ernfors, M., von Arnold, K., Stendahl, J., Olsson, M. & Klemedtsson, L: Nitrous oxide emissions from drained organic forest soils—an up-scaling based on C:N ratios. *Biogeochemistry* 89 (2008), pp. 29–41. http://www.eubia.org/ European Biomass Industry Association: Source: Kaltschmitt, 2001.

FAO: Major food and agricultural commodities and producers — countries by commodity. Food and Agriculture Organization of the United Nations, Rome, Italy, 2010a, Fao.org. http://faostat.fao.org/site/567/DesktopDefault.aspx?PageID=567#ancor (accessed July 2012).

FAO: *State of the world fisheries and aquaculture 2010*. Fisheries and Aquaculture Department, Food and Agriculture Organization of the United Nations, Rome, Italy, 2010b.

Gautam, P., Gustafson, D.M. & Wicks III, Z.: Phosphorus concentration, uptake and dry matter yield of corn hybrids. *World J. Agri. Sci.* 7:4 (2011), pp. 418–424.

Heilman, P.E. & Stettler, R.F.: Genetic variation and productivity of populustrichocarpa and its hybrids. II. Biomass production in a 4-year plantation. *Can. J. Forest Res.* 15:2 (1985), pp. 384–388.

Huber, G.W. & Dale, B.E.: Grassoline at the pump. *Sci. Amer.* (July 2009), pp. 40–47.

IEA (International Energy Agency): Ttracking industrial energy efficiency and CO_2 emissions — Executive summary. Paris, France, 2007, http://www.iea.org/Textbase/npsum/tracking2007SUM.pdf (accessed July 2012).

IEA (International Energy Agency): *Key world energy statistics 2010*. Paris, France, 2010, www.iea.org (accessed July 2012).

Karlsson, R. & Carlsson-Kanyama, A.: Food losses and their prevention strategies in food service institutions. *Food Policy* 29:3 (2004), pp. 203–213.

Kukkonen, C.: Perennial grass as a dedicated energy crop. *11th International Conference on Clean Energy*, 2–5 Nov 2011, Taichung, Taiwan, 2011, www.VIASPACE.com (accessed July 2011).

Lambers, H., Chapin, F.S. & Pons, T.L.: *Plant physiological ecology*. Springer Verlag, 1998.

Larsson, M. & Granstedt, A.: Sustainable governance of the agriculture and the Baltic Sea — agricultural reforms, food production and curbed eutrophication. *J. Ecol. Econom.* 2009.

Li, S.-H.: The priorities and goals of the development of China's biomass resources. *Workshop between Chinese Academy of Engineering and Swedish Royal Academy of Engineering*, 17 June 2008, Stockholm, Sweden, 2008.

Longman, K.A.: *Preparing to plant tropical trees*. Blakton Hall, 2002.

McLeod, A. (ed): *World Livestock 2011. Livestock in food security*. ISBN 978-92-5-107013-0. Rome, Italy, 2011.

Monsanto Imagine: Corn stalk lodging. http://www.dekalb.ca/content/pdf/corn_stalk_lodging.pdf (accessed October 2008).

Moorthy, S.N.: Physicochemical and functional properties of tropical tuber starches: a review. *Starch – Stärke* 54:12 (2002), pp. 559–592.

NationMaster: Agricultural statistics. http://www.nationmaster.com/graph/agr_gra_whe_pro-agriculture-grains-XXX-production XXX is the crop type asked for, wheat, corn, rice etc. (accessed July 2012).

Nyström, J.: Insects for dinner. *Research and Progress* (March 2012), pp. 66–68.

Odlare, M., Arthurson, V., Pell, M., Nehrenheim, E., Abubaker, J. & Svensson, K.: Land application of organic waste: effects on the soil ecosystem. *Appl. Energy* 88:6 (2011), pp. 2210–2218.

Paz, A., Starfelt, F., Dahlquist , E., Thorin, E. & Yan, J.: How to achieve a fossil fuel free Malardalen region. *Conference proceedings of 3rd IGEC-2007*, 18–20 June 2007, Västerås, Sweden, 2007.

Quorn: http://www.quorn.com/ (accessed March 2012).

Salisbury, F.B & Ross, C.W.: *Plant physiology*. 4th edn, Wadsworth Publishing Co., Belmont, CA, and internal data from SLU, dept of Plant Physiology. 1992.

Schmer, M.R. , Vogel, K.P., Mitchell, R.B. & Perrin, R.K.: Net energy of cellulosic ethanol from switchgrass. *Proc Natl. Acad. Sci. USA* 105:2 (2008), pp. 464–469.

Singh, S.S.: Wheat production in India and future prospects. *8th International Wheat Conference*, 1–4 June 2010, St. Petersburg, Russia, 2010.

Snaprud, P.: Good new quorn. *Forskning och Framsteg*, March 2012, pp. 20–27 (in Swedish).

Twenneboah, C.K.: *Modern agriculture in the tropics – food crops*. Co-wood Publishers, 2000.

USDA: Oilseeds: World market and trade. United States Department of Agriculture, Foreign Agricultural Service, Circular Series FOP 1-09, 2009, http://www.fas.usda.gov/oilseeds/circular/2009/January/Oilseedsfull0109.pdf. (accessed July 2010).

Weidow, B.: The basics of crop farming (Växtodlingens grunder). LTs förlag, Stockholm, Sweden, 1998 (in Swedish).

Widen, J.: *System studies and simulations of distributed photovoltaics in Sweden*. PhD Thesis, Uppsala University, Uppsala, Sweden, 2010, http://urn.kb.se/resolve?urn=urn:nbn:se:uu:diva-132907 (accessed July 2012).

World Bank: World Development Indicators & Global Development Finance. The World Bank, Washington, DC, 28 July, 2011.

Zalesny, R.S., Hall, R.B., Zalesny, J.A., McMahon, B.G., Berguson, W.E. & Stanosz, G.R.: Biomass and genotype × environment interactions of *Populus* energy crops in the Midwestern United States. *Bioenergy Res.* 2:3 (2009), pp. 106–122.

CHAPTER 2

Chemical composition of biomass

Torbjörn A. Lestander

2.1 INTRODUCTION

Biomass is here defined as biological material derived from living (or recently living) plants that is used as a source of energy; the term also encompasses peat. It is assumed that any biomass used is harvested sustainably without endangering food security, habitats, or soil conservation.

Biomass is a complex fuel. Its primary components are cellulose, hemicellulose and lignin, which are the main constituents of the plant cell wall. These species also account for the bulk of its energy content (calorific value). The non-uniformity and varied sources of biofuels make it necessary to convert them to standardized solid, liquid, or gaseous fuels.

2.1.1 *A new biocarbon era*

The world's population has recently passed seven billion (UNFPA 2011) and atmospheric CO_2 levels continue to increase, reaching new all-time peaks, also in 2011 (NOAA 2012). Over the last decade, cumulative global anthropogenic carbon dioxide (CO_2) emissions have increased by more than 3% annually (Raupach *et al.*, 2007). More than half a trillion tonnes of CO_2 have been added to the atmosphere since the onset of industrialization (Allen *et al.*, 2009). This increases the risk of rapid changes in the Earth's atmospheric system followed by irreversible climate and ecosystem change (Solomon *et al.*, 2009).

Energy conversion plays a key role in sustainable development. As stated in the Kyoto protocol (UNFCC, 1998), there is a great need for renewable sources of energy to prevent the worst effects of climate change, which is largely driven by non-neutral carbon dioxide (CO_2) emissions to the atmosphere. Today, renewable energy sources supply only around 16% of the world's total energy demand (REN21, 2011), even if the use of firewood in the Third World is included. The development of renewable energy sources also offers a way to reduce dependence on fossil fuels. As such, there is increasing pressure to shift from fossil fuels to renewable energy sources as a means of meeting social energy demands. Achieving a shift of this kind will require the rapid development of novel technologies and methods for the large-scale exploitation of sustainable and CO_2-neutral energy sources such as bioenergy.

The amount of biomass used in energy conversion processes such as combustion, co-firing, torrefaction, pyrolysis and gasification is expected to increase rapidly. Renewables will become the world's fastest-growing source of energy, causing bioenergy to become one of the fastest growing energy sectors. For example, the amount of electricity generated from biomass is expected to triple within the first quarter of the 21st century (EIA, 2011).

We are currently at the beginning of the industrialized biocarbon era and must develop novel know-how concerning the pre-treatment, handling, and transportation of biomass from farmland to industrial facilities, in addition to the new scientific and technical knowledge that will be needed to develop large-scale biomass-based industrial processes. The development of complex and integrated processes for exploiting biomass will turn yesterday's unwanted waste into a crucial resource for sustainable development. To achieve this, it will be necessary to reduce the time spent between cropping, pretreatment and processing so as to avoid degradation of the organic matter, i.e. the development of moulds, bio-contamination, and oxidation. Conservation methods will have to be developed for species that can only be cropped at certain times of year in order to overcome microbial degradation of stored biomass and to minimize the spreading of microbes that are harmful to health.

Industrial usage of bioenergy has increased significantly in recent years and continues to rise. Sweden and Finland are both global leaders in terms of the extent of their adoption of bioenergy. Notably, bioenergy recently surpassed oil in Sweden to become the country's largest source of energy; the government has set a target of having almost 50% of the nation's energy requirements met by renewable sources, including biomass (which is primarily to be obtained from forest trees; Energimydigheten, 2011).

In addition to the increased production and usage of chopped or chipped raw biomass for energy conversion, there has been a significant increase in demand for refined solid biofuels such as pellets. Europe has around 650 pellet mills with a combined annual output of 9.8 million tonnes of fuel pellets in 2009. The size of the EU-27 market for fuel pellets is expected to increase between 10- and 30-fold by 2020 (Sikkema *et al.*, 2011). Pelletizing of torrefied materials is of special interest for co-combustion in coal-based combined heat and power plants. The predicted increases in demand for biofuel will lead to regional shortages of by-products such as sawdust and shavings that serve as feedstock for pellet production. Thus, if the expansion of the pellet industry is to be maintained, it will be necessary to identify alternative raw materials for their production.

2.1.2   *The potential of biomass for energy conversion*

The exploitation of bioenergy will play a key role in the transition away from fossil fuels, which currently account for 81% of the world's annual energy turnover (REN21, 2011). Bioenergy is the renewable energy source whose usage is increasing the most rapidly because it can replace fossil fuels for heating applications, in power production, and as fuel for vehicles (AEBIOM, 2011; BP, 2011). The world's yearly gross primary production of plants is about 150 Pg C/y (Still *et al.*, 2003) but only 0.9–1.5 Pg C/y is currently used for the production of bioenergy (Andres *et al.*, 1999; Ludwig *et al.*, 2003). Calculations performed by Tao *et al.* (2012a) using the residue-to-product ratio outlined by Koopmans (1997) and data from the FAOSTAT (2009) suggest that each year around 2.7–3.5 Pg C/y of agro-crop residues are generated in addition to the 'waste' biomass produced by forest industries (see Table 2.1). This corresponds to nearly one tonne of dry biomass per person on the planet. As such, there is considerable potential for exploiting residual biomass from the agricultural sector for energy production, along with the current usage of residues and byproducts from forestry.

It is estimated that China and the USA generate around 0.74 and 0.5 Pg (Gt, billion dry tonnes), respectively, of residual biomass suitable for energy production each year (Lal, 2005; Liao *et al.*, 2004; Xie *et al.*, 2010). However, there is some uncertainty about the total capacity for biomass production. The US Department of Energy (2011) estimates that the United States' annual production of energy crops could potentially reach 1.0–1.3 Pg (Gt, billion dry tonnes) by the year 2035.

Approaches that only consider the energy that can be obtained from biomass are shortsighted, since it can be used to produce other useful things such as bio-based chemicals and products. In addition, agro-biomass can often be used for other useful purposes, e.g. as a feed, fertilizer or feedstock for pulp production. According to Liao *et al.* (2004), nearly 60% of the biomass residuals in China could potentially be used in biofuel production. Today, about 2.5 billion people around the world use biomass as their primary fuel for cooking, especially in developing countries (IEA, 2009; Mishra and Retherford, 2007). However, there is also substantial on-going misuse of biomass, with large quantities of material being burned in open fields to remove waste in order to facilitate the growth of the next season's crop and to clear land for agricultural use, as well as in so-called back yard burning. In addition to energy losses, these practices also cause air pollution.

Table 2.1. Global production of agro-crop residues as calculated by Tao *et al.* (2012a) using FAOSTAT statistics (2009) and the residue–product ratio of Koopmans (1997).

Agrocrop	Residue	Tg*	Tg**	% of total*	% of total**	% accumulated
Maize			2020.7		30.2	30.2
	Stalks	1634.2		24.4		
	Cob	223.1		3.3		
	Husks	163.4		2.4		
Rice			1373.7		20.5	50.7
	Straw	1192.5		17.8		
	Husk	181.2		2.7		
Wheat			1193.4		17.8	68.5
	Straw	1193.4		17.8		
Sugar cane			992.7		14.8	83.4
	Tops	504.8		7.5		
	Bagasse	487.9		7.3		
Soybeans			777.9		11.6	95.0
	Straw, pods	777.9		11.6		
Groundnut			98.6		1.5	96.5
	Straw	81.7		1.2		
	Husks	16.9		0.3		
Oil palm			90.2		1.3	97.8
	Bunches	47.7		0.7		
	Fibre	29.0		0.4		
	Shell	13.5		0.2		
Millet			56.1		0.8	98.6
	Stalks	56.1		0.8		
Coconut			32.3		0.5	99.1
	Husks	25.1		0.4		
	Shells	7.2		0.1		
Coffee			17.3		0.3	99.4
	Husks	17.3		0.3		
Cassava			14.9		0.2	99.6
	Stalks	14.9		0.2		
Tobacco			13.8		0.2	99.8
	Stalks, etc.	13.8		0.2		
Jute			8.5		0.1	99.9
	Stalks	8.5		0.1		
Cocoa			4.2		0.1	100.0
	Pods	4.2		0.1		
Sum		6,694.3	6,694.3	100.0	100.0	

*million tonnes; 1 Pg = 1000 Tg = 1000 million tonnes. **accumulated.

Including forestry residuals and manure, Woolf *et al.* (2010) calculated that around 5 Pg of biomass is produced around the world annually that is suitable for energy conversion and could be exploited without endangering food security, habitats, or soil conservation. Table 2.2 presents a breakdown of this available biomass in terms of the feedstocks from which it derives.

There are some limitations that restrict the scope for growing biomass crops as feedstocks for biofuel production. Some of these stem from the need for specific ecosystem services to maintain crop growth (e.g. the availability of fresh water). Others relate to the need to conserve existing habitats (i.e. biological conservation). Finally, most existing ecosystems have some quantity of stored carbon that may be released into the atmosphere as CO_2 or methane (CH_4) in response to land use changes, which would result in increased greenhouse gas emissions. Consequently,

Table 2.2.   Global availability of biomass suitable for energy conversion according to Woolf *et al.* (2010).

Biomass	Maximum sustainable technical potential* [Pg/y]
Agroforestry	1.28
Biomass crops	1.25
Rice	0.67
Manure	0.59
Cereals excluding rice	0.42
Forestry residues	0.29
Sugar cane	0.27
Harvested wood	0.21
Green waste	0.07
Total	5.10

*1 Pg = 1000 Tg = 1000 million tonnes.

multiple utilization of existing agro- and forest-based crops is likely to be the preferred source of biomass for energy conversion in most cases.

2.2   MAJOR COMPONENTS OF BIOMASS

The major components of biomass are shown in Figure 2.1. The composition of biomass is more complex than other energy sources such as oil or coal. For example, the water content of natural biomass can be more than 50% on a wet basis, and both its organic and inorganic components are highly variable. Biomass from different plant species (and from different structural elements of individual species) can have very different relative contents of cellulose, hemicellulose and lignin, as well as extractives (fats, oils etc.), sugars, starch, and proteins. This results in differences in the relative abundance of carbon (C), hydrogen (H), oxygen (O), nitrogen (N) and sulfur (S), which affect the resulting fuel's energy value. Volatile matter accounts for a significant fraction of all biomass. The amount of volatile matter in a sample is determined using a standardized method involving measuring the amount of organic material lost when the dry sample is briefly subjected to a high temperature (typically 900°C for seven minutes). The organic material that remains following such treatment is referred to as fixed carbon.

The biochemical composition of biomass is expressed in terms of the percentage of the weight of the dry material that is attributable to each constituent. However, the amount of water in a sample is often expressed in terms of the percentage of the mass of the wet material that is due to water.

The inorganic content of biomass is usually defined in terms of the mass of ash remaining after its combustion. This is typically lowest for pure wood (accounting for 0.3–0.5% of the original sample mass) but may exceed 10% for bark or other structural tissues. At a more detailed level, ash is defined by its elemental composition. One problem when determining the ash content of biomass is that biomass often is contaminated, e.g. by soil particles.

There are two major groups of analytical procedures used when studying biomass: proximate and ultimate analyses. Proximate analyses are used to determine the sample's moisture content, the weight percentage of volatile matter and fixed carbon in dry biomass (wt% d.b.), the heating value (gross calorific value), and the ash content. Ultimate analyses are those that determine the wt% of carbon (C), hydrogen (H), oxygen (O), nitrogen (N) and sulfur (S) in dry biomass, as well as those of chlorine (Cl), fluorine (F) and/or bromine (Br) in some cases.

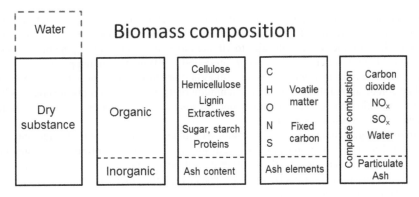

Figure 2.1. The different constituents of biomass.

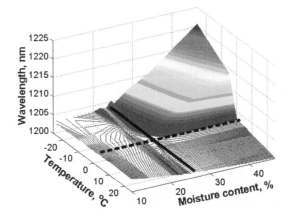

Figure 2.2. Peak wavelength shifts around 1200 nm at different moisture contents (10–45% weight at wet basis) and temperatures between –20 and 20°C (from Lestander *et al.*, 2008). The solid line is the fiber saturation point and the dashed line is the freezing point of free water in the biomass.

During complete combustion, all of the elemental constituents of biomass are oxidized. This generates gases, namely carbon dioxide, nitrogen (NO_x) and sulfur (SO_x) oxides, and water (vapor), as well as solids in the form of fly ash, bottom ash, and particles of various sizes.

2.2.1 *Water in biomass*

While it may seem strange to treat water as a component of biomass, moisture content is a major issue in most processes where biomass is used as a source of energy. The 'separation' of water and dry substances in natural moist biomass, i.e. drying, is an energy-consuming process.

The nature of the moisture in biomass at different moisture contents and temperatures is illustrated in Figure 2.2, which shows the overtones of water molecule vibrations at one of three peaks in the near-infrared region; the slower the molecular vibrations, the longer the wavelength (in the z-axis). The peak in the plot at high moisture contents for temperatures below 0°C shows that the overtones are shifted towards longer wavelengths under these conditions. This is due to the formation of ice crystals within the biomass, which only occurs if its moisture content is above the fiber saturation point. Below this point, water molecules are attached to binding sites on the internal surfaces of the biomass and so the water molecules in this state do not undergo these drastic changes with temperature. Notably, the 'floor' is somewhat tilted. Ongoing from higher to lower temperatures the 'floor' tilts towards longer wavelengths at lower temperatures. As the moisture content increases, the binding energy for the interaction between water and biomass decreases, allowing the water molecules to vibrate more freely and shifting the peak to shorter wavelengths. In pure water, this peak occurs at around 1190 nm at room temperature.

The presence of water in biomass is also important because it influences its susceptibility to microbial colonization followed by the consumption of its nutrients. At moisture contents below the fiber saturation point, there is little scope for microbial degradation, which is completely inhibited at lower moisture contents.

Biomass is hygroscopic, and dried biomass will therefore equilibrate to the humidity of the ambient air. This is an exothermic process. The adsorption of water vapor from the surrounding air thus releases heat that can potentially set off other self-heating processes such as auto-oxidation, leading to the emission of toxic and hazardous substances. If the process continues and the material heats up even further, it may catch fire.

2.2.2   *Dry matter content*

The dry matter content of biomass is usually defined using oven-based methods involving drying at 105°C. At this temperature, volatile organic compounds (VOCs) may vaporize, causing the real dry matter content to be underestimated. In such cases, lower drying temperatures can be used. The dry substance consists of both organic matter and inorganic substances. If the biomass is contaminated by inorganic dust or soil particles, these will be included in the measured dry matter content.

The relative abundance of different organic and inorganic compounds in biomass varies between plant genera, species, age, site of harvesting (including genotype × environment interactions), management (e.g. liming and fertilization regimes), season, structural origin (tissue type), cell type, and location in cell wall. This makes it difficult to provide a full and detailed review of the composition of biomass. The brief discussion in this chapter provides only a rough overview of the organic and inorganic contents of dry plant biomass. The cells of most plant species have both a primary and secondary cell wall. While the cell is growing, it has only a primary wall, but when growth stops, the secondary wall is formed inside the boundaries of the first, i.e. between the primary cell wall and plasma membrane. The primary wall consists of densely packed cellulose chains that are bound to one another by hydrogen bonds and arranged in micro fibrils, which are arranged in a network to increase the strength of the cell wall. Conceptually these micro fibrils are like a cross-linked network of girders, with the crosslinking units being embedded hemicelluloses and/or pectin, as well as proteins in some cases. The cell walls are laid down in stages. First, the primary wall and the middle lamella that occupies the space between cells are laid down. These are rich in pectic material. When the primary cell wall has been established, a secondary wall is formed inside the cell. This secondary cell wall consists mainly of cellulose, hemicellulose and lignin; it includes no proteins but pectin may also be absent. The cellulose micro fibrils of the secondary wall are cross-linked by hemicellulose. The secondary cell wall strengthens the cell against dynamic forces and static forces such as gravity, and helps to hold the plant up. It is therefore thicker and less pliant than the primary wall. It often has three distinct layers that are known as s1, s2 and s3. Typically, s2 is the thickest and accounts for about 90% of the fibrous mass of plants (Batchelor *et al.*, 2000). In subsequent phases of cellular development, lignin is deposited as a filler material around the primary wall, in the cell corners and the intercellular space, and also in the secondary wall.

2.3   ORGANIC MATTER

The main organic components of biomass are cellulose (the most abundant organic polymer on earth), hemicelluloses, lignin, pectin, extractives, starches, sugars, and proteins. Proteins

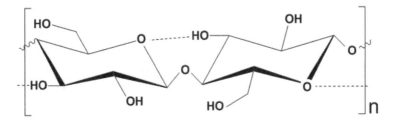

Figure 2.3.   Structure of cellulose.

contain N, which generates NO_x when biomass is combusted. In addition, the S in biomass can cause problems during gasification and may contaminate catalysts.

Extractives have the highest energy values of all the organic components. The high oxygen content of cellulose reduces its energy content per unit mass.

2.3.1   *Cellulose*

Cellulose is a polymer that accounts for a large proportion of plant biomass. It is the most abundant natural polymer; having crystalline and amorphous regions. Pure cellulose consists of linear chains of 7000 to 15,000 linked glucose ($C_6H_{10}O_5$) monomers. Its molar H/C and O/C ratios are 1.67 and 0.83, respectively. The bond between the glucose molecules is formed between the first carbon atom of one ring and the oxygen atom bound to the fourth carbon atom of the next, as shown in Figure 2.3.

These bonds are stabilized by hydrogen bonding between a hydroxyl group on one ring and the oxygen atom within the adjacent ring. This hydrogen bonding causes cellulose to form micro fibrils with a flat ribbon-like conformation. These micro fibrils are relatively stable and have high axial stiffness. Networks of stacked cellulose fibrils in the S2 cell wall structure provide the bulk of the plants' structural reinforcement. To withstand bending forces, cellulose fibrils have regions where the cellulose chains are disordered (amorphous) while others are highly ordered and crystalline.

Oxygen accounts for nearly half the total mass of cellulose, which therefore has a relatively low energy content; its calculated gross calorific value is 17.68 kJ/g (Gaur and Reed, 1998) and according to Walawender *et al.* (1985) (α)-cellulose has a gross calorific value of 17.492 kJ/g on a dry ash-free basis.

2.3.2   *Hemicellulose*

In contrast to cellulose, which consists exclusively of glucose and forms a long un-branched polymer, hemicellulose contains many different sugar monomers. That is to say, it is a mixed polymer including rings with five carbon atoms (C5; pentose) and also those with six (C6; hexoses). Hemicellulose has short side chains and some varieties are heavily branched, such as arabinogalactan, which is frequently found in larch wood. Individual hemicellulose molecules typically consist of 80–200 hexose and pentose monomers. While individual cellulose molecules stack to form crystalline micro fibrils and networks, hemicelluloses are characterized by their non-bonding steric repulsion, which disfavors their close packing. As a result, hemicellulose is amorphous in its native form and is therefore more reactive than cellulose. In addition, the more extensive the branching in the hemicellulose chain, the more soluble it is. The solubility of hemicellulose is further enhanced by the availability of its hydroxyl (O–H) groups for hydrogen bonding with water.

The term hemicellulose encompasses a very wide range of different molecules. The individual hemicelluloses in a biomass sample along with their structures, composition, and relative abundance vary depending on the genus, species, structural origin (tissue type, cell type) and location in the cell wall. In the secondary cell wall, hemicellulose acts as a bridge between micro fibrils,

p-coumaryl alcohol   coniferyl alcohol   sinapyl alcohol

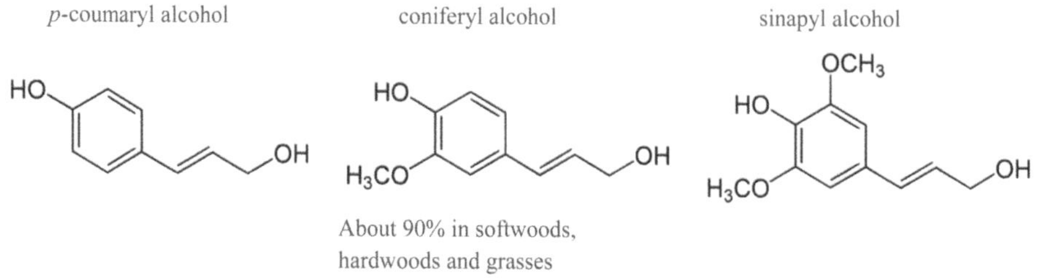

About 90% in softwoods,
hardwoods and grasses

Figure 2.4.   Basic building blocks of lignin.

keeping them separate and also preventing them from sliding past one another and causing the fibril networks to collapse.

It has been proposed that hemicellulose absorbs energy when the plant is subjected to different dynamic forces (e.g. wind) and static forces (e.g. snow, ice) since it serves as a link between cellulose and lignin and can thus be used to redistribute stresses. Because some hemicellulose remains associated with cellulose even after prolonged cooking of chemical pulps, it has been suggested that at least one end of the branched hemicellulose chain may be adsorbed onto the cellulose micro fibrils. This is observed for approximately one third of the short-sided hemicelluloses. In softwoods and hardwoods, these tend to be glucomannans and xylans, respectively, for which both ends of the branched chain can bind to micro fibrils, forming snakelike patterns of relatively inflexible chains (Walker, 2006).

2.3.3   Lignin

Lignin is an unusual aromatic biopolymer because it has a random, complex and cross-linked network structure, i.e. its structure is not well defined. It is also difficult to study in its unmodified form as it readily undergoes oxidation and condensation reactions. Cell wall lignins are always associated with hemicelluloses and they form covalent bonds with each other.

Lignin is the second most abundant natural polymer in plant biomass: it accounts for 17–24, 18–25 and 27–33% of the total mass of grasses, hardwoods and softwoods, respectively. It is a natural amorphous polymer that effectively functions as "glue" maintaining the plant's structural integrity. The basic building blocks of in lignin are three monolignols: p-coumaryl alcohol, coniferyl alcohol, and sinapyl alcohol (Fig. 2.4). They form, respectively, p-hydroxyphenyl, guaiacyl and syringyl phenylpropanoid units when incorporated into the lignin polymer (Boerjan et al., 2003). Lignins exist as a complex branched and cross-linked network of phenylpropenyl subunits.

2.3.4   Extractives

The extractives are a wide range of compounds that can be extracted using polar (e.g. water) or non-polar (e.g. ether) solvents. Plant extractives include waxes, oils, fats, lipids, resins etc. and typically have high-energy values; fats and oils can yield as much as 38 kJ/g. For example, palm fruits contain approximately 30% oil by mass and resin acids account for up to 45% of the dry weight of the knot wood from certain softwood trees (Boutelje, 1966).

2.3.5   Sugars

When discussing biomass, "sugars" is typically taken to mean sucrose, which consists of glucose and fructose. Sugar cane is the most well-known example of a plant containing high concentrations of sugar. A mature stalk of sugar cane typically contains 12–16% (wt%, dw) soluble sugars.

Another well-known example is sugar beet, whose sugar content ranges from 12 to 17% of its dry weight (Zhou *et al.*, 2011).

2.3.6 *Starch*

Starch is often found in all the living tissues of a plant. High amounts of starch are found in many fruit tissues e.g. in grains of maize (corn), wheat and rice and also in underground tubers such as potato and cassava. Starch generally contains 20 to 25% amylose and 75 to 80% amylopectin (Brown and Poon, 2004).

2.3.7 *Proteins*

In contrast to other major compounds present in biomass, proteins contain high amounts of nitrogen (N); their average ratio of N to C is 0.25 (Perrett, 2007). Some plant species (especially legumes) are known for their high protein contents. In such cases the highest protein contents are generally found in fruit tissues, as with beans and lentils. However, all living plant cells contain proteins that play crucial roles in various life-supporting processes. In dead cells, proteins are found in the cell walls and the remnants of the protoplast. Wood has extremely low nitrogen to carbon (N/C) molar ratios of 0.003 or below.

2.4 INORGANIC SUBSTANCES

The reason for studying ash elements in biomass is that they can cause significant operational problems during energy conversion (for instance by combustion), including deposition, corrosion, and particle emissions.

The term biomass encompasses a wide range of very heterogeneous substances with widely varying properties. This is also true of the content and abundance of ash-forming matter, which both vary depending on the species from which the biomass was obtained and the specific tissue harvested. In general, woody biomass contains relatively little inorganic matter. For instance, ash-forming materials (i.e. non-combustible inorganic matter) account for only 0.3–0.4% of the dry mass of stem wood from pines, spruces, birches and aspen. Other parts of the tree, such as the bark, branches, twigs, shoots and leaves/needles, contain higher levels of inorganic matter, rising to approximately 7 wt% for aspen shoots.

Plants take up macronutrients, micronutrients and trace elements from the soil. The inorganic content of biomass comprises many metals as well as some non-metals. Specifically, plants tend to contain at least some amount of Si, Ca, Mg, K, Na, P, S, Cl, Al, Fe, Mn, N Cu, Zn, Co, Mo, As, Ni, Cr, Pb, Cd, V, and Hg. Most of these elements act as nutrients and have essential biological functions in the living plant.

Some plant species and varieties may hyper-accumulate metals. At present, there are about ten species that are known to hyper-accumulate Mn, most of which are woody trees and shrubs. Plants that accumulate above-normal metal concentrations in their leaves do so mainly by primary sequestration in the non-photosynthetic part of the leaves. This is the case for *Salix*, which is an important crop for biofuel production in some countries. It has been shown that the pectin-rich layers of the collenchyma cell walls of leaf veins are an important Cd sink in Cd-tolerant *Salix viminalis*, and it has been suggested that some varieties of this species could be used as soil photoremediators. Reed canary grass is a non-food biomass crop that accumulates Si when grown on soils containing clays but has a low Si content when grown on organogenic soils (Burvall, 1997).

Another important problem is the *in situ* contamination of biomass crops with inorganic materials during growth, harvesting, handling, or processing. This problem is particularly severe when using low-value and residual biomass components such as straw, branches, tops, stumps, or lingocellulosic waste in bio-based industrial processes. The contaminants are usually a complex mixture of substances that include geogenic materials (soil particles, etc.) and atmospheric pollutants.

The elemental composition of ash from biofuels tends to differ significantly from that of coal ash (Vassilev *et al.*, 2010) and to be much more variable. It therefore exhibits different and more complex chemical behavior at high temperatures during combustion and gasification. Together with the considerable potential for contamination, this will necessitate the development of systems for classifying and characterizing ashes in biomass feedstocks.

In a review of the elemental composition of biofuel ash, Vasssilev *et al.* (2010) defined seven different biomass groups using a ternary diagram with the following elements (clusters) in each corner of the triangular panel: (i) Si + Al + Fe + Na + Ti; (ii) Ca + Mg + Mn; and finally, (iii) K + P + S + Cl. Biomass samples with high levels of elements from the first cluster are described as "high acid" (S-type) samples, while those with high levels of elements from the second two clusters are regarded as "low acid." Within the "low acid" group, those with high levels of Ca + Mg + Mn are defined as C-type biomasses, while those with high levels of K + P + S + Cl are referred to as K-type biomasses. There is also a group of CK-type biomasses whose ash elemental composition is intermediate between that of the C and K types. The seven groups are defined by abundance boundaries set at 30% for either of the low acid oxide clusters, and at 40% and 60% for the high acid cluster. This scheme defines three of the seven groups as having intermediate acidity. The authors also stated that there are strong associations between the abundances of the individual elements of the N-S-Cl, Si-Al-Fe-Na-Ti, Ca-Mg-Mn and K-P-S-Cl groups.

Such classifications are very important in fully understanding the complex mixtures of substance in biofuel ashes. However, the number of elements involved and their various interactions during ash formation make it difficult to provide a detailed overview of the processes involved. This problem is often addressed by using ternary diagrams, but simple combinatorics state that if each ternary diagram features 3 ash elements ($k = 3$) and there are (for example) a total of 10 elements to consider ($n = 10$) then a full overview would require the use of 120 ternary diagrams. The expression for the number of possible permutations when making k choices of n possible objects is:

$$\binom{n}{k} = \frac{n!}{k!(n-k)!}$$

An illustrative four-dimensional ternary diagram is shown in Figure 2.5.

It would be necessary to use a very large number of diagrams to cover all of the inorganic elements in ash. In order to reduce the number that have to be considered, Tao *et al.* (2012b) used principal component analysis to gain an overview of the ash elements in biomass and the relative abundance of SiO_2, K_2O, CaO, MgO, Na_2O, P_2O_5, Al_2O_3, Fe_2O_3 and SO_2 in biofuel ash. Principal component analysis was used to identify a small set of orthogonal components that account for the greatest possible amount of the variation in the observed data and could thus be used to briefly describe the properties of a given ash. The studied dataset comprised 367 different ash composition analyses, and a number of interesting conclusions were drawn. The first was that the abundance of Si increases and that of Ca decreases gradually in the following order: woody groups, herbaceous dicots, C4 graminoids and C3 graminoids. C4 and C3 plants differ in the way they handle the first steps of photosynthesis; C4 plants are mostly found in tropical and subtropical regions. Examples of C4 graminoids are sugarcane, maize, millet, sweet sorghum, switchgrass and miscanthus, while wheat, oat, barley, rye, rice, alfalfa, sunflower and reed canary grass are C3 plants as are most woody species.

Second, it was found that levels of K, P, Mg and S were higher in herbaceous plants than in graminoids and dicots (grasses). Woody species were generally associated with high levels of Ca-Mg-P-K-S and herbaceous ones with Si; of all the elements involved, Ca and Si made the

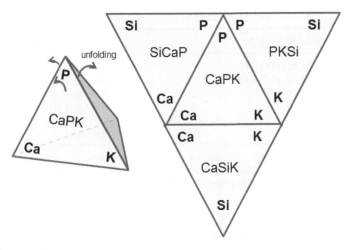

Figure 2.5.   Ternary diagram for the elements Ca, P, K and Si in biomass, depicted as the unfolding of a three-dimensional tetrahedron into two dimensions.

greatest contributions to this trend. When all of the woody species were treated as a single group, the variation in their ash composition was greater than that for the herbaceous group.

According to Vassilev *et al.* (2010), the most abundant ash elements (in decreasing order) in biomass are Ca, K, Si, Mg, Al, S, Fe, P, Cl, Na, Mn and Ti. In the earth's crust, their abundance decreases in the following order: Si>Al>Na>K>Ca>Fe>Mg>Ti>P>Mn>S>Cl. All of these species are rock-forming elements. Consequently, plants are enriched in Ca, K, Mg, S, and P and have comparatively low levels of Si, Al, Fe and Ti relative to their abundance in the earth's crust. It is possible that these 12 most abundant ash elements in biomass also account for 95% or more of the total amount of mineral oxides that are formed during combustion and end up in the ashes.

Boström *et al.* (2012) have constructed a simplified model to describe the ash-forming reactions by first considering the relative thermodynamic stabilities of the oxides of the different elements involved in ash formation and then considering how they might react in the secondary reactions. First, they reduced the number of significant ash elements to eight: Ca, P, Mg, S, K, Si, Na and Cl, omitting Fe, Mn, Al and Ti. Fe and Mn were excluded because they often form individual oxides that do not interact significantly with other ash elements during combustion. Al and Ti were excluded because they are not essential metals for most plants, but certain soil conditions and/or contamination can make them relatively abundant in ash; this is especially true for Al.

The oxides formed after the initial stages of combustion and the main oxides originating from the organic components of biomass (i.e. H_2O and CO_2) are divided into two categories; basic oxides and acid oxides; see Table 2.3 (after Boström *et al.*, 2012). The thermodynamic stability of all initial ash element oxides decreases with increasing temperature between 200 and 1600°C. However, the order of stability remains unchanged throughout this range, with a few exceptions. The most notable exception is that the oxides of K and Na exhibit intermediate thermodynamic stability at the lower end of this temperature range but are the least stable oxides of those mentioned in Table 2.3 at the higher end of the range.

According to the model of Boström *et al.* (2012), Ca interacts with P, and two Ca phosphates that are commonly found in woody biomass ashes were identified: apatite ($Ca_5(PO_4)_3OH$) and whitlockite ($Ca_3(PO_4)_2$). If there is still an surplus of calcium oxide, the next most acidic oxide, $SO_2(g)$ and/or $SO_3(g)$ will react to form Ca-sulfates. On the other hand, if there are P-based initial oxides left over after all of the K, Na and Ca oxides have been consumed, Mg-phosphate oxides will be formed. The same is true for the other basic and acid oxides. While this is a simplified model, it gives a good overview of how complex ash mixtures are formed during biomass combustion.

Table 2.3.   Initial basic and acid oxides in primary ash formation.

Initial basic oxides	Initial acid oxides
KOH (l, g) (K_2O)	P_2O_5 (g)
NaOH (l, g) (Na_2O)	SO_2 (g)/SO_3 (g)
CaO (s)	SiO_2 (s)
MgO (s)	HCl (g) (Cl_2)
H_2O (g)	CO_2 (g)
	H_2O (g)

The states within parenthesis indicate whether the primary oxide is refractory or volatile. For example, K_2O is a solid but is assumed to combine with water to form volatile KOH (l, g).

The reaction zones will also influence the outcome of the secondary reactions in ash formation. If for example the feedstock consists of two different biomasses then there may be three sets of ash reactions proceeding at once (one for each of the two biomass types and one arising where the two interact) if they are not sufficiently well-separated to prevent interaction. For the gaseous species in Table 2.3, such separation is probably impossible and so gaseous species can be assumed to always interact with every biomass type present. The model is further complicated by the differing availability of various species for participation in secondary reactions. For example, CaO (s) is a refractory initial oxide that may form nano- or sub-micron particles whereas KOH (g) is gaseous and is therefore much more available for reaction with other initial oxides. Volatile compounds may therefore have higher effective concentrations than refractory ones even if their true abundance is equal or lower. Furthermore, the volatile gases and particles of the refractory initial oxides may be fractionated by the movement of the flue gases, depending on their speed. This fractionation could prevent some expected chemical reactions from taking place. The kinetics of the reactions involved will of course also influence the outcome of the ash transformation processes. These classifications and pathways of chemical reactions will be important in predicting operational problems.

For more complex and realistic situations, Boström *et al.* (2012) write: "To transfer these general concepts to a realistic situation, the physical characteristics of the specific energy conversion facility, such as process temperature, residence time, air supply, and flue gas velocities, have to be taken in account. Thus, the practical consequences of the ash transformation reactions for a certain fuel may be quite different depending upon if the fuel is used in a fluidized bed, on a grate, or within a powder burner."

2.5   ENERGY CONTENT

The most important characteristic of a biofuel is its energy content. The energy value or heating value of biomass is the amount of heat released per unit mass during complete combustion. The energy released during combustion is that tied up in the covalent bonds for C, H, O, N and S atoms and their valence numbers in the biomass. The calorific value of different covalent bonds in gases increases in the following order: O–H < C–O < C=O < N–H < C–N < C–C < C–H < S–H < C≡C (Eberson, 1969).

Heating values for biomass-based fuels are measured in units of energy per unit mass of fuel, i.e. kJ/g (or alternatively, MJ/kg). The energy content of a dry substance is expressed in terms of the gross calorific value (*GCV*), which is also known as the higher heating value, the gross energy, the upper heating value, or the higher calorific value. By definition, the *GCV* is the quotient of the thermal energy of complete combustion and the mass of solid fuel, whereby the water formed during combustion is liquid and the temperature of the fuel (before combustion) and the combustion products are of the same specified value. This thermal energy is commonly determined by subjecting dry biomass samples to complete combustion to CO_2, NO_2, SO_2 and H_2O in a bomb calorimeter under an excess of oxygen with a starting temperature of 25°C. Thus, the heat released between the initial and final temperatures (which are both usually 25°C) is recorded as the gross calorific value. The heat of vaporization of any water vapor produced during combustion is therefore included in the *GCV*.

The gross calorific value (*GCV*) takes into account the latent heat of vaporization of water in the combustion products, and is useful in calculating heating values for fuels where condensation of the reaction products is practical (e.g., in CHP plants connected to a district heating system). In other words, *GCV* assumes all that all of the water in the fuel will be in the liquid state at the end of combustion (and will be mixed with the other combustion products).The net calorific value (*NCV*) does not include this heat of condensation of water vapor and is measured using a heating cycle whose final temperature is above the boiling point of water, e.g. at 150°C for smoke gases. It is also known as the net energy, the lower heating value, or the lower calorific value. Thus, the *NCV* is the energy obtainable if the water produced during combustion remains in the form of vapor, while the *GCV* is the energy obtainable if the water is in the liquid state. The relationship between *NCV* and *GCV* is outlined in a CEN standard (EN 14918:2009):

$$NCV = (GCV - 21.22 \times H - 0.08 \times (O + N)) \times (1 - m) - 2.443 \times m$$

Here, the *NCV* and *GCV* are measured in kJ/g. H, O and N are the mass concentrations on a dry basis (db) of hydrogen, oxygen and nitrogen in the dry biomass, and m is the mass concentration of water (on a wet basis, w.b.) in the wet biomass.

Two of the major contributors to the variation in the energy value of different dry biomass samples are their moisture and ash contents. There is little variation in the energy contents of ash- and water-free biomass samples, so when the H content is known, the *NCV* can be estimated according to Stockinger and Obernberger (1998) as follows:

$$NCV = GCV \times (1 - m) - 2.447 \times m - 0.5 \times H \times 18.02 \times 2.447 \times (1 - m)$$

where *GCV* is the gross calorific value (kJ/g, dry basis), m is the mass concentration of water in the wet biomass, and H is the mass concentration of hydrogen in the dry biomass.

The net calorific value (*NCV*) on a wet basis (wb) and a dry basis (db) can also be calculated using the equations of van Loo and Koppejan (2008):

$$NCV_{wb} = GCV \times (1 - m) - 2.444 \times m - 2.444 \times 8.936 \times H \times (1 - m)$$

$$NCV_{db} = GCV - 2.444 \times m \times (1 - m)^{-1} - 2.444 \times 8.936 \times H$$

where the constant 2.444 is the enthalpy difference between gaseous and liquid water at 25°C in KJ/g (or MJ/kg) and the constant 8.936 is the ratio of the molar masses of water (H_2O) and hydrogen (H_2).

The variation in the calorific value of fresh biomass is highly dependent on the composition of the material as well as its moisture and ash content. It is shown that these parameters can be predicted accurately by using multivariate calibration modeling based on near-infrared spectra from biomass samples (Lestander and Rhén, 2005). The same reference gives the information that the energy contents of cellulose, lignin and extractives range from 17.5–19.5, 23.3–26.8 and 32.3–39.4 kJ g in wood, respectively, assuming that the dry matter contains 6% H.

If the mass concentrations of C, H, O, N, S and ash are known, the higher energy content of biomass can be estimated using the expression developed by Gaur and Reed (1995):

$$P_{GCV} = 0.3491 \times C + 1.1783 \times H - 0.1034 \times O - 0.0151 \times N + 0.1005 \times S - 0.0211 \times ash$$

Here, P_{GCV} is the calculated gross calorific value in kJ/g and C, H, O, N, S and ash are the mass concentrations of the corresponding substances in the dry biomass. This calculated heating

Table 2.4.   Predicted gross calorific values (P_{GCV}) for biomass from different plant groups.

Material	Number of observations	Predicted GCV* P_{GCV} [kJ/g]	Standard deviation [kJ/g]
WN bark	18	21.04	1.42
WN wood	56	20.10	1.50
WB bark	9	19.68	1.82
WB fruit-residue	104	19.68	2.06
Bamboo	13	19.57	0.41
WN log-residuals	6	19.56	0.64
HD fruit-residue	27	19.23	1.71
HG bagasse	28	18.73	1.46
HG stems	3	18.61	0.37
HG cob	14	18.14	1.08
HD stems	66	17.88	1.87
HG straw	183	17.82	1.57
HG leaves	7	17.46	1.55
HG husk	37	16.06	1.33

*Note that these predicted energy values based on C, H, O, N, S and ash levels may be biased and overestimated by 1.8% on average; W: woody; N: needle trees; B: broad-leaf; H: herbaceous; G: grass; D: dicot.

value may be biased and it is claimed that this expression overestimates the *GCV* by 1.8% on average (Obernberger and Thek, 2010).

For woody biomass (including bark), the gross calorific value is generally around 20.0 kJ/g (dry basis), with a slightly lower value for herbaceous biomass (about 18.8 kJ/g). Tao *et al.* (2012a) studied literature data on analyzed biomass samples and found that the first two principal components explained about 70% of the variation in the observed C, H, O, N, S, and ash concentrations. The first component was spanned by the C and H group and ash; the second was spanned by O on one side and the N and S group on the other. Groups on opposite sides of a component are negatively correlated with one-another. However, the correlations between the concentrations of elements within individual groups, such as the C and H group or the N and S group were strong and positive.

A dataset consisting of 775 observations for various biomasses was collected by Tao *et al.* (2012a). Of these observations, 571 provided enough information to predict the corresponding gross calorific value according to Gaur and Reed (1995); see Table 2.4. The average predicted gross calorific value (P_{GCV}) for this dataset was 18.78 kJ/g, with a standard deviation of 1.89 kJ/g.

2.6   CHEMICAL COMPOUNDS AND BIOMASS PROCESSING

2.6.1   *Drying*

One drawback of many processes involving biomass relates to its water content. The water content of freshly harvested green biomass is often as high as 50–60% water (wet basis). Drying is an energy consuming process but can be done at low temperature, meaning that surplus low-value and low-temperature (<95°C) heat can be used. At industrial facilities and combined heat and power (CHP) plants, heat generated by the condensation of vapor produced during drying at high temperatures (>100°C) is often reused to increase the energy efficiency of the drying processes.

Patzek and Pimentel (2005) illustrated the importance of moisture content in biomass. These authors discussed a case in which raw biomass with a natural moisture content of around 55% was upgraded by drying it to yield a fuel with a moisture content of 10% by weight. Because the heat of condensation was not recycled during the drying process, the net gain in available energy was low.

In the interval of 20–60% moisture content (wet basis) the weight of raw biomass needed, to produce heat in order to dry 1 kg of upgraded fuel to 10% moisture content, is more than doubled for each 10% going from the bottom of the interval to the top. The higher moisture content the lower net gain. At high moisture contents the concept without recovery of condensation heat starts to break down completely.

On the other hand, Wahlund *et al.* (2002) reported that biomass drying could be achieved with low energy input at a bio-based CHP facility. In this case, the biomass was dried using pressurized steam and a heat exchanger was used to transfer heat from the 'dirty' steam generated by moisture in the wood to 'clean' steam that could be used to generate power. In addition, the heat released during the condensation of the 'dirty' steam was recycled and reused in the drying process. Additional heat exchange below the vapor condensation point could further reduce the specific energy of drying and increase the net output of bioenergy in this process. Thus, the most effective drying process accounts only for heat exchange losses in the system.

These examples illustrate that controlling the moisture content of biomass is crucial for its industrial use and that the heat needed for drying should be reused in as many ways as possible in order to continuously increase the net output of upgraded fuel. This requires the development of integrated processes that allow for the efficient use of surplus heat; many current industrial and energy conversion processes produce a lot of surplus low-value, low-temperature heat that could be exploited for drying biomass.

There are of course other ways to reduce or overcome problems arising from the moisture content of biomass. One is delayed harvesting, which has been demonstrated for the rhizome grass, reed canary grass (*Phalaris arundinacea* L.), and for industrial hemp, by cropping in spring instead of autumn (Xiong *et al.*, 2009). During late fall and winter, the above-ground biomass of these grasses dies, and over the course of the early spring, it dries out, at which point it is harvested. A notable advantage of such delayed harvesting is the low ash content of the biomass obtained and the favorable composition of the ash that is formed. Other systems used in forestry involve storing covered piles of logging residues in such a way that they self-dry over time in the air. A drawback of these outdoor storage systems is that some biomass is lost on standing due to biological and other degradation. In addition, infestations of mould and other bio-contaminants are hard to avoid and may be harmful downstream in the value chain.

2.6.2 *Wet processing*

Novel systems have been developed to avoid having to dry biomass completely. These typically focus on carrying wet biomass through all the steps in energy generation. Thus, the biomass is stored after pretreatment at low temperature with bio-preservatives such as polysaccharide-degrading enzymes (which can be added by treatment with yeasts) that maintain its condition and facilitate subsequent enzyme treatments (Passoth *et al.*, 2009). It can then be subjected to fermentation e.g. ethanol and biogas production in sequence (Dererie *et al.*, 2011). It is also possible use hydrothermal upgrading processes e.g. hydrothermal carbonization of lignocellulosic biomass (Hoekman *et al.*, 2011). Gasification of wet biomass feedstocks using supercritical water oxidation has been studied and may become more practical following the discovery of new catalysts that improve reaction efficiency and product yield (Azadi and Farnood, 2011; Robbins *et al.*, 2012).

2.6.3 *Health aspects*

Biomass in and of itself can be hazardous and cause health problems. Notably, it releases volatile species such as terpenes, aldehydes, and ketones that are collectively known as volatile organic compounds (VOCs). Banerjee (2001) has shown that during softwood drying, there are three mechanisms involved in the release of α-pinene and other terpenes. Moreover, dissolved terpenes accounted for approximately 0.1% of the total quantity of vapor released during softwood drying. Wajas *et al.* (2007) used headspace solid-phase micro extraction to analyze VOC emissions from biomass. They found that vapors from different species had unique chemical compositions and that 64–98% of the eluted compounds were monoterpenes. Arshadi *et al.* (2009) have studied the emission of volatile aldehydes and ketones from wood pellets. Long-term exposure to high levels of VOCs may cause health problems. Thus, gaseous substances such as VOCs from biomass should also be monitored regularly, especially in industries that handle biomass. In addition, good ventilation should be maintained.

It is possible that stored biomass may undergo self-heating, which presents a risk of explosions and fires. If the moisture content of the biomass is sufficiently high, this initial self-heating often has biological causes, as thermophilic microorganisms increase the temperature of the fuel while decomposing it. At low moisture levels, which tend to suppress microbial growth, biomass may adsorb vapor from the ambient air, which can trigger self-heating due to differential rates of adsorption and vapor condensation. These initial heating processes can trigger auto-oxidation of the biomass, raising its temperature to the point that increasing quantities of VOCs and non-condensable gases like CO, CO_2 and CH_4 are released, greatly increasing the risk of explosions and/or fires. The temperature of stored biomass should therefore be monitored continuously and if it starts to rise, preventative measures should be taken to reduce the risk of fire.

Of the non-condensable gases, CO is the most hazardous because it prevents oxygen uptake by humans even at low atmospheric concentrations. Unprotected people exposed to an atmosphere containing around 800 ppm of CO will lose consciousness within 2 hours. At 12,800 ppm, death occurs within 1–3 minutes. Svedberg *et al.* (2004, 2008 and 2009) have studied VOCs and also cases where people were killed by emissions of non-condensable gases from biomass. These typically occur in environments where biomass is stored in rooms with no or very little ventilation, i.e. limited exchange of air. Such places are found in ships transporting biomass, in silos and in storage areas, etc. The solution is simple: increase the amount of ventilation and fresh air. People working in such environments must ensure that the levels of both CO and O_2 are safe (it is not enough to measure O_2 alone) before entering any space where biomass is stored in an enclosed area with little or no ventilation. Because CO can leak out to other rooms, the air in adjacent spaces should also be checked before entering. The safety measures for fuel pellets discussed in the *Pellet Handbook* (Obernberger and Thek, 2010) are valid for most biomass and should be implemented as a matter of course.

Another problem associated with handling biomass is the presence of dust as well as moulds and other microorganisms that grow on biomass. These are also hazardous. Explosions can occur if the concentration of dust in the air becomes too high, and prolonged inhalation of dust is harmful. Furthermore, microorganisms can be pathogenic and produce toxins and allergenic spores; exposure to these for even relatively short periods may result in sickness.

2.6.4   *Bulk handling*

It is well known that industrial processes involving solid biomass are significantly more problem-atic than those involving only gases or liquids. This point was highlighted three decades ago by the RAND Corporation (Merrow *et al.*, 1981) and is still valid. The general conclusion of these reports was that the start-up costs and processing times for fluid processing plants exhibited little variation, with most being within 20% of the mean, whereas those for plants based on solids were 200 to 300% greater. After completion, fluid-processing plants typically operated at around 90% of the designed throughput whereas solid-based plants tended to operate at around 50%. This is mainly due to the variable flow ability of solids compared to fluids. One should also keep in mind that viscoelastic solid biomass is harder to handle than hard solids such as gravel.

Feeding problems are a common reason for delayed start-ups and failure to sustain the designed-for production capacity. In comparison to fluid-based indusial processes, relatively little is known about biomass flow ability in different industrial processes and during internal transportation. As such, it will be important to study the tribology and rheology of solid biomass-derived materials in order to further their widespread adoption in industrial processes.

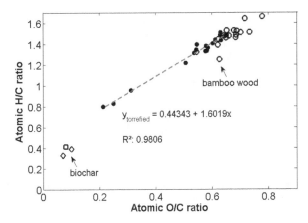

Figure 2.6.   A van Krevelen diagram for un-treated biomass (open circles); Norway spruce wood chips torrefied at 260–300°C for 8 to 25 minutes and commercial charcoal from softwood produced at about 450°C (filled circles); and biochar produced at 600°C (Lehmann *et al.*, 2011) (diamonds). The material with the highest average ratios is leaves from switch grass and sugarcane. Data on biomass and torrefied wood were kindly provided by Tao *et al.* (2012a) and Nordwaeger *et al.* (2013 a and b), respectively.

Biomass is a visco-elastic and fibrous material and its bulk properties are generally difficult to predict because of its irregular shapes, wide particle size distributions, low bulk densities, and often high moisture contents. Further, biomass can have particles with very inconvenient shapes in which one or two dimensions are very much smaller than the third (e.g, straw, fibers and flakes); this causes phenomena such as nesting. There are various mechanical problems associated with the bulk handling and flow characteristics of biomass, including bridging, compaction, and unwanted separation in silos and transporters. All of these are exacerbated by the presence of fine particles and dust, which also increase the risk of dust explosions.

The chemical composition of the material also affects its flow ability and processing. Samuelsson *et al.* (2009) and Nielsen *et al.* (2009) have shown that biomass samples with high concentrations of extractives in biomass will generate comparatively low wall friction at high pressures in systems such as pellet presses. High levels of organic acids and inorganic components in biomass will increase the wear on steel in transport canals, transporters, mills, presses, and so on.

2.6.5   *Heat treatment of biomass*

Pyrolysis is thermal treatment under anaerobic conditions, usually at temperatures of 450–550°C or more. Torrefaction is a mild form of pyrolysis conducted at about 250 to 330°C. In general, compared to raw and absolutely dried biomass, torrefaction produces material that has lost more of its mass (as a percentage) than its energy content. The oxygen-containing constituents of biomass have lower energy contents than other constituents. Consequently, the gases released during torrefaction typically have higher oxygen contents than the remaining biomass, as demonstrated by analyses of C, H and O levels in torrefied materials. A good overview of the torrefaction process can be obtained by constructing van Krevelen diagrams (van Krevelen, 1950) with the atomic O/C ratio on the horizontal axis and the H/C ratio on the vertical axis. A linear trend was observed in a series of experiments conducted using different torrefaction temperatures and treatment times (see Fig. 2.6).

Van Krevelen diagrams for other materials such as coal and anthracite have steeply sloping H/C trend lines when the O/C ratio is lower than 0.1, as shown by van der Stelt (2011). As such, it may be hard to obtain oxygen-free materials that still contain hydrogen by pyrolysis of biomass.

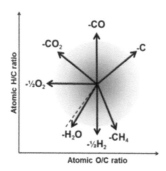

Figure 2.7.   Changes in relative biomass atomic (molar) ratios after losses of C, H, O, CO CO$_2$, H$_2$O and CH$_4$ in processes such as fermentation (biogas formation), torrefaction, pyrolysis, and gasification.

Furthermore, as indicated by the somewhat anomalous average value for bamboo (see Fig. 2.6), the starting atomic ratios in the untreated biomass may influence the product O/C and H/C ratios after treatment.

Charcoal, which is also known as biochar, is a carbon-rich product produced by pyrolysis of plant biomass at about 350–600°C. The storage of biochar in soil has been suggested as a method for sequestering carbon from CO$_2$ in the atmosphere and thus mitigating climate change. The attractiveness of this idea is increased by the fact that biochar enhances plant growth in many soil types, facilitating bioenergy production and increasing crop yields. Woolf *et al.* (2010) studied the technical potential for using pyrolysis on a global scale to produce biochar for storage in soils. They estimate that the maximum potential offset is 12% of current anthropogenic CO$_2$-C-equivalent emissions without endangering soil conservation habitat or food security. That is to say, a sustainable global implementation of biochar has the potential to negate approximately 1.8 Pg CO$_2$-C$_e$ of the 15.4 Pg CO$_2$-C$_e$ emitted annually.

In gasification, all of the hydrogen in the substrate can potentially be extracted for syngas production and for forming more complex products such as dimethylether (DME), which has the empirical formula CH$_3$OCH$_3$, an H/C ratio of 3 and an O/C ratio of 0.5. If hydrogen is trapped in the residual carbon after gasification, this directly reduces the efficiency of the process. This is especially true if the raw material has lower H/C ratios than the gasification products, which is the case for polymers (bio-plastics) consisting of C$_n$H$_{(1.5-2)n}$ and other elements. As such, pretreatments such as torrefaction that are used to make biomass more suitable for gasification could reduce its H/C ratio. This in turn would reduce the product's potential value. On the other hand, torrefaction could increase the scope for producing more suitable substrates that leave smaller quantities of carbon-containing residues during gasification. New pretreatments and approaches to gasification should therefore be developed to keep as much of the substrate's hydrogen as possible in a reduced state and minimize the presence of hydrogen-containing biochar residues.

The results presented in Figure 2.6 and 2.7 can be combined to estimate the net effects of different processes on the basis of C, H and O analysis of samples. The dashed line in Figure 2.7 shows the linear relationship for torrefied samples presented in Figure 2.6. For these samples, both the H/C and O/C ratios decrease at higher temperatures and with prolonged treatment. The direction of this linear relationship indicates that torrefaction under anaerobic conditions induces a major overall reaction whereby the oxygen in biomass reacts with its hydrogen to produce water. The torrefaction reaction is described in more detail in Chapter 7 of this book and was recently reviewed by van der Stelt (2011). Likewise it can be postulated from Figure 2.7 that the remaining material after biogas production, during which the main species lost from the biomass is CH$_4$, will have a lower H/C ratio but probably an elevated O/C ratio, especially if the losses of CO,

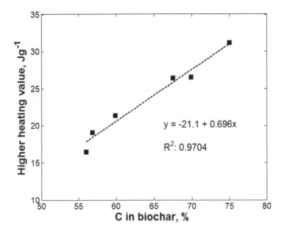

Figure 2.8.   The relationship between energy content and carbon content (C%) in biochar produced by pyrolysis (data from Woolf *et al.*, 2010).

CO_2 and H_2O are low in comparison to that of methane (CH_4). This fermentation process also occurs in anaerobic environments.

It is indicated by van der Stelt (2011) that like natural peat produced in mires and mosses, natural lignite also has a somewhat higher H/C ratio than other thermally treated biomasses with comparable O/C ratios. This may also indicate that peat and lignite formation are both characterized by a relatively more rapid loss of oxygen than hydrogen. However, these formation processes are slow, occur at low temperatures, and may have a biological component. The data presented in Figure 2.7 suggest that the higher H/C ratio may be due to losses of CO_2 from the biomass.

Carbonization of biomass by different thermal and biochemical processes increases its gross calorific value. Figure 2.8 shows the linear relationship between energy and carbon content in biochar produced by pyrolysis.

The carbonization process for converting fresh biomass to biochar seems to be linear according to the van Krevelen diagram (van der Stelt, 2011), but the corresponding progression of "natural" materials (lignite – coal – anthracite) is non-linear. Thus, thermal carbonization of biomass may be a natural first step that will facilitate the large-scale industrial use of biomass in existing coal-based plants as part of a broader effort to reduce the risk of irreversible climate change while avoiding damage to ecosystems and maintaining sustainable development that does not endanger food security, habitats, or soil conservation.

2.7   CONCLUSION

In conclusion, the chemical composition of biomass is complex and variable. Moreover, non-refined solid biomass has low flow ability and a low energy density whereas fossil oil and gas have good flow properties and oil and coal both have high energy densities. These factors, together with the health risks involved in handling raw and refined biomass, cause many problems that must be solved if biofuels are to replace fossil fuels on a large scale. It will be particularly important to monitor and control the inorganic contents of biomass, i.e. the ash-forming elements, since unfavorable compositions may cause operational problems during energy conversion. Thermal pretreatment of the organic components of biomass can homogenize feedstock quality, making it more like lignite or fossil coal.

We are currently on the doorstep of the industrialized biocarbon era. To cross the threshold, we will have to develop new knowledge and know-how in many sectors of society. We will also have to be patient and accept that there will initially be some drawbacks with the new systems because the problems to be solved along the way are more challenging than those encountered during the adoption of fossil fuels in the last century. However, we have a great advantage compared to the pioneers who first introduced widespread use of fossil fuels in that our technological sophistication and understanding of science have increased enormously. This puts us in a strong position as we strive to develop sustainably and to develop efficient and non-harmful methods for energy conversion.

2.8   QUESTIONS FOR DISCUSSION

- Why is biomass a suitable chemical resource for production of so many different types of products?
- How can biomass compete with oil and natural gas for production of plastics, textiles and other chemical products?
- What is the difference between cellulose, hemicellulose, starch and lignin?
- What do you believe biomass will be used for in the future? Any totally new applications?

REFERENCES

AEBIOM: Annual statistical report on the contribution of biomass to the energy systems in the EU27. AEBIOM – European Biomass Association, Brussels, Belgium, 2011.

Allen, M.R., David, J., Frame, D.J., Huntingford, C., Jones, C.D., Lowe, J.A., Meinshausen, M. & Meinshausen, N.: Warming caused by cumulative carbon emissions towards the trillionth tonne. *Nature* 458 (2009), pp. 1163–1166.

Andres, R.J., Fielding, D.J., Marland, G., Boden, T.A. & Kumar, N.: Carbon dioxide emissions from fossil-fuel use. 1751–1950. *Tellus* 51B (1999), pp. 759–765.

Arshadi, M., Geladi, P., Gref, R. & Fjallstrom, P.: Emission of volatile aldehydes and ketones from wood pellets under controlled conditions. *Ann. Occup. Hyg.* 53:8 (2009), pp. 797–805.

Azadi, P. & Farnood, R.: Review of heterogeneous catalysts for sub- and supercritical water gasification of biomass and wastes. *Int. J. Hydrogen Energy* 36:16 (2011), pp. 9529–9541.

Banerjee, S.: Mechanisms of terpene release during sawdust and flake drying. *Holzforschung* 55:4 (2001), pp. 413–416.

Batchelor, W.J., Conn, A.B. & Parker, I.H.: Comparison of techniques to measure the fibril angle. *Appita J.* 53:6 (2000), pp. 432–437.

Boerjan, W., Ralph, J. & Baucher, M.: Lignin biosynthesis. *Annu. Rev. Plant Biol.* 54 (2003), pp. 519–546.

Boström, D., Skoglund, N., Grimm, A., Boman, C., Öhman, M., Broström, M. & Backman, R.: Ash transformation chemistry during combustion of biomass. *Energy Fuels* 26:1 (2012), pp. 85–93.

Boutelje, J.B.: On the anatomical structure, moisture content, density, shrinkage, and resin content of the wood in and around knots in Swedish pine (*Pinus sylvestris* L.), and in Swedish spruce (*Picea abies* Karst.). *Svensk Papper* 69:1 (1966), pp. 1–10.

BP: BP Statistical review of world energy June 2011. British Petroleum, London, UK, 2011.

Brown, W. H. & Poon, T.: *Introduction to organic chemistry*. 3rd edn, Wiley, Hoboken, NJ, 2004.

Burvall, J.: Influence of harvest time and soil type on fuel quality in reed canary grass (*Phalaris arundinacea* L). *Biomass Bioenergy* 12:3 (1997), pp. 149–154.

Dererie, D.Y., Trobro, S., Momeni, M.H., Hansson, H., Blomqvist, J., Passoth, V., Schnurer, A., Sandgren, M. & Stahlberg, J.: Improved bio-energy yields via sequential ethanol fermentation and biogas digestion of steam exploded oat straw. *Bioresour. Technol.* 102:6 (2011), pp. 4449–4455.

Eberson L.: *Organisk kemi* (Eng: *Organic chemistry*). Almquist & Wiksell, Stockholm, Sweden, 1969, pp. 568.

EIA: *Annual energy outlook 2011 with projections to 2035.* US Department of Energy, Energy Information Administration, Office of Integrated and International Energy Analysis, Washington, DC, USA, 2011, www.eia.gov/forecasts/aeo/ (accessed 28 March 2012).

EN 14918:2009. Solid biofuels — Determination of the content of volatile matter. European Committee for Standardization, Brussels, Belgium, 2009.

Energimydigheten: Energiläget 2011. ET 2011:42, Statens Energimyndighet, Eskilstuna, Sweden, 2011. ISSN 1403-1892.

FAOSTAT: FAO Statistics, Food and Agriculture Organization of the United Nations, Rome, Italy, 2009, http://www.fao.org (accessed March 2012).

Gaur, S. & Reed, T.: *Thermal data for natural and synthetic fuels*. Marcel Dekker, New York, NY, 1998.

Gaur, S. & Reed, T.B.: An atlas of thermal data for biomass and other fuels. NREL/TB-433-7965, UC Category:1310, DE95009212, National Renewable Energy Laboratory, Golden, CO, USA. 1995.

Hoekman, S.K., Broch, A. & Robbins, C.: Hydrothermal carbonization (HTC) of lignocellulosic biomass. *Energy Fuels* 25:4 (2011), pp. 1802–1810.

IEA (International Energy Agency): World energy outlook 2009. IEA, Paris France, 2009.

Koopmans, A. & Koppejan, J.: Agricultural and forest residues generation, utilization and availability. Regional Consultation on Modern Applications of Biomass Energy, Kuala Lumpur, Malaysia, 1997.

Lal, R.: World crop residues production and implication of its use as a biofuel. *Environ. Int.* 31:4 (2005), pp. 575–584.

Lehmann, J., Rillig, M.C., Thies, J., Masiello, C.A., Hockaday, W.C. & Crowley, D.: Biochar effects on soil biota — A review. *Soil Biol. Biochem.* 43:9 (2011), pp. 1812–1836.

Lestander, T.A. & Rhén, C.: Multivariate NIR spectroscopy models for moisture, ash and calorific content in biofuels using bi-orthogonal partial least squares regression. *Analyst* 130 (2005), pp. 1182–1189.

Lestander, T.A., Hedman. B., Funkquist, J., Lennartsson, A. & Svanberg, M.: On-line NIR-fukthaltsmätning för styrning av panna i värmekraftverk (In Swedish: Measuring of moisture content in biofuel with on-line NIR, for forward control of a fluidised boiler). Report I6-605, Värmeforsk Service AB, Stockholm, Sweden, 2008.

Liao, C.P., Yan, Y. J., Wu, C.Z. & Huang, H.B.: Study on the distribution and quantity of biomass residues resource in China. *Biomass Bioenergy* 27:2 (2004), pp. 111–117.

Ludwig, J., Marufu, L.T., Huber, B., Andreae, M.O. & Helas, G.: Domestic combustion of biomass fuels in developing countries: a major source of atmospheric pollutants. *J. Atmos. Chem.* 44:1 (2003), pp. 23–37.

Merrow, E.W., Philips, K.E. & Myers, C.W.: Understanding cost growth and performance shortfalls in pioneer process plants. R-2569-DOE, Rand Corporation, Santa Monica, CA, 1981.

Mishra, V. & Retherford, R.D.: Does biofuel smoke contribute to anaemia and stunting in early childhood? *Int. J. Epidemiol.* 36:1 (2007), pp. 117–129.

Nielsen, N.P.K., Norgaard, L., Strobel, B.W. & Felby, C.: Effect of storage on extractives from particle surfaces of softwood and hardwood raw materials for wood pellets. *Eur. J. Wood Prod.* 67:1 (2009), pp. 19–26.

NOAA: US Department of Commerce, National Oceanic & Atmospheric Administration, Earth System Research Laboratory, Boulder, CO, ftp.cmdl.noaa.gov/ccg/co2/trends/co2_annmean_mlo.txt (accessed March 2012).

Nordwaeger, M., Olofsson, I., Pommer, L., Wiklund-Lindström, S. & Nordin, A.: Parametric study on torrefaction of spruce wood. Unpublished manuscript, Energy Technology and Thermal Process Chemistry, Umeå University, Sweden, 2013a.

Nordwaeger, M., Olofsson, I., Pommer, L., Wiklund-Lindström, S. & Nordin, A.: Parametric study on torrefaction of logging residues. Unpublished manuscript, Energy Technology and Thermal Process Chemistry, Umeå University, Sweden, 2013b.

Obernberger, I. & Thek, G. (eds): *The pellet handbook*. Earthscan Limited, London, UK, 2010.

Passoth, V., Eriksson, A., Sandgren, M., Ståhlberg, J., Piens, K. & Schnürer, J.: Airtight storage of moist wheat grain improves bioethanol yields. *Biotechnol. Biofuels* 2:16 (2009).

Patzek, T.W. & Pimentel, D.: Thermodynamics of energy production from biomass. *Crit. Rev. Plant Sci.* 24:5–6 (2005), pp. 327–364.

Perrett, D.: From 'protein' to the beginnings of clinical proteomics. *Proteomics – Clin. Applicat.* 1:8 (2007), pp. 720–738.

Raupach, M.R., Marland, G., Ciais, P., Le Quéré, C., Canadell, J.G., Klepper, G. & Field C.B.: Global and regional drivers of accelerating CO_2 emissions. *PNAS* 104:24 (2007), pp. 10,288–10,293.

REN21: Renewables 2011 — Global status report. REN21 Secretariat, Paris, France, 2011.

Robbins, M.P., Evans, G., Valentine J., Donnison, I.S. & Allison, G.G.: New opportunities for the exploitation of energy crops by thermochemical conversion in Northern Europe and the UK. *Prog. Energy Combust. Sci.* 38:2 (2012), pp.138–155.

Samuelsson, R., Thyrel, M., Sjöström, M. & Lestander, T.A.: Effect of biomaterial characteristics on pelletizing properties and biofuel pellet quality. *Fuel Process. Technol.* 90:9 (2009), pp. 1129–1134.

Sikkema, R., Steiner, M., Junginger, M., Hiegl, W., Hansen, M.T. & Faaij, A.: The European wood pellet markets: current status and prospects for 2020. *Biofuels Bioprod. Biorefin.* 5:3 (2011), pp. 250–278.

Solomon, S., Plattner, G., Knutti, R. & Friedlingstein, P.: Irreversible climate change due to carbon dioxide emissions. *PNAS* 106:6 (2009), pp. 1704–1709.

Still, C.J., Berry, J.A., Collatz, G.J. & DeFries, R.S.: The global distribution of C3 and C4 vegetation: carbon cycle implications. *Global Biogeochem. Cy.* 17:1 (2003), p. 1006.

Stockinger, H. & Obernberger, I.: Systemanalyse der Nahwärmeversorgung mit Biomasse. Book series *Thermische Biomassenutzung*, vol. 2. DBV – Publisher of Graz University of Technology, Graz, Austria, 1998.

Svedberg, U., Högberg, H.-E., Högberg. J. & Galle, B.: Emission of hexanal and carbon monoxide from storage of wood pellets, a potential occupational and domestic health hazard. *Ann. Occup. Hyg.* 48:4 (2004), pp. 339–349.

Svedberg, U., Samuelsson, J. & Melin, S.: Hazardous off-gassing of carbon monoxide and oxygen depletion during ocean transportation of wood pellets. *Ann. Occup. Hyg.* 52:4 (2008), pp. 259–266.

Svedberg, U., Petrini, C. & Johanson, G.: Oxygen depletion and formation of toxic gases following sea transportation of logs and wood chips. *Ann. Occup. Hyg.* 53:8 (2009), pp. 779–787.

Tao, G.C., Lestander, T.A., Geladi, P. & Xiong, S.J.: Biomass properties in association with plant species and assortments I: a synthesis based on literature data of energy properties. *Renew. Sust. Energy Rev.* 16:5 (2012a), pp. 3481–3506.

Tao, G.C., Geladi, P., Lestander, T.A. & Xiong S.J.: Biomass properties in association with plant species and assortments II: a synthesis based on literature data for ash elements. *Renew. Sust. Energy Rev.* 16:5 (2012b), pp. 3481–3506.

UNFCC: Kyoto protocol to the United Nations framework convention on climate change. United Nations, New York, 1998.

UNFPA: State of world population. United Nations Population Fund, New York, 2011.

US Department of Energy: Annual energy outlook 2011 with projections to 2035. DOE/EIA-0383, US Department of Energy Washington, DC, 2011.

van der Stelt, M.J.C., Gerhauser, H., Kiel, J.H.A. & Ptasinski, K.J.: Biomass upgrading by torrefaction for the production of biofuels: a review. *Biomass Bioenergy* 35:9 (2011), pp. 3748–3762.

van Krevelen, D.W.: Graphical-statistical method for the study of structure and reaction processes of coal. *Fuel* 29 (1950), pp. 269–284.

van Loo, S. & Koppejan, J. (eds): *The handbook of biomass combustion and co-firing.* Earthscan, Sterling, VA, 2008.

Vassilev, S.V., Baxter, D., Andersen, L. & Vassileva, C.G.: An overview of the chemical composition of biomass. *Fuel* 89: 5 (2010), pp. 913–933.

Wahlund, B., Yan, J. & Westermark, M.: A total energy system of fuel upgrading by drying biomass feedstock for cogeneration: a case study of Skellefteå bioenergy combine. *Biomass Bioenergy* 23 (2002), pp. 271–281.

Wajs, A., Pranovich, A., Reunanen, M., Willför, S. & Holmbom, B.: Headspace-SPME analysis of the sapwood and heartwood of *Picea abies, Pinus sylvestris* and *Larix decidua. J. Essential Oil Res.* 19:2 (2007), pp. 125–133.

Walawender, W.P., Hoveland, D.A. & Fan, L.T.: Steam gasification of pure cellulose. 1 a uniform temperature profile. *Ind. Eng. Chem. Process Des. Dev.* 24:3 (1985), pp. 813–817.

Walker, J.: Basic wood chemistry and cell wall ultrastructure. In: J. Walker (ed): *Primary wood processing – principles and practice.* Springer, Dordrecht, The Nederlands, 2006.

Woolf, D., Amonette, J.E., Street-Perrott, F.A., Lehmann, J. & Joseph, S.: Sustainable biochar to mitigate global climate change. *Nat. Commun.* 1:56 (2010), doi: 10.1038/ncomms1053.

Xie, G., Wang, X. & Ren, L.: China's crop residues resources evaluation. *Chinese J. Biotechnol.* 26:7 (2010), pp. 855–863 (in Chinese with English abstract).

Xiong, S., Landstrom, S. & Olsson, R.: Delayed harvest of reed canary grass translocates more nutrients in rhizomes. *Acta Agr. Scand.* B *Soil Plant Sci.* 59:4 (2009), pp. 306–316.

Zhou, G.-Q., Zhang, G.-F. & Qi, D.-M.: A new method of producing bio-energy by using sugar-beets. *Energy Procedia* 12 (2011), pp. 873–877.

CHAPTER 3

Characterization of biomass using instruments – Measurement of forest and crop residues

Robert Aulin

3.1 INTRODUCTION

In 2009, renewable energy made up for 47% of the entire energy consumption in Sweden where two major sources of energy are biofuel and hydropower. In 2010, they contributed with 141.5 TWh and 66.8 TWh respectively. The corresponding shares of the entire 616.5 TWh energy supply in Sweden were 23% and 11%, respectively.

Due to the large forest areas in Sweden, wood and forest residue have been an important source of energy for centuries. Since the early 1980s, they have increasingly been used as an energy source in the combined heat and power plants (CHP's) that constitute a backbone in municipal heating. According to the Swedish Energy Agency, wood, tops and branches extracted directly from the forest contributed to 26 TWh in 2010. In addition to this figure, e.g. stubs, residue from pulp and paper plants, sawmills and recycled wood are used for energy production.

Since production capacity increases, there is a debate over a potential fuel shortage in the future. Except for rising prices, another implication would be that the use of less desirable fuel sources such as stubs and recycled wood is likely to increase. In this case, one might expect larger quality fluctuations and problems with fuel shipments that do not meet up to the specifications agreed upon by the buyer and seller. This, in turn, increases demand for quality control. Increasing quality fluctuations will also have a negative impact on the plants themselves. With real-time quality assessment of the fuel, adequate countermeasures may be taken through improved process control.

This text is written from a Scandinavian perspective with a large demand for heat in the winter and where the installed capacity of a single CHP often exceeds 100 MW. As a result, the implications of this text might be less applicable for countries with a warm climate and smaller plants. However, these lines of reasoning may be useful since economies of scale suggest that plant sizes will rise generally. Recent directives from the European Union will also enhance this development.

This chapter discusses sources of moisture content variation, their magnitude and techniques to measure moisture content in large material flows. Focus is put on forest residue such as tops, branches, sawdust and bark. Many aspects and solutions might well apply for other materials such as peat, recycled wood and household waste. Today, there is a fast development in the technology for moisture measurement. Any detailed description of such technology is likely to become outdated within a few years. Some examples of commercially available measurement equipment are given but the emphasis of this chapter is on sources of quality fluctuation and strategies for using the measurement results.

3.2 QUALITY ASPECTS AND SOURCES OF VARIATION

3.2.1 *Volume, weight and moisture content*

There are several parameters of interest with respect to fuel characteristics. In Sweden, volume, weight and moisture content are three important parameters that are often measured for shipments as small as the containers of a single truck.

Volume measurement has a long tradition. A major advantage is that volume is simple and easy to measure and verify by all parties. A buyer or seller can easily challenge an existing measurement by replicating it. However, the measurement result depends e.g. on how compressed the material is. A shipment of fresh branches and tops will not have the same volume is if the material is collected after three months in the forest. Further, the density is also likely to increase during transport. During the last decade, the industry has lesser attention on measuring the volume of forest residue. Today, weight is the dominating measure.

Weight is a more objective measure in the sense that weight is less affected by transport. Therefore, weight is increasingly accepted for shipment control as the scale technology has improved. Further, weight is used as the basis for calculation of transportation costs. Assuming that the moisture content is constant, the weight would also give a good indication of the amount of dry biomass. However, reasons cited in section 3.3 indicate that this assumption should not be trusted.

Volume and weight might be combined in order to calculate density. For a given type of material, density might be used in order to calculate approximate moisture content. Unfortunately, the result is affected by the type of material, the degree of compression and of its fractional distribution. However, a scale mounted on a front-end loader might e.g. provide useful information when biomass is to be mixed according to a recipe before it enters the boiler.

A third parameter of great importance is the moisture content (*MC*) since it determines the amount of dry mass in a certain shipment. Moisture content is important because it affects the energy content, price and combustion properties.

$$\text{dry mass} = \text{total mass} \times (1 - MC)$$

3.2.2   *Calorific value*

The above measures may all be used in order to estimate the energy content. Assuming that the energy content in dry biomass is constant, information about weight and moisture content can be used in order to calculate the energy content of the biomass. However, the energy content of dry biomass varies depending on the type of material and on its storage time. Therefore, it would be desirable to measure the energy content itself.

Energy content is traditionally determined through manual sampling followed by the use of a calorimeter in order to estimate the energy content. Since this process is tedious and time-consuming, it is impractical for everyday measurement.

3.2.3   *Other parameters*

Ash content and fractional distribution are other parameters with an economic relevance. The term ash content often refers to the weight ratio of organically bound elements such as Na or K. The term is also used as a general expression for various impurities such as dirt. While the presence of dirt also affects the energy content, high sodium and potassium levels will have an impact on the plant and its wearing.

Fractional distribution refers to the size distribution of the biomass. Two extremes are sawdust and branches. The parameter is of interest because the size distribution of the biomass affects the combustion process. Large differences or changes in size complicates boiler control and decreases the overall efficiency of a boiler. An example is that fine fraction might be carried away by the exhaust gases and burn in the upper part of the boiler while large pieces of fuel might not burn through entirely.

Ash content and fractional distribution might be determined in a laboratory but these methods are less practical for everyday measurement. Since determination of such parameters requires manual labor, the sampling intensity will be a trade-off between utility and measurement costs. There are several types of strategies depending on preferences of the buyer. One might for example take reference samples from the different biomass types of each respective supplier and then assume that the measured properties will be representative for a large number of shipments during a contract period. In the case of large shipments, such as 5000 m.t. vessel, the size of the shipment often motivates specific sampling. In other cases, determination of ash content or fractional distribution might be done on a regular basis such as with a weekly or monthly interval.

3.3 THE FUEL CHAIN AND ITS IMPACT ON THE MOISTURE CONTENT

3.3.1 *The fuel chain*

There are several routes for forest biomass to the end user. Forest residue is either transported directly to the end user or stored at a terminal. Sawmills and pulp factories deliver a part of their by-products to the energy producers. However, pulp factories use part of the residue for their own energy productions while saw mills deliver saw dust for the production of pellets rather than use by CHPs.

Wood for burning, and felling residue such as tops and branches, are often stored for some time in the forest. Often, the material is left to dry and its extraction coordinated with other activities in the forest. Chipping or crushing might take place on site in the forest, at a terminal or at the power plant (Fig. 3.1). Due to economies of scale, it is desirable to perform this task in large quantities. However, it is often made in the forest if the receiving plant has no chipping equipment or when transport costs make it too expensive to send the forest residue to a terminal for processing. In Sweden, processed residue is usually collected in metal containers in the forest. This way, the material can be stored on several forest sites and be collected by a vehicle typically being able to carry three containers.

Part of the material is transported straight to the energy producer while other fuel might be stored at a terminal in order to build up a supply before the cold season, i.e. the winter. The fuel is often transported in side-tipping trucks (Fig. 3.2). The volume is somewhat larger than for container trucks since it is desirable to distribute overhead on the largest possible volume. Side-tipping trucks are also common for transportation of residue from pulp factories and saw mills. However, since the containers are fixed to the truck, this type of vehicle is not suitable for collection of material in the forest.

3.3.2 *Sources of variation in moisture content*

3.3.2.1 *The forest*
While the moisture content in the respective parts of a tree or a fresh log is relatively stable, the measured moisture content in the resulting biomass tends to vary depending on storage and transportation.

In the forest, the moisture content of the forest residue is affected by the season of the year, the storage time and by the choice of storage area. The traditional pattern is that the moisture content of the delivered forest residue is higher during the winter than in the summer. In some cases, forest residue is left on the forest floor in order to dry during the late spring and summer. However, several factors affect the moisture content of a certain shipment. Examples of actions that affect moisture content include if the material is put in a high and dry place, or in a low place where rainwater collects. There will also be a difference in moisture content depending on whether the forest residue has been protected by a covering material or not. Further, when biomass is extracted in the winter, rain and snow might settle in the open metal containers when they are stored in the forest. Another example from the Scandinavian winter is that roads often have to be opened from snow in order to access the biomass. To facilitate transport, the forest residue is stored adjacent to the roads. Therefore, snow walls are pushed into the piles that will be partially filled with snow. This snow will then be transported along with the biomass to the energy producer.

3.3.2.2 *Terminal storage*
When forest residue is stored at a terminal in order to build up supplies, material is gathered in large piles. These piles tend to be filled with biomass from several origins and with different humidities. Further, the moisture content will change during storage.

Figure 3.1.   Container truck.

Figure 3.2.   Side-tipping truck.

Storage will also affect the moisture content. The humidity of the surface area will be affected by the recent weather. Rain or snow will remain close to the surface during precipitatious weather while the surface dries in warm and sunny weather. Further, material close to the ground tends to contain some amount of dirt that is scraped up when the fuel is picked up and loaded in a truck.

In addition to this, biological activity will cause a degradation of the biomass. Bacteria and fungi will consume part of the biomass in order to produce water and heat. In the center of the pile, the moisture content is likely to increase since water is a by-product of the biological activity. Accordingly, the calorific value decreases as the energy is consumed by the bacteria and fungi. However, the moisture content in adjacent biomass might decrease because the heat evolved dries the material. The calorific value of terminal fuel will therefore differ a lot depending on the storage time and its position in the pile.

The previous examples do not constitute an exclusive list. However, they illustrate some of the many causes of why the moisture content of a shipment will fluctuate within a large range.

For this reason, manual sampling will yield different measures of moisture content depending on where the sample is taken in a shipment.

3.3.2.3 *Transport*

Transportation is unlikely to cause major changes in the moisture content. However, a slight change in the moisture profile might occur so that the bulk of a shipment will have slightly different moisture content than material close to the upper surface.

An analysis of data from Bestwood customers suggests that this phenomenon is due to factors that may be divided into three groups. First, when forest residue is stored in an open container it is subject to weather that will affect the moisture content near the surface. When rain falls, precipitation will make the surface area more wet then the bulk. The contrary occurs during a warm and sunny day. This effect, which might take place during storage as well as during transport, is seasonal. Further, if an open container is transported during warm weather, the speed wind is likely to dry the surface area. Finally, transport vibrations are supposed to make smaller particles propagate to the lower part of the container. With the smaller fraction removed from the surface, the upper region of the container is more likely to contain material of a slightly larger size and somewhat lower moisture content. A similar effect is anticipated to occur in the winter when transport vibrations cause snow particles to move towards the bottom of the containers.

As a rule of thumb, one might assume that the surface region is somewhat drier than the bulk. The magnitude and direction of this bias depends on the season and weather. When sampling is done close to the upper surface, this bias will affect the results of the moisture determination and the price settled for a shipment.

3.3.2.4 *Site storage and fuel handling*

Storage at the CHP is subject to similar effects as in the terminal although bacterial activity tends to be less of a problem since the average storage time tends to be far lower due to limitations in storage capacity.

While the moisture content itself is unlikely to change during storage, several actions may be taken in order to stabilize the moisture content of the fuel that is fed to the boiler. First, it is common to mix different types of forest residue in order to compensate for anticipated difference in humidity. Then, the fuel storage silo is used for further homogenization due to additional mixing. Finally, using screws that extract fuel from an intersection of different layers of material, the variation might be decreased further.

3.4 MOISTURE MEASUREMENT

3.4.1 *Gravimetric moisture measurement*

3.4.1.1 *The gravimetric method*

Moisture measurement traditionally uses manual sampling and gravimetric determination. Moisture determination is often performed according to the Swedish standard SS 187 113 (1998) but the actual procedures are carried out in a multitude of ways. The samples are combined into a single sample, where after a smaller subsample is chosen and weighed in a tray (Fig. 3.3). The tray is dried in an oven for 20–24 hours. After weighing the dried sample, and knowing the weight of the empty tray, the moisture content is calculated.

An indicative precision of the gravimetric method is approx. 2%-units of moisture when two trays from the same sample are compared. The performance depends on the heterogeneity of the material, since the precision is likely to deteriorate if the sample is less homogeneous.

The oven method is not flawless. Volatile organic compounds with high heating value may be evaporated during drying. This flaw can be compensated for by using default calorific values that refer to the respective types of biomass. Further, the degree of evaporation differs somewhat depending on the position of a certain tray in the oven and on the overall level of moisture in the

Figure 3.3.   Laboratory trays for use with oven.

samples (Anerud and Jirjis, 2011). Changes in drying time and thermostat settings would also affect the result. A comparison of results over time would therefore benefit from ring analyses where sub-samples of a certain sample are processed in different ovens at different laboratories.

The performance of the gravimetric method also has implications for the calibration and validation of the instrumental methods. Regardless of technique, they are usually calibrated against a reference method that is likely to be the gravimetric method. Further, instrumental methods may require additional reference values over time in order to make certain that their calibration is stable. Deviations in performance may include changes in the hardware. Further, the biomass mix might change and no longer be covered by the calibration model.

More crucial is that discrete shifts in the reference method would have a severe impact on the evaluation of an instrumental method. If a discrete shift is introduced after a certain calibration has been set, one will erroneously conclude that the result of the instrumental method suffers from a bias. Further, erroneous reference samples are likely to be entered into the subsequent, revised calibration models and decrease model performance because the error in reference values is no longer normally distributed.

Another error source is caused by inappropriate storage of samples. Ideally, they ought to be dried immediately. In practice, samples are dried in batches of, say, 30–60 trays depending on oven size. A sample of wet fuel might dry significantly if taken in the morning, not weighed immediately and stored in a warm room during the day before it is put in the oven. A similar effect would occur if a truck driver, for convenience, takes a sample to be analyzed the next day and stores it overnight in a non-sealed container in a warm room.

Sampling errors constitute the main error source. The need for representative sampling depends on the intended use. When the sampling error and the measurement error are uncorrelated, the total error is computed as the sum of the variances of the sampling error and the measurement error:

$$s_{tot}^2 = s_{sampling}^2 + s_{measurement}^2$$

3.4.1.2 *Sampling*

Today, moisture measurement is mainly done for shipment control and subsequent price settlement. The Swedish wood surveying organization VMF (1998) has a recommendation stating that moisture measurement for price settlement should be unbiased and be based on a sample size that yields a standard deviation of 1%-unit of moisture. A rising number of samples will yield a more correct price settled between buyers and sellers, but the cited precision requirement implies a very large number of samples. The industry seems to have made a tradeoff against what is practically feasible and, instead of 20–30 samples taken from all parts of a shipment, some 2–3 samples are typically taken.

A key issue is to differ between random errors and systematic errors. Although the gravimetric moisture measurement method is quite reliable in itself, the practical implementation deserves special attention. Representative sampling is also of great importance since the objective is to obtaining a measurement result that represents the bulk moisture content of a certain shipment. A common sampling method is to use a shovel to take samples in 2–3 positions in a shipment.

The two main approaches are to collect the samples close to the surface layer of the shipment or to collect samples from material tipped on the ground. The first alternative might be the only option when sampling on tipped fuel is either inconvenient or not possible because the biomass is tipped directly to an entry pocket of a silo or the boiler feed. Another reason for the procedure is that the reception station might have an inspection bridge adjacent to the lab ovens so that samples can be taken easily.

In another section, a number of factors that might cause a systematic difference between the bulk humidity and the moisture close to the surface are discussed. Therefore, a more representative sampling can be expected if the shipment is unloaded on the ground and sampling can be made in randomly chosen positions in the bulk.

Another rationale for sampling on tipped material is that the supplier has less potential for manipulation of the measurement result. One way to manipulate the price would e.g. be to load wet material in the lower section of the container if sampling is made close to the surface. A key issue to remember from the above is that the choice of sampling method, and the number of samples taken, will have an impact on the measurement results.

3.4.1.3   *Practical illustration*

When the price is based upon a large number of containers, individual errors tend to cancel out. This is also likely to be the case if similar material is delivered from the same supply. In these cases, the parties may agree on a lower sampling intensity. In other cases, the number of containers from a certain supplier might be small. Then, individual errors will have a larger impact on the price paid to each supplier. This is a reason why high precision is desirable.

Figure 3.4 shows a histogram of the variation in moisture content in a container with chipped branches and tops sent from a terminal to a Swedish plant in February 2012. The material has been tipped on the ground. Sampling was made in 20 evenly distributed squares. Then, the moisture content was determined using the gravimetric method. The *y*-axis shows the number of samples that fall within the respective 2%-unit intervals indicated on the *x*-axis. Assuming that sampling is made at one point per container it is clear that manual sampling does not give a reliable result for individual containers. This conclusion is supported in a study by Aulin *et al.* (2008) which also indicates that Swedish industry participants are well aware that manual sampling might lead to erroneous results and that the total error is far larger than the standard deviation for gravimetric method itself.

3.4.2   *Instrumental methods*

3.4.2.1   *Introduction*

Moisture determination through loss of drying has been described earlier in this chapter. It is a reliable method with a precision of approx. 2%-units of moisture. A clear disadvantage is that the method is subject to a sampling error and that results usually are obtained after 20–24 hours.

While sawdust or woodchips may be analyzed with, for example, impedance measurement, it is far more difficult to determine the moisture content of complex mixtures. Today, new technologies such as NIR spectroscopy, RF spectroscopy and X-ray methods have emerged for more advanced applications such as process control.

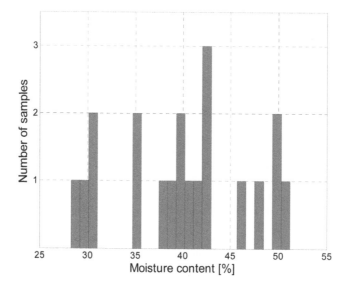

Figure 3.4.   Example of heterogeneity in a single container.

It is difficult to compare the performance of commercially available measurement systems since published results depend on a large number of factors. It is easy to achieve high precision and accuracy for a homogenous material such as sawdust while it is far harder to measure moisture content in complex mixtures of different types of biomass. Other factors that affect the outcome of a comparison are the moisture content range itself, temperature differences or the presence of frozen water. Finally, the performance of a certain instrumental technique depends on the calibration skills of the supplier. Rather than viewing a certain technology as superior or inferior, evaluation should be made with the actual performance requirements of the intended user in mind.

It can also be noted that the sampling techniques in the below examples range from point wise sampling to sampling along intersections of the bulk and analysis of a part-stream of the flow on a conveyor belt. Therefore, a comparison of performance between methods should also consider the impact of the respective sampling errors.

3.4.2.2   *Near-infrared spectroscopy (NIR)*

Near-infrared spectroscopy (NIR) uses the near-infrared region of the electromagnetic spectrum in the range of approx. 800–2500 nm. The method uses information in molecular overtone and combination vibrations. There is no immediate correlation between the elementary composition and spectrum. Material properties are usually calculated using multivariate methods. Another complication is that moisture peaks are temperature sensitive.

The penetration depth varies with the material but an indicative depth is 10 mm for forest residue. Therefore, analysis of bulk properties has to be made analyzing a part stream or an intersection in the material. Figure 3.5 shows a system for on-line analysis on a conveyor belt while Figure 3.6 displays equipment for bulk analysis of truck containers.

A clear advantage of NIR is that it provides information about the organic matrix. NIR can e.g. be used in order to determine the mix of different fuels, estimate ash content or approximate the energy content. Since various types of biomass differ in energy content such estimation of the calorific value will be more reliable than if the moisture content is determined and a fixed calorific value is used is used in order to estimate the energy content. A study by Lestander and Rhén (2005) confirms that near-infrared spectroscopy might be used for estimation of ash content as well as the gross calorific value in ground samples from stem and wood branches.

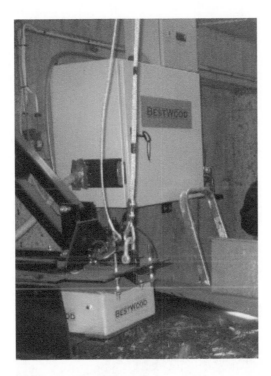

Figure 3.5.   Bestwod BAS-600 system for on-line analysis.

3.4.2.3   *Microwave spectroscopy (RF)*

Microwave spectroscopy is often referred to as radio frequency spectroscopy (RF) which analyses the absorption of microwaves. Depending on the moisture content, the microwave pulse will be subject to different attenuation and time delay. This relationship is represented in a calibration model that is used for moisture estimation. There is no immediate correlation between the elementary composition and spectra. Among the error sources in RF measurement of biomass are fluctuations in density and the concentration of Na and K ions. This technique is described in detail by Nyström (2006) and Paz (2010).

One advantage of RF spectroscopy is that it might analyze the bulk properties of a material in larger, closed compartments. The sampling error is reduced because the radio waves travel through the entire part of the material to be analyzed. This is important for analysis of very heterogeneous materials, for example. Examples of RF-based equipment are the desktop and on-line analyzers from the Finnish supplier Senfit and the Bestwood BAS-800 system that is used for on-line moisture determination of household waste and boiler control.

3.4.2.4   *X-ray spectroscopy*

X-ray spectroscopy analyses photons emitted when inner shell electrons are excited and replaced with outer shell atoms. The wavelengths of the photons are characteristic for the respective elements in the sample. Data about the elemental composition of the sample is used in order to calculate the moisture content. X-ray spectroscopy can analyze a large part of the bulk in a sample or material flow.

Figure 3.7 shows a desktop scanner from the Swedish supplier Mantex. It uses photon absorptiometry to register the effective atomic numbers and density of the material analyzed. By using two different energy levels, this technique may be used to identify different materials.

Figure 3.6.   Bestwod BAS-700 system for shipment control.

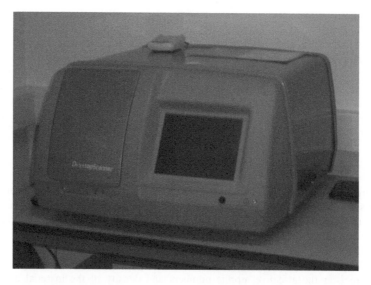

Figure 3.7.   Mantex desktop instrument.

Figure 3.8.   Inray fuel analyzer.

The absorption and scattering of photons are due to the photoelectric effect and Compton scattering. Figure 3.8 shows an on-line system from the Finnish manufacturer Inray.

3.4.2.5   *Method selection*
Although a large number of instrumental methods allow for indicative measurement of moisture content, the actual choice of an instrumental method is likely to depend on several factors. Rather than recommending a certain technology, the choice depends on the performance of the systems commercially available. There is also a trade-off between measurement performance and the cost of a certain system. Further, due to economies of scale a large plant can distribute the cost over large volumes while a small actor may not afford measurement at all.

A central issue is the type of material to be analyzed and the properties to be determined. The requirements may range from moisture analysis of a homogeneous material such as sawdust to determination of several properties of a complex mixture of various types of forest residue. A special consideration in the Nordic countries is whether the system must be able to analyze frozen material or not. Further, there are requirements concerning accuracy, precision and the need for representative sampling. These requirements vary depending on application and are discussed in more detail in section 3.5.

3.5   PRACTICAL APPLICATIONS FOR MOISTURE DATA

3.5.1   *Real-time measurement*

During the last decade, several new methods for non-destructive, real-time moisture measurement have become commercially available. A number of actual, and potential, applications are listed in the section below.

3.5.2   *Price settlement*

Price settlement has been discussed in the section about sampling. Three important requisites for a reliable method are accuracy, unbiasedness and representative sampling. Price settlement might not require a high precision if the measurement result is used for a large number of shipments. However, whereas a small bias has lesser importance for logistic purposes, systematic measurement errors would cause a corresponding error in the prices paid.

A large number of instrumental methods deliver indicative measurements of moisture content. Possibly, buyers and sellers could agree on any measurement method that they are comfortable with it. However, since unbiasedness is an important for price settlement, performance requirements are likely to high for systems that will determine the price of a shipment. To facilitate trade, methods for large-scale use are likely to be subject to accreditation The Bestwood BAS-700 was the first method to receive such an accreditation in Sweden.

3.5.3 *Logistics*

Access to real-time moisture information facilitates a large number of applications. One obvious application is the rejection of shipments that do not meet the specified quality requirements. However, a fully automatic system for price settlement will also have economic benefits in terms of avoiding manual labor. Other applications include automatic handling of invoices, real-time inventory control and order systems that adapt to changes in the production targets.

Access to real-time feedback also has benefits for the supplier since it will be easier to adapt to deviations in terms of contracted energy content or changes in production forecasts. On-site measurement at a terminal would also allow for delivery of a biomass that is more optimal with respect to the moisture content and variation required by the energy producer. In this case, measurement costs have to be weighed against the price premium paid by the customer. Another aspect is that rising demand is likely to increase the use of biomass that is considered inferior today. With an adequate measurement system, it is possible to deliver a more stable product despite the lower quality of the biomass.

3.5.4 *Fuel mixing*

There are several applications for fuel mixing on the site of the energy producer. The general theme is to direct material flows in a way that reduces the variation in moisture content and to reach an average moisture content that is optimal for the configuration of the boiler and the production targets with respect to power and heat.

The potential in this respect will depend on the possibilities in mixing fuel at the respective plants. In Sweden, some plants have large areas to store fuel. It is then possible to store biomass in different piles, depending on moisture content, and mix an appropriate fraction when the fuel is to be used. However, the majority of the plants have a small storage area or no storage area at all. If there are several entry points where the biomass is tipped into a silo, one strategy is to direct trucks to appropriate entry points depending on the moisture content. This strategy is used by a large producer on the Swedish West coast.

3.5.5 *Boiler control*

Boiler control utilizes information about the moisture content of the fuel on its way to the boiler. Being able to compensate for fluctuations in fuel moisture, it is possible to stabilize the fuel feed and combustion process and slightly increase capacity and overall efficiency.

Precision is more important than accuracy since the control loops act on relative changes where the presence of a systematic and constant bias has less importance. An indicative precision requirement for on-line control might be approx. 2%-units of moisture, which is of the same magnitude as the gravimetric method in a lab.

3.6 FUTURE PERSPECTIVES

Today, the measurement technology is developing rapidly. Some examples of hardware were shown in section 3.4. New systems are likely to enter the market and the range of products might change in the near future. Certain trends are already more or less visible.

First, precision and accuracy will improve as calibration techniques and hardware develop further. The price performance ratios of the analysis systems will then decline with time so that the systems will be increasingly attractive also to smaller plants. Further, the sampling errors are likely to decline as the systems will sample from a larger part of the bulk of containers and material flows. Then, additional parameters such as ash content and calorific value are likely to be presented by a large array of analysis systems.

Proper action has to be taken in order to use the new technology and increase efficiency in various parts of the chain from forest to boiler. Therefore, we might expect a large number of new applications such as decision support systems and optimization platforms. Concerning price settlement, we may expect standardization efforts since international trade would benefit, and require, an objective and practically relevant quality assessment.

ACKNOWLEDGEMENTS

The author thanks Mikael Karlsson (Bestwood AB, Stockholm, Sweden) and Torbjörn A. Lestander (Swedish University of Agricultural Sciences, Umeå, Sweden) for valuable comments and reviews of this chapter.

REFERENCES

Allmänna och särskilda bestämmelser för mätning av biobränslen. Industry recommendation, Swedish Wood Measuring Association (VMF), 1998 (in Swedish).

Anerud, E. & Jirjis, R.: Fuel quality of Norway spruce stumps – influence of harvesting technique and storage method. *Scand. J. Forest Res.* 26:3 (2011), pp. 257–266.

Aulin, R. & Karlsson, M.: Standardisation of moisture measurement using NIR spectroscopy for delivery control. Värmeforsk Report I6-604, ISSN 1653-1248, 2008 (in Swedish).

Dahlquist, E., Axrup, L., Nyström, J., Thorin, E. & de la Paz, A.: Automatic moiosture measurment on biomass fuel using NIR and Radio frequency spectrometri. Swedish Heat and Power Research Report (Värmeforskrapport), October 2005 (in Swedish).

Lestander, T.A. & Rhén, C.: Multivariate NIR spectroscopy models for moisture, ash and calorific content in biofuels using bi-orthogonal partial least squares regression. *Analyst* 130:8 (August 2005), pp. 1182–1189.

Nyström, J.: *Rapid measurements of the moisture content in biofuel.* PhD Thesis, No. 24, Mälardalen University, Sweden. 2006.

Nyström, J. & Dahquist, E.: Methods for determination of moisture content in wood chips for power plants – a review. *Fuel* 83 (2004), pp. 773–779.

Nyström, J., Thorin, E., Backa, S. & Dahlquist, E.: Filling level measurement in woodchips bins with radio frequency spectroscopy. Nordic Energy Research Program Biomass Combustion, Thermochemical Conversion of Biofuels, Norway, Trondheim, Nov. 2002, 2003.

Nystrom, J., Thorin, E., Backa, S. & Dahlquist, E.: Moisture content measurements on sawdust with radio frequency spectroscopy. *Proceedings of PWR2005.* ASME Power, 5–7 April 2005, Chicago, Illinois, 2005.

Paz, A.M.: *The dielectric properties of solid bofuels.* PhD Thesis, No. 90, Mälardalen University, Sweden, 2010.

Paz, A., Thorin, E. & Dahlquist, E.: A new method for bulk measurments of water content in woody biomass. *Ecowood 2008, 3rd Conference on Environmental Compatible Forest Materials Proceedings,* Portro, Portugal, 2008.

Swedish Standard, SS 187113, Swedish Standards Institute, 1998.

CHAPTER 4

Bioenergy in Brazil – from traditional to modern systems

Semida Silveira

4.1   FROM DEVELOPING COUNTRY TO LEADING ECONOMY

In the past decades, Brazil has moved from a position among developing countries to a key position as a BRIC[1] country, member of the G20[2], and soon the sixth largest economy in the world.[3] Behind the Brazilian economic miracle, a number of specific efforts have been particularly important. This includes the efforts towards industrialization started in the 1930s, modernization of agriculture started mainly after the 1960s, and stronger integration of the country with the global economy after the mid-1980s. Political stability and economic growth has led to unprecedented poverty reduction particularly after the turn of the last century (OECD, 2011). More recently, Brazil has also emerged as a major scientific nation, a development that is gaining speed with increasing efforts towards more innovation (Adams and Christopher, 2009). In addition, energy provision and use, and the transformation of the Brazilian energy system have served as an important pillar for the great economic transformation that the country has undergone.

In energy terms, Brazil has gone from strong dependency on traditional biomass to a diversified and modern energy matrix within less than half a century. In 1950, Brazil was still largely dependent on traditional biomass for energy purposes, and only larger cities had access to electricity (Fig. 4.1). In fact, some 80% of the energy used in the country was based on traditional biomass technologies. By the turn of the century, the country's energy consumption had increased manifold, but then relying on modern and innovative energy systems largely based on renewable energy sources. Brazil is now rapidly approaching full electricity coverage thanks to recent efforts prioritizing universal coverage (MME, 2012; Gomez and Silveira, 2011).

The strong reliance on renewable energy sources observed in the Brazilian energy matrix differentiates it from most countries. The common rule in the past decades has been that, as countries became richer and more industrialized, they increased their dependency on fossil fuels. In developing countries, such a development has gone hand in hand with the abandonment of wood fuel in favor of more efficient fuels and technologies (Silveira, 2005). As a result, many countries are trapped in imports of costly fossil fuels and unsustainable energy paths. Unfortunately, this path is still the most common despite an increasing understanding about the need to deploy sustainable alternatives. Also rich countries such as the G7 which, given their wealth, could have chosen to evolve in a more sustainable direction have not done so in the past decades, and remain largely dependent on fossil fuels.

The Brazilian experience shows that it is possible to evolve in a different way, also when starting from low levels of economic development. Although Brazil did not manage to prevent continuous increase in oil demand due to the rapid expansion of its economy and the development priorities chosen, it has managed to modernize the energy sector and diversify the energy sources of the

---

[1] BRIC is an acronym refering to the countries of Brazil, Russia, India and China, which are considered as being in an advanced stage of economic development.
[2] G20 Refers to the group of 20 major economies: 19 countries plus the European Union. Together, the G-20 economies comprise more than 80% of the global gross national product (GNP).
[3] http://www.bbc.co.uk/news/business-17272716.

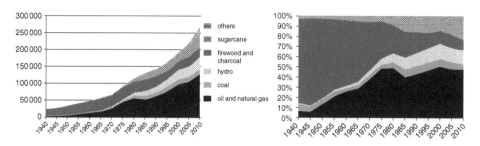

Figure 4.1.   The development of the Brazilian energy matrix 1940–2010, in 10^3 toe (EPE, 2011).

country in innovative ways. This has led to a reduction in the country's relative dependency on oil and improved security of supply. Bioenergy has played a key role in this process as is discussed in this chapter.

My purpose here is not to give a full account of the efforts made to transform the Brazilian energy matrix but to look at the evolving role played by bioenergy in the country. I start providing the broad context in which bioenergy efforts have taken shape in Brazil in the last few decades. I continue with a description of the present conditions of bioenergy in Brazil with particular attention to (i) forest-based bioenergy in the form of traditional fuel wood and charcoal, and (ii) biofuels for transport in the form of ethanol and biodiesel. Biomass utilization in connection with urban waste management is not covered in this chapter, but it is worth mentioning that this is also an evolving area of bioenergy expansion in urban areas of Brazil.

I look at how bioenergy applications have evolved in the past decades and what perspectives they have in the next ten years. I rely on scientific literature, official data provided in consolidated energy balances for the country (EPE, 2011a) and other Brazilian statistics (IBGE), information provided by business association reports, and my own research and extensive on-the-ground experience of working with bioenergy in Brazil. For the analysis of perspectives in the next ten years, I particular refer to the Brazilian energy expansion plan PDE 2020 and conjuncture analysis made by EPE (2011b; 2011c). In addition, ongoing international processes, particularly the negotiations under the climate convention (UNFCCC) and formation of biofuel markets are taken into account. I finalize with reflections about the development of bioenergy in Brazil, and the importance of the Brazilian experience for climate change mitigation, and other countries as they develop their bioenergy potential.

4.2   FROM TRADITIONAL FUELWOOD TO MULTIPLE BIOENERGY SYSTEMS

The economic development of Brazil in the past decades has relied on the expansion and trans-formation of the country's infrastructure, including the energy base. In the middle of the past century, Brazil was in a position quite similar to that of many developing countries in Africa today (see Fig. 4.1). Not only was industrialization still at an early stage, but agriculture had low productivity and most of the country relied on wood fuel for energy purposes. Electricity was only available in larger urban areas – and only some 35% of the population was urban, compared to 85% today (EPE, 2011a).

The period of the late sixties and early seventies is often referred to as the period of the so-called *Brazilian miracle*, a time when the Brazilian economy really took off (Baer, 2008). This is reflected not only in the growth of energy demand in the country, but also in the initiation of new developments in the energy sector, including deployment of new energy sources and the shift towards modern technologies. The seeds sown in this period served as the basis for defining new directions when oil price shocks forced the government to look for other energy options. Notably,

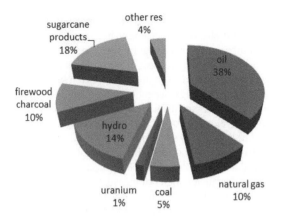

Figure 4.2.   Primary energy supply in Brazil, by source 2010 (EPE, 2011a).

the expansion of the energy supply in Brazil continued at high rates in the 1980s, despite the fact that this is often referred to as the lost decade, due to the low economic growth of the period.

Figure 4.1 shows the development of the Brazilian energy matrix between 1940 and 2010. It is interesting to notice that biomass still accounted for more than half of Brazil's energy supply in the early 1970s, mainly in the form of firewood and charcoal. Sugarcane-based bioenergy was limited to internal uses of residues in the sugar production. Biomass was the most important source of energy in Brazil until 1973 when it was surpassed by oil. Oil became gradually more important particularly as road transport was expanding rapidly with the construction of road infrastructure that was to connect huge extensions of land. The use of gas has only become more significant in the last decade. Hydropower became the backbone of the Brazilian electricity system, a position that it still holds with as much as 74% of the country's electricity being generated in hydropower plants (EPE, 2011a).

Brazil had a total energy supply of 269 million toe in 2010. Figure 4.2 shows the total primary energy supply by source in 2010. Biomass accounted for 28% of the total energy supply, being the second largest energy source in the country after oil. Two thirds of that, or 18% of the total supply, was sugarcane-based. This can be taken as an indicator for the modernization of the bioenergy segment in Brazil since most of the ethanol production and use in the country is connected to rather modern supply-and-use chains from agriculture to industrialization all the way to fuel distribution and utilization. More recently, Brazil has been developing bio-diesel production, an industry that is also based on modern technologies and applications.

Notably, bioenergy continued growing in the past decades as an important source of energy in Brazil, with multiple carriers and end-uses. However, there has been a revolutionary change in the role played by biomass as energy source, and the technologies used. Sugarcane-based energy has grown as a result of the ethanol program launched in the mid-1970s aimed at gasoline substitution, and become an important development engine in the Brazilian economy. Meanwhile, the use of firewood decreased in importance over time as LPG entered the domestic markets for cooking and the population became more urbanized. Charcoal remained important for metallurgical industries, although the integration of plants have led to a gradual shift towards imported coal. Still charcoal is important in the supply chain of the sector, and the traditional character of charcoal production contrasts with the modern character of the iron and steel mills operating in the country. In absolute terms, the use of firewood and charcoal decreased by 18% between 1970 and 2010, while the sugarcane-based energy increased manifold in the same period.

Certainly, the oil price shocks of the 1970s served as incentive to the development of domestic energy alternatives in Brazil. Oil prices went up fourfold in 1973 at a time when Brazil's total energy dependency was around 35% (70% for oil), putting a strong pressure on the Brazilian economy. This economic pressure resulted in increasing borrowing, economic recession, inflation and a serious debt crisis in the 1980s. Brazil became an net oil exporter in 2006 and is now heading towards becoming a major oil exporter in the near future.[4] Since peak-oil is approaching and most oil is coming from the Middle East, the prospect of becoming the 5th largest oil producer in the world within this decade gives Brazil a very strategic position when it comes to global energy security. As of 2010, the Brazilian energy dependency amounted to 7.8%, mostly in the form of imported coal for metallurgical uses and a small amount of electricity (IBGE, 2011).

Brazil's strategies for the development of energy supply after the 1970s included oil prospection and research, deployment of the country's hydropower potential, and development of alternative energy sources such as ethanol from sugarcane. The successful achievements were many but there have been also many social and environmental costs. In particular, deforestation has continued throughout the past decades despite a number of measures to curtail environmental degradation. Environmental and social liability in the charcoal segment is one of the most serious to be tackled by the Brazilian industries. As climate change issues become more pressing, renewed attention is put on the Brazilian deforestation problems, the largest source of emissions in the country. As much as 61% of the Brazilian emissions are related to land and forest use change while energy only accounts for 15% of the total emissions.[5]

The next two sections offer a brief description of main sources and uses of bioenergy in Brazil with focus on forest-based biomass and transport fuels. It is not an exhaustive description but does provide an overview of how Brazil has systematically developed major segments of bioenergy, and how biomass has become a modern energy alternative in the country also serving the objectives of sustainable development.

4.3   FOREST-BASED BIOMASS IN BRAZIL

Forests have tradionally been important sources of food, timber, pulp and paper, medicines and energy. Biomass has been used in the early stages of industrial development, but resource depletion and availability of other alternatives have gradually led to energy source substitution and improved forest management as a way to protect standing forests. Still native forests are being used as sources of energy particularly in developing countries. According to the FAO's recent global forest assessment, deforestation continues throughout the world even if the rate of clearance has slowed down somewhat (FAO, 2011). The prospects are quite good for increasing the amount of planted forests in the next two decades, although deforestation and degradation is most likely to proceed in regions where environmental control is weak and poverty is widespread. Planted forests are becoming increasingly important but still only account for less than 2% of the land use globally (Carle and Holmgren, 2008).

In Latin America, most of the deforestation is presently related to the expansion of agriculture and urbanization but the causes vary around the continent (FAO, 2011). In some regions, the demand for round wood is still a major source of deforestation. Worldwide, there is expectation that natural forests will become more attractive to preserve as eco-systems are better valued, this leading to improved management and market value creation for natural stands. In fact, the process of valuing natural resources, also contemplated in the REDD[6] mechanism of the climate convention, could help change the use of natural forests from single purposes such as energy or round wood to multiple services including, for example, product extraction together with

---

[4]The Economist, Nov 5, 2011. Filling up the future, available at http://www.economist.com/node/21536570.
[5]Information on the Brazilian inventory available at http://www.brasil.gov.br/cop-english/overview/what-brazil-is-doing/emissions-inventory.
[6]**UN-REDD** is an initiative launched in 2008 under the climate convention aimed at Reducing Emissions from Deforestation and forest Degradation (REDD) in developing countries. More info available at http://www.un-redd.org/ and http://www.redd-monitor.org/.

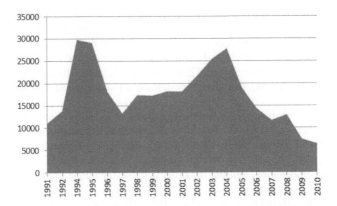

Figure 4.3. Deforestation in the Brazilian Amazon 1991–2010, in km^2 (IBGE, 2011).

leisure and tourism. In this context, guaranteeing local access to the forest and the continuation of non-cash activities is a major issue for local populations (Dhakal *et al.*, 2012).

Brazil is one of the five richest countries in the world when it comes to forest coverage, and the number one in continuous coverage of tropical forest. In fact, 22% of the world's forests are located in Brazil (FAO, 2011). Deforestation is still a major problem and the largest source of greenhouse gas emissions in the country. In general, deforestation has been performed to expand agriculture and cattle raising activities, to produce charcoal aimed at the iron and steel industry, to supply wood for construction purposes or markets for exotic species, to name some of the major causes. After the start of industrialization in the 1930s, the causes of deforestation have varied significantly among regions. In the Southeast, the development of the iron and steel industry triggered the demand for charcoal after the 1940s. In the west and north, policies to occupy and explore the region fostered deforestation particularly after the 1960s. The development of a metallurgical pole in the Amazon in the last two decades has also pushed deforestation in the region. The use of fuel wood for cooking has had a decreasing importance when it comes to threats to Brazilian forests.

Internationally, threats on the Amazon region have caught particular attention. Figure 4.3 shows deforestation in the Brazilian Amazon, where more than half of the Brazilian deforestation occurs, between 1991 and 2010. The decreasing rate of deforestation observed is a result of a number of measures to contain forest clearance in the region including continuous monitoring with satellite images. However, Brazil has other important biomes, which have been under continued pressure, particularly the Atlantic forest, and *cerrado* areas, which have been systematically used for charcoal production or cleared to accommodate agricultural expansion in the past. The remaining areas of Atlantic forest are protected at present and also the *cerrado* to some extent, but illegal activities still occur particularly in the latter.

There is expectation that the national and international demand for wood will increase fast in the near future, reaching 21 million m^3 per year. This will require 36 million ha of forests in a 30-year cycle. This too could lead to increased pressure on native forests, and has led to the proposition to open part of the public forests for sustainable management as a way to avoid uncontrolled and illegal deforestation. Although planted forests have been substituting natural forests in a number of industrial applications, the expected increase in demand could pose new threats to the Brazilian forests. At present, 11 million ha of public forests are managed on a productive basis. There are still significant extensions of forests under federal and state jurisdiction, which could be used for economic purposes on a sustainable basis, thus avoiding illegal pressure on natural forests – this excludes natural reserves and indigenous land. If properly implemented, an increase in the economic value of forests might help towards more preservation (SFB and IPAM, 2011).

Increasing pressure to explore forests has also led to a discussion about the national forest code dating from the 1960s. The code is considered quite progressive, but has often been illegally challenged due to widespread practices of deforestation. The Congress approved a number of revisions in 2011. The code needs also to be approved by the senate and President Dilma Rousseff. Environmental groups in Brazil and abroad have strongly condemned the proposed changes in the Brazilian forest code[7] claiming that, if approved, there is risk for significant reduction in forest preservation.

Despite continued threats to natural forests, significant progress has been achieved in Brazil in the past decades when it comes to the control of deforestation and establishment of reforestation practices. The area of planted forests has increased successively in the country since the mid-1960s when a new forest code was established by law, generous fiscal incentives were put in place to promote reforestation, and the Brazilian Institute for Forest Development (IBDF) was established to implement forest policies. FAO (2003) estimates that some 4 million hectares of forests were planted in Brazil between 1967 and 1987. The incentives were removed in 1988 and forest policies became the responsibility of the Brazilian Institute for Environment and Renewable Resources (IBAMA), which is under the Ministry of Environment and the Legal Amazon. For many years, the forest industry in Brazil has claimed that this shift in responsibility also marks the point when forests started being treated as an environmental issue, disfavoring the development of the forestry industry.

Truly, in the past two decades, forests have not received the same type of attention as agriculture when it comes to economic policies. Nevertheless, decades of practice and forest expansion have set the basis for a modern forest industry in the country, which is expanding rapidly at present. Brazil has become well known for its fast-growing forests. The average yield of a Brazilian forest is around 450 GJ/ha/y, while high yields of commercial forests in countries such as the US, Finland or Sweden will not reach more than one fifth of that (IPCC, 2001). Best-performing eucalyptus plantations in Brazil can reach up to 1000 GJ/ha/y, indicating the enormous potential still to be explored through continued research in this area. Development of research in silviculture has helped guarantee a competitive position for Brazil in the global context of the pulp and paper industry. Forest productivity has practically doubled in Brazil since 1980, growth rates now reaching an average of 44 m^3/ha/y in eucalyptus plantations using seven rotation years, and 38 m^3/ha/year in planted pine forests using 15 rotation years. However, a level of 70 m^3/ha/y is achievable in eucalyptus plantations (BRACELPA, 2011).

There are approximately 6.5 million hectares of planted forests in Brazil today (0.7% of the country's territory), two thirds using eucalyptus and one-third pine (ABRAF, 2011). The pulp and paper industry in Brazil, one of the world's major producers, is totally based on planted forests. According to the Brazilian pulp and paper association, BRACELPA, this sector alone had 2.2 million ha of planted forests throughout Brazil in 2009 (BRACELPA, 2011). In addition, the steel producers have traditionally planted forests for the production of charcoal. All in all, the area of planted forests is increasing in Brazil, having also more recently expanded towards the western and northern parts of the country.

In short, markets for forest products including pulp and paper, sawn wood, fuel wood and charcoal are well established and the comparative advantages of Brazil have made the country a major player among forest-based industries. There are well-established companies in the country that have mastered the know-how of reforestation. In fact, companies are largely involved in silviculture research in Brazil. The potential is quite large for expansion of planted forests, particularly eucalyptus. Nevertheless, many challenges lie ahead as the demand for forest products increases and continues putting pressure on natural stands. Transforming policies for environmental protection into policies for sustainable and productive use of natural forests, while providing the necessary incentives for reforestation remains a challenge in Brazil despite the progress made in the past few decades.

---

[7]http://www.sosflorestas.com.br/

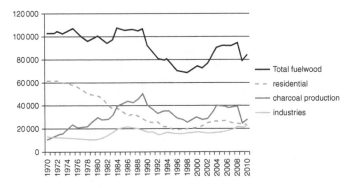

Figure 4.4.   Total fuel wood in Brazil and final uses, 1970–2010, in 10^3 tonnes (EPE, 2011a).

4.3.1   *Fuel wood and charcoal – traditional uses of biomass in Brazil*

In the shadow of on-going modernization of forestry activities, which particularly involves extensive plantations aimed at pulp and paper, traditional energy uses of forests remain significant in Brazil. In fact, forests are an important source of biomass for energy, accounting for approximately 10% of Brazil's total energy supply. The most common is the use of forest residues as firewood and charcoal production but there are also large extensions of eucalyptus plantations dedicated to charcoal production for industrial use.

The production of fuel wood in Brazil amounted to 84 million tonnes in 2010. This indicates an absolute decrease by almost one fifth in the use of fuel wood since 1970 (Fig. 4.4). One third of the fuel wood is used for charcoal production, another third in industries such as ceramics and the pulp and paper, and the other third is used domestically. Between 1970 and 2010, the Brazilian population grew from 94.5 million to 190.8 million. During the same period, the demand for firewood for domestic use went down two thirds. This was possible due to the shift towards other cooking fuels at the same time that the population also became more urbanized. The most common fuel used for cooking in Brazil today is LPG but other uses may co-exist, for example the use of charcoal for grilling which is a tradition in the country. Meanwhile, the use of fuel wood has decreased significantly.

Almost twenty million tonnes of fuel wood were used to produce approximately 7 million tonnes of charcoal in 2010 (EPE, 2011a). The use of charcoal in Brazil is closely related to industrial uses. Although charcoal is also used domestically and in agrarian applications, these uses have been rather stable in recent years. Still, this translates into more than 2 million people depending on charcoal for cooking, mainly in the northern parts of the country (Nogueira *et al.*, 2007). The most important use of this energy source is found in metallurgical industries, particularly the production of pig-iron, steel, and iron alloys. The use of charcoal is a unique feature of the Brazilian iron and steel industry when compared internationally. Although charcoal has been historically used in many countries for iron reduction, it has been gradually substituted by coke. This substitution has also partially taken place in Brazil where integrated plants have shifted towards imported coke particularly as a result of more recent expansion. Nevertheless, given the fact that Brazil does not have coke and the fact that environmental control on native forests has been lax in the past, charcoal was kept as a major energy input in non-integrated metallurgical industries. Charcoal-based furnaces account for 25–30% of the total Brazilian production in the sector (CGEE, 2010).

Figure 4.5 shows how the use of charcoal has developed in Brazil between 1970 and 2010 based on official statistics and the consolidated energy balance (EPE, 2011a). Until 1985, the use of charcoal increased rapidly in line with the expansion of the industry and exports of metallurgical products. Brazil had a period of recession in the late 1980s and early 1990s during which production and exports went down. There was significant recuperation after the turn of the century

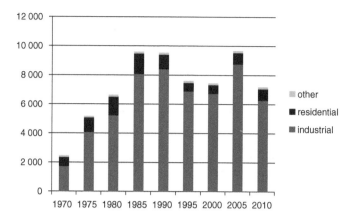

Figure 4.5.   Charcoal use in Brazil 1970–2010 in 10^3 tonnes (EPE, 2011a).

but then, again exports were affected by the international economic recession. Thus, the use of charcoal in Brazil varies in line with the conjunctures that affect the demand for metallurgical products nationally and internationally, while other uses of charcoal have been quite stable.

Brazil is the ninth largest steel producer in the world. The production has increased continuously in parallel with the industrialization of the country and demand in world markets. Today, Brazil has installed capacity for production of more than 40 million tonnes of steel in both integrated and semi-integrated plants – all certified according to ISO 14001 (IAB, 2010). Integrated and semi-integrated plants are 27 in total, controlled by eight national and international business groups, and located in ten different states in Brazil from north to south. In 2011, production reached 32.5 million tonnes of which approximately two thirds were consumed internally and the rest exported. Since 2009, these industries have developed greenhouse gas emissions balance. According to IAB, the Brazilian steel association, only 5% of the iron reduction in these plants is made with charcoal (IAB, 2011). However, these plants receive pig-iron from independent producers where the use of charcoal is the most common. There is also an international market for pig-iron.

The modern segments of the metallurgical industry in Brazil in the form of integrated and semi-integrated plants contrasts with the often lower efficiency of pig-iron plants and the charcoal production on which it is based. Large losses have been observed along the charcoal supply chain. Some of the losses are gradually being used for the production of briquettes but there is still a significant amount of waste. Research to develop charcoal kilns has been conducted continuously but the implementation of more efficient technologies has been slow and uneven.

Reforestation aimed at charcoal production does take place, particularly articulated by the large steel companies, which also use more efficient technologies. Nevertheless, charcoal production has been traditionally decentralized in Brazil and largely based on the use of native forests and traditional kilns. Large companies both produce and buy charcoal from small producers. The *cerrado* area, the second largest biome of the country, has been highly affected by this process. Violation of environmental rules is still a problem also among the large international companies as also indicated by the industry's sustainability report (IAB, 2010). Thus, forests are being planted, but native forests are still being lost.

In the past years, forest-based industries including charcoal have been continuously discussed in Brazil. Internationally, however, there has been very little interest in scrutinizing charcoal production and use in Brazil. While there are companies organizing charcoal production in large scale based on planted eucalyptus, a significant part of the production is decentralized, coming from many small producers. The conditions of traditional charcoal production are socially appalling. Approximately half of the wood source is illegal thus linked to natural forests. The modern character of the metallurgical industry stands in strong contrast with the conditions of charcoal production in many regions of the country.

4.4   BIOFUELS FOR TRANSPORT

4.4.1   *The development of modern bioethanol production*

The history of modern bioenergy in Brazil is strongly linked to the development of the ethanol industry after the 1970s. The oil price shock of 1973 caught the world by surprise, revealing not only the strong reliability of global energy systems on oil but also the vulnerability of many economies to oil price variations. In Brazil, oil imports amounted to approximately 70% of the national needs in the 1970s and actually increased to a proportion of 85% by 1980 (EPE, 2011a). Oil price increases implied very large and increasing energy import costs, strongly affecting the Brazilian trade balance. However, between 1980 and 2005, Brazil not only managed to revert the situation, but actually became an oil exporter. The ethanol development in Brazil is now well known for being the most successful of the global attempts to substitute oil in transport, both when it comes to scope and scale. It is well documented and, therefore, there is no need to review the whole process here (Moreira and Goldemberg, 1999; Hira and Oliveira, 2009). However, it is worth looking into what has been achieved in the past decades and putting it in perspective to understand the transition that is being accomplished in Brazil, from traditional to modern bioenergy.

Like many other countries, when oil prices went up, Brazil had to search for new energy sources particularly to substitute oil. Sugar production was well established in the country – it was actually Brazil's oldest industry. However, despite its export capacity, the industry was still rather traditional and had low productivity. Nevertheless, the government identified the sugar industry as a good starting point for building an ethanol industry. Using the existing agricultural structure and the potential for coordinating sugar and ethanol production, a set of supply and demand measures were put in place to boost the segment. It made sense in the context of rising oil prices, the rapidly growing car ownership in the country, increasing transport needs, and the ambition to develop agriculture and industry, generate jobs and economic development. In addition, sugar prices declined significantly in 1974 and this served to mobilize producers to modernize the industry and develop new products. Finally, Brazil had previous experience with using ethanol in transport particularly from the world war periods (Hira and Oliveira, 2009).

The Brazilian ethanol program Proalcool was launched in 1975 with the objective to reduce oil dependency, promote the development of ethanol fuel and strengthen the sugarcane and sugar-producing sector (GoB, 1975). It included both expansion of sugarcane production and distilleries, as well as development and modernization of the whole supply chain from agriculture to distribution. Initially, also manioc was contemplated as a potential ethanol crop, but sugarcane crops and sugar production offered a synergy of higher economic value, creating impulse for the so-much-needed modernization of agriculture and the sugar industry (Moreira and Goldemberg, 1999; Hira and Oliveira, 2009). The triggering effect that the ethanol program had on the industry is illustrated by the rapid expansion of the sugarcane and sugar-ethanol production. Figure 4.6 shows the development of sugarcane crops between 1970 and 2010.

The Proalcool program combined a number of elements that, together with the exogenous shock represented by increasing oil prices, contributed to the successful development of the sugar-ethanol industry in Brazil. The policies put in place included incentives to expand ethanol production, mandatory targets for ethanol mix in the gasoline, agreement with car manufacturers to produce ethanol-cars, procurement to create government car fleets driven by ethanol, research and development to improve crops and yields of sugarcane.

The Brazilian move was innovative and bold, and the results came relatively fast. There was significant improvement in yields on the agricultural side and design of distilleries, and successful expansion of the distribution infrastructure throughout the country. In the first ten years, the

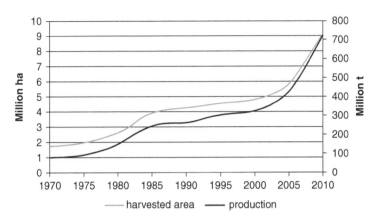

Figure 4.6.   Sugarcane production expansion in Brazil 1970–2010 – harvested area in million hectares and production in million tonnes (IBGE, 2011).

harvested area doubled and the production almost tripled reflecting significant increase in yields. The expansion of sugarcane has been constant since 1975 albeit much slower between 1985 and 2000 (Fig. 4.6). Cane-based energy increased by 120% in the first five years of the ethanol program, and then doubled between 1980 and 1985. The initial target of 3.5 billion liters set for 1980 was achieved already in 1982. In 1985, 95% of the light vehicles produced and sold in Brazil were ethanol driven.

The rapid fall in oil prices in the second half of the 1980s put a strong pressure on the Brazilian ethanol program. The government had difficulties justifying its focus on the fuel shift at a time when oil prices were low and climate change efforts had not started. There was not yet broad common understanding about the multiple benefits of the sugar industry for the country. At the same time, the Brazilian economy was being gradually opened up, and ethanol-related policies moved towards removing price regulations and the incentives previously provided to the sector. As a consequence, research and development on sugarcane production slowed down, while producers shifted to sugar, where prices had recovered and offered a better return in export markets. The situation culminated in a shortage of ethanol in the pumps leading to ethanol and methanol imports, undermined consumer confidence and collapse of the market for ethanol cars (Rosillo-Calle and Cortez, 1998).

However, some positive effects came out from the challenges of the late 1980s as well. In particular, the efforts made by producers to adapt to the new policy and market context led to significant efficiency improvements. The average ethanol production cost declined twice as fast in the late 1980s compared with the previous ten years (Goldemberg et al., 2004; Moreira and Goldemberg, 1999). The high blend of ethanol in gasoline guaranteed a market which, given the expansion of demand, still offered potential to absorb production. Nevertheless, production increased much slower in this period than before. The introduction of the flex-fuel technology in 2003 boosted a new phase in favor of ethanol markets. With a well-established infrastructure for ethanol distribution, and flexibility for consumers to choose their fuel, the flex-fuel cars gained the new sales market very fast. In 2010, flex-fuel cars comprised 86% of new car sales in Brazil. Today, 44% of the Brazilian fleet of light vehicles is composed of flex-fuel vehicles. The production of ethanol doubled since the introduction of flex-fuel cars and is now at 28 billion liters of ethanol per year (EPE, 2011c).

Since 2005, the area planted with sugarcane doubled and the production almost tripled (see Fig. 4.6). In 2010, the area harvested reached 9.0 million hectares, while production reached 717 million tonnes. The area planted is equivalent to 2.7% of Brazil's total arable land. Yields increased by 30% between 1990 and 2010, reaching 80 tonnes/ha in average. In the southeast region, however, yields have reached considerably higher levels. Sugarcane amounts to 18.4% of the agricultural production value in Brazil, only second to soya, which amounts to 24.2% of the total value (IBGE, 2012).

Another important contribution of the sugar-ethanol industry to the overall energy supply in Brazil refers to the generation of heat and power using sugarcane residues. Traditionally, the *bagasse* was used to generate heat and power and meet internal process needs. In 2009, however, sugar-ethanol producers exported almost 6 TWh of power to the grid. The opening of electricity markets and government efforts to diversify the sources of electricity generation have provided an incentive to upgrade cogeneration facilities using *bagasse* (Khatiwada *et al.*, 2012). The CDM[8] (Clean Development Mechanism of the climate convention) also helped boost investments on efficiency improvements since bioelectricity is eligible to generate tradable certificates. In fact, this type of project represents one third of the CDM renewable energy projects realized in Brazil.

All in all, the ethanol program started in 1975 has contributed to a reduction of more than 600 million tonnes of CO_2 emissions. Increased climate benefits are expected as the production increases further and fuel substitution proceeds both nationally and internationally due to the formation of global ethanol markets. Despite preoccupation about the competition between food and fuel, the Brazilian food production has increased continuously side by side with the expansion of sugarcane for sugar and ethanol over the past decades. In fact, Brazil is the world's largest exporter of grain. To a great extent, the productivity gains accrued from the modernization of the agriculture have benefited both food and fuel production.

In the next decade, ethanol production is expected to increase further in Brazil to meet both national and international demand for transport fuel. EPE (2011c) projects continued expansion of sugarcane planted areas and production capacity. Further efficiency improvements shall also be achieved along the supply chain, including increasing production and importance of bioelectricity. In this way, ethanol shall remain as a competitive option in the market. As a result of the agro-ecological zoning developed by the Brazilian government, it is not possible to plant sugarcane in the sensitive biomes of the Amazon and Pantanal (wetlands in the center-western part of Brazil). Neither is sugarcane expansion allowed in areas where there is native vegetation, for example in the *cerrado*. Nevertheless, some 65 million hectares can still be used for sugarcane (Leite *et al.*, 2009).

4.4.2   *The development of biodiesel production*

Although the use of vegetable oils in engines dates from the experiments made by Rudolf Diesel as long ago as 1900, Expedito Parente, a Brazilian researcher, was the first to patent biodiesel in the world as late as 1980. In Brazil, the first attempts to use vegetable oils in transport date from the 1940s. Vegetable oils had been used in emergency situations but were not considered suitable as heavy vehicle fuel, in contrast with biodiesel, which can be mixed with the fossil diesel in varied proportions (Yusuf *et al.*, 2011). More recently, biodiesel has attracted global interest and many countries are investing in research in this field. Brazil has an intermediary position in this new context but is reviewing strategies, given the recent success of the biodiesel program so far and potential for market development.

The Brazilian National Energy Commission created the Pro-Oleo program in 1980 aiming at 30% fossil diesel substitution initially (Pousa *et al.*, 2011). Like in the case of the Proalcool, the Pro-Oleo program also included research efforts, including the development of technologies for biodiesel production, and engines, which were done in cooperation with motor manufacturers in the country. However, the biodiesel program did not get the same government engagement and public support as the alcohol program. Eventually, production costs were judged too high to be competitive, and the program had already been abandoned in the mid-1980s when oil prices came down.

---

[8]http://cdm.unfccc.int/

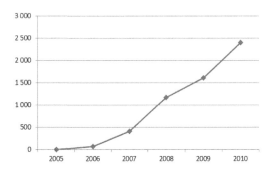

Figure 4.7.   Biodiesel production in Brazil 2005–2010 – in 10^6 liters (IBGE, 2011).

At the end of the 20th century, the issue of biodiesel was on the government's agenda once again, and in 2002 a biodiesel program was launched aimed at the substitution of 5% of the fossil diesel consumed in the country by 2013. Production of biodiesel was started in 2005 when a mixture of 2% biodiesel was also authorized. Later, the mix became compulsory, and was successively raised to 5%. Figure 4.7 shows the rapid development of biodiesel production in Brazil, which reached 2.4 billion liters in 2010, thus allowing the 5% mix of biodiesel in the total diesel consumption of the country, and meeting the mandatory target long before the target year (EPE, 2011a). Meanwhile, the installed capacity is already in the order of 6 billion liters and the government is being asked to set new targets and authorize increases in the mix, initially to 7% and gradually up to 20% (IPEA, 2012).

Notably, Brazil was able to establish production to meet a 5% biodiesel mixture in only 5 years. There is here a clear distinction from the efforts made in the 1980s and the successful policies implemented in the past few years. First, there is now a global conducive environment to develop biofuels for transport in face of high oil prices and climate change mitigation challenges. Second, Brazil had a long and successful experience developing the ethanol industry when the recent biodiesel program was launched. This includes the modernization achieved in agriculture, and technological development favored by R&D, both very important for a competitive biodiesel production.

One point to notice, however, is that the biodiesel program was launched to be more socially inclusive and diverse than the ethanol program. Given the variety of species that can be used for biodiesel production in Brazil, the idea was to build upon regional vocations and promote different species. However, although production goals and fuel mixture were reached, the dominance of two major sources indicates that more attention is needed if the distributive social and regional benefits are to be fully explored. Most of the production came from soya (85%) and a small part from meat fats (13%) in 2010, indicating that the most established and export-oriented segment of the Brazilian agribusiness (soya) was able to very rapidly capitalize on the synergies that the market incentives for the biodiesel program made possible. Notably, this has resulted in significant participation of international capital in the development of biodiesel production, and higher capital concentration than in the ethanol sector (IPEA, 2012).

The social certification created early in the program to favor small producers has had limited impact. Only some 109,000 families have benefited directly from the program in southern regions, while the target was 245,000 families with particular focus on the northeast (IPEA, 2012). This goes hand in hand with the fact that soya is often organized in large-scale production and regionally concentrated. Likewise, the biodiesel production is mainly concentrated in four states of the center-south regions while the northeast would better contribute with other crops. For other crops to become more important, technological improvements are needed to establish cost-efficient industrial processes for inputs other than soya. Since at least 70% of the production costs of the biodiesel are in the oil source, there is a need for coordinating R&D efforts with distributive policies to achieve the full social benefits of the biodiesel development.

Given the rapid development of the biodiesel industry, idle capacity awaiting new legal frameworks to increase production, and on-going discussions for a new regulatory framework for the sector, it is surprising that the government's ten-year plan for the energy sector indicates no expectation to increase the percentage of biodiesel mix in the near future. This is justified by the fact that input prices are expected to increase in the next years, putting biodiesel in strong competition with other end-uses for relevant crops. In the ten-year plan report, EPE (2011b) indicates an increase from 2.4 billion liters in 2010 to 3.8 billion liters in 2020, which is needed to continue fulfilling the compulsory mix of 5% biodiesel. This is definitely a very conservative view given that installed capacity already allows production of 6 billion liters and the industry is putting pressure on the government to increase the compulsory mix. In any case, a new legal framework for biodiesel is needed to strengthen the social and regional development dimensions of the program, as well as research and technological innovation in the sector.

4.5 BIOENERGY – OPPORTUNITIES FOR SUSTAINABLE DEVELOPMENT

The Brazilian economic development of the past decades is reflected in the expansion and transformation of the country's energy base. Not only has the Brazilian energy consumption increased rapidly in line with major development efforts, but the diversification of the energy sources has been quite significant. In this process, bioenergy not only remained an important source of energy in the country but is increasingly characterized by modern energy transformation, multiple carriers and end-uses. Synergies have been established between bioenergy and various sectors of the economy including agriculture and forestry. To promote these synergies, innovative cross-sectorial policies and multiple actions have been implemented, and significant progress has been achieved. Still, many challenges lie ahead.

Today, Brazil is the world leader in the production and use of sugarcane ethanol – the first large scale alternative for substitution of fossil fuel in the transport sector. The positive results of the efforts made have been gradually recognized internationally not least due to the climate change mitigation benefits achieved. The benefits of a more sustainable energy path are also strongly felt in the Brazilian economy, and this is triggering new efforts to continue on track. A study by Weidenmier *et al.* (2008) found empirical evidence that diversification of energy supply resulted in large macroeconomic benefits to Brazil. According to their study, the GDP was almost 35% higher in 2008 compared to 1980 due to reduction of oil imports, increased domestic oil production and development of sugarcane ethanol. In addition, diversification of the energy system helped reduce business-cycle volatility in the range of 14 to 22% particularly in the past decade. Three quarters of the welfare benefits are related to reduced oil imports and development of national oil production. However, sugarcane ethanol had a major role in the other part of the benefits not least in protecting the economy from oil price shocks.

Fuel wood and charcoal together still account for 10% of the energy supply in Brazil. Domestic uses of fuel wood for cooking are common in rural areas although gradually going down in importance as the economy grows and energy modernization reaches the countryside. Meanwhile, the increasing demand for round wood, pressure to expand agriculture and pasture, added to the urgency to generate income continue acting as strong counterforces to preservation of Brazilian forests, in spite of stringent environmental control. Charcoal production is still to a great extent based on natural forests, and this leads to significant impacts on sensitive biomes. Given the role of charcoal in the metallurgical industries, there are strong reasons to upgrade this part of the supply chain and transform it into a sustainable proposition to mitigate emissions and generate socio-economic development. Such a discussion has already been triggered in Brazil due to the emissions reductions potential implied.

In fact, new strategies are needed to change the role of forest-based bioenergy and make it an integrated part of the modernization of forestry activities in Brazil. For example, the potential of synergetic exchanges between the modern segment represented by the pulp and paper industry and the traditional segments of bioenergy has not yet been fully explored. Neither has this been fully contemplated in policies for the forest sector. Brazil's pulp and paper industry is expanding rapidly and is based on eucalyptus plantations. Residues from these plantations comprise a significant amount of fuel wood that is commonly used in small industries including pizzerias, ceramics and charcoal production. Significant amounts of eucalyptus plantations are also dedicated to charcoal production. Technological improvements in charcoal production are on the way and may lead to modernization of the segment. This implies higher efficiency and extraction of by-products, often with higher value than the charcoal itself, paving the way for improved sustainability in the metallurgical industry as a whole.

Furthermore, research and past experiences with eucalyptus and pine forest plantations provide a strong knowledge basis for further expansion of forests. A better exploration of the synergies between the pulp and paper industry, forest residues and bioenergy will lead to improved efficiency and sustainability of forestry-related industries. A systems approach in policy design and close cooperation between public and private entities will be needed together with the engagement of the civil society to make this a successful proposition. Improvements are needed in forest management so that the large environmental value of Brazilian forests can be properly combined with higher economic value on a sustainable basis – this will help promote the preservation of natural forests.

In the past decades, increasing environmental awareness has led to a number of efforts to deal with deforestation including strong environmental laws, community projects for local management of forest resources, and increasing control of forest activities and trade. Causes and figures on deforestation vary considerably from region to region, but Brazilian forests remain under strong pressure. Brazil's environmental laws provide a framework to protect natural forests and indigenous land but stronger enforcement of these laws, together with policies to promote afforestation and reforestation activities, and higher economic value for forests are necessary to guarantee sustainable development. Climate change mitigation efforts may help catalyze innovation in forestry not only in technology, but also in management practices, helping to mobilize the financial resources needed to enhance the sustainability of forest-based industries and the protection of native forests.

The Brazilian experience offers important lessons to other countries particularly developing countries aiming at orchestrating their own bioenergy transition. This is the case of many poor countries in Africa that are fully dependent on oil imports, which in turn imply a real drain in their economies. Their agriculture needs modernization, not least to produce more food, their industries wait for a dynamic push of markets, and their populations need jobs, income and electricity. Bioenergy can provide a catalytic role in mobilizing efforts to create new impulse in national economies. Certainly, coordination of national and international efforts, if achieved, can speed up the process of disseminating sustainable practices for bioenergy development. The climate change agenda can be used to support these national efforts, particularly among the poorest countries as they still depend on external financial support to pursue development goals.

The energy development of the last decades in Brazil reflects the shift from traditional to modern bioenergy, but also the increasing integration of energy in development strategies. The Brazilian experience shows that a lot can be realized at national level, and the modern bioenergy transition achieved in a few decades only. Through national mobilization of multiple stakeholders, and the capacity to define goals, catalyze industries, market forces and investments, together with coordinated efforts on the supply and demand sides, and R&D, Brazil achieved excellent success with the ethanol program. There is great promise when it comes to the biodiesel program and forest sector as well. As international competition intensifies, it will be important to devise new strategic policy frameworks to guarantee continued leadership in the ethanol segment (Souza and Macedo, 2011). Likewise, sectorial and innovative policies will be needed in the other bioenergy segments to fully explore their national and global benefits.

The future development of bioenergy is not given. Well-trodden unsustainable paths, established infrastructure and economic interests, lack of political systematic and cross-sectorial support, the conflict of fuel *versus* food, and public opinion are some of the barriers to be addressed in Brazil and in the world at large. Planning for social, economic, spatial and environmental balance simultaneously will be crucial for correcting distorted processes of environmental degradation, capitalizing on the bioenergy benefits and potential sector synergies, turning the present dynamics into processes of sustainable development.

ACKNOWLEDGEMENTS

This chapter has been written in the scope of bioenergy research carried out at the Division of Energy and Climate Studies. The division is partly funded by the Swedish Energy Agency.

REFERENCES

ABRAF: *Statistical Yearbook*. 2011, http://www.abraflor.org.br/estatisticas/ABRAF11/ABRAF11-EN.pdf (accessed July 2012).

Adams, J. & Christopher, K.: *Global Research Report* – Brazil. Thomson Reuters, 2009.

Baer, W.: *The Brazilian economy – growth and development*. Lynne Rienner Publishers, 2008, https://www.rienner.com/uploads/47e29b0ea1787.pdf (accessed July 2012).

BRACELPA (Brazil Pulp and Paper Association): Brazilian Pulp And Paper Industry. 2011, http://www.bracelpa.org.br/eng/estatisticas/pdf/booklet/booklet.pdf (accessed July 2012).

Carle, J. & Holgrem, P.: Wood from planted forests – a global outlook 2005–2030. *Forest Prod. J.* 58:12 (2008), pp. 6–18.

CGEE: Siderurgia no Brasil 2010–2025. Ministry of Science and Technology, Brazil, 2010.

Dhakal, B., Bigsby, H. & Cullen, R.: Socioeconomic impacts of public forest policies on heterogeneous agricultural households. *Environ. Resour. Econ.* 2012. Accepted for publication.

EPE: BEN – Brazilian Energy Balance. Ministry of Mines and Energy (MME), Brazilian Government, 2011a, https://ben.epe.gov.br/BENSeriesCompletas.aspx (accessed July 2012).

EPE: Plano Decenal de Expansão de Energia 2020. Ministry of Mines and Energy (MME), Brazil, 2011b, http://www.cogen.com.br/paper/2011/PDE_2020.pdf (accessed July 2012).

EPE: Análise de conjuntura dos biocombustiveis. Jan 2010–Dez2010. Ministry of Mines and Energy (MME), Brazilian Government, 2011c.

FAO: State of the world forests. Food and Agriculture Organization of the United Nations, Rome, Italy, 2011, http://www.fao.org/docrep/013/i2000e/i2000e.pdf (accessed July 2012).

FAO: Forests and the forestry sector – Brazil. Food and Agriculture Organization of the United Nations, Rome, Italy, 2003, http://www.fao.org/forestry/country/57478/en/bra/ (accessed July 2012).

Ferreira, N.C., Ferreira, L.G. & Miziara, F.: Deforestation hotspots in the Brazilian Amazon: evidence and causes assessed from remote sensing and census data. *Earth Interactions*, vol. 11, paper 1, 2007, pp. 1–16.

Furtado, A.T., Scandiffio, M.I.G. & Cortez, L.A.B.: The Brazilian sugarcane innovation system. *Energy Policy* 39:1 (2011), pp. 156–166.

GoB (Government of Brazil): Decreto No. 76.593, de 14 de novembro do 1975, D.O.U. de 20.02.1976, Government of Brazil, 1975.

Goldemberg, J.: Ethanol for a sustainable energy future. *Science* 315:5813 (2007), pp. 808–810.

Goldemberg, J. & Coelho, S.T.: How adequate policies can push renewables. *Energy Policy* 32:9 (2005), pp. 1141–1146.

Goldemberg, J., Coelho, S.T., Plinio, M.N. & Lucond, O.: Ethanol learning curve – the Brazilian experience. *Biomass Bioenergy* 26 (2004), pp. 301–304.

Gomez, M. & Silveira, S.: Rural electrification of the Brazilian Amazon – achievements and lessons. *Energy Policy* 38:10 (2010), pp. 6251–6260.

Hira, A. & de Oliveira, L.G.: No substitute for oil? – How Brazil developed its ethanol industry. *Energy Policy* 37:6 (2009), pp. 2450–2456.

IAB (Instituto Aço Brasil): Relatorio de Sustentabilidade. 2010, www.acobrasil.org.br/site/portugues/sustentabilidade/downloads/relatorio08_2010.pdf (accessed July 2012).

IAB (Instituto Aço Brasil): Aço Brasil Informa. 2011, http://www.acobrasil.org.br/site/portugues/biblioteca/pdf/acobrasilinformanovo.pdf (accessed July 2012).

IBGE: PAM – Produção Agricola Municipal 2010. 2010, http://www.ibge.gov.br/home/estatistica/economia/pam/2010/default.shtm (accessed July 2012).

IBGE: 2011, http://www.ibge.gov.br (accessed July 2012).

IPCC: Summary assessment report. 2001.

IPEA: Biodiesel no Brasil: desafios das políticas públicas para a dinamização da produção. Comunicado do IPEA no 137. Secretary for strategic affairs of the Brazilian government, 2012.

Khatiwada, D., Seabra, J., Silveira, S. & Walter, A.: Power generation from sugarcane biomass – a complementary option to hydroelectricithy in Nepal and Brazil. *Energy* 48:1 (2012), pp. 241–254.

Kuchler, M.: Unravelling the argument for bioenergy production in developing countries: a world-economy perspective. *Ecol. Econ.* 69 (2010), pp. 1336–1343.

Leite, R.C. de C., Lima Verde Leal, M.R., Barbosa Cortez, L.A., Griffin, W.M. & Gaya Scandiffio: Can Brazil replace 5% of the 2025 gasoline world demand with ethanol? *Energy* 34:5 (2009), pp. 655–661.

MME (Ministry of Mines and Energy) (2012). Programa Luz para Todos. Brazilian Government, http://luzparatodos.mme.gov.br/luzparatodos/Asp/o_programa.asp (accessed July 2012).

Moreira, J.R. & Goldemberg, J.: The alcohol program. *Energy Policy* 27 (1999), pp. 229–245.

Nogueira, L.A.H., Coelho, S.T. & Uhlig, A.: Sustainable charcoal production in Brazil, Food and Agriculture Organization of the United Nations, Rome, Italy, 2007, http://www.fao.org/docrep/012/i1321e/i1321e04.pdf (accessed July 2012).

OECD: OECD economic surveys – Brazil. 2011, http://www.oecd.org/dataoecd/12/37/48930900.pdf (accessed July 2012).

Pousa, G.P.A.G., Santos, A.L.F. & Suarez, P.A.Z.: History and policy of biodiesel in Brazil. *Energy Policy* 35:11 (2011), pp. 5393–5398.

Santos, M.A.: A brief history of energy biomass in Brazil. COPPE/UFRJ, Rio de Janeiro, Brazil, not dated.

SFB and IPAM: Florestas nativas de produção brasileiras – relatório técnico. 2011.

Silveira, S.: *Bioenergy – realizing the potential*. Elsevier, 2005.

Sousa, E.L.de & Macedo, I. de C.: Ethanol and bioelectricity – sugarcane in the future of the energy matrix. UNICA, Sao Paulo, Brazil, 2011 (Portuguese version dated 2010).

Uhlig, A., Goldemberg, J. & Coelho, S.T.: O uso de carvão vegetal na indústria siderúrgica brasileira e o impacto sobre as mudanças climáticas. *Revista Brasileira de Energia* 14:2, 2° sem (2008), pp. 67–85.

Weidenmier, M.D., Davis, J.H. & Aliaga-Diaz, R.: Is sugar sweeter at the pump? The macroeconomic impact of Brazil's alternative energy program. NBER Working Paper No. 14362, 2008, http://www.nber.org/papers/w14362.pdf (accessed July 2012).

Yusuf, N.N.A.N., Kamarudin, S.K. & Yaakub, Z.: Overview on the current trends in biodiesel production. *Energy Convers. Manage.* 52 (2011), pp. 2741–2751.

CHAPTER 5

Biomass in different biotopes – an extensive resource

Erik Dahlquist & Jochen Bundschuh

In this chapter, an overview will be made of the available biomass resources, primarily in Europe, covering different climate zones from north to south, from arctic over boreal into Mediterranean temperate or subtropical climates. Europe can be seen as an example of what is actually very similar around the globe in Russia, China and North America. Some other climates and biotopes are also covered as comparison, especially the tropics in West Africa (Twenneboah, 2000) and sub-tropics in South Africa. The intention is to give more details than the chapter on the global resources, where the coverage was more general. Much of the data come from Eurostat (2010) and World Development Indicators (2011). Concerning specific data on selected crop sources are among others Salisbury and Ross (1992) and Weidow (1998).

5.1 BIOENERGY IN NORTHERN EUROPE

One country in Europe has really utilized biomass energy in a very extensive way. This is Sweden, where 132 TWh biomass was used 2010 of a total 400 TWh energy. This is more than the amount of fossil oil used although the applications are different. Of the total energy use approximately 55–65 TWh/y comes from Nuclear power, and 60–70 TWh/y from hydropower. About 12 TWh electricity comes from bioenergy converted after combustion and steam production to electricity in steam turbines. In addition, 6 TWh wind power was produced 2011. Of course, it may be unfair to compare the total use of biomass to electric power made in other ways, but biomass is mostly used for heating buildings and for use in process industries. Thus, it is reasonable. Most of the fossil fuels thus are utilized for transportation, where 99 out of 103 TWh comes from fossil oil products in Sweden. The rest is electricity, ethanol and biogas. In the future, we can foresee that the gas and diesel fuel will be replaced with bio-fuels, as well as by electricity. The electricity then would be produced in CHP, combined heat and power plants, in energy combines, where different chemicals and electricity are produced aside of heating and cooling, as well as with hydropower, wind power and solar power. There should be no technical problem for Sweden to be independent of fossil fuels by the year 2050. The problem is more to find enough economic incentives for new technologies. In Dahlquist *et al.* (2007) and Paz *et al.* (2007), pathways to *fossil-fuel free energy system* are discussed on a regional level in the Stockholm–Mälardalen region of Sweden.

There is generally an issue if we can achieve a fossil-fuel-free society without causing harm to the environment. Different aspects of this are covered in many papers. Some examples like Bassam (1998) are discussing the use of different energy crops and their impact on environment and development. Lambers *et al.* (1998) are covering ecological aspects related to plant physiology, while Larsson and Granstedt (2009) are presenting how a sustainable agriculture could be achieved. One issue is to reduce the negative effect of agriculture on the Baltic Sea, which is today suffering strongly from eutrophication. Their conclusion is that we should halve the fertilizer dosage and not use more than what can be produced by animals in a balanced farming. They also say that this will lead to no further leakage of phosphorus to the Baltic Sea, but also the production of food will go down by some 25%. This can be compensated by reducing the amount of meat we eat to give a better efficiency from the given farmland area we have in the region to final food supply. Ernfors *et al.* (2008) have studied how emission of greenhouse gases is correlating to agriculture of different kind. In northern Europe, we can see a major emission of N_2O during springtime

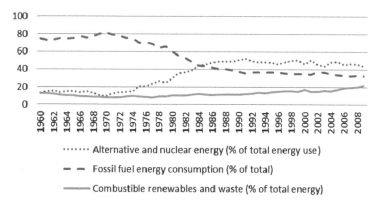

Figure 5.1.   Development of primary energy sources in Sweden from 1960 to 2008.

when the ice in the ground is melting. This is causing anaerobic conditions, which also cause the production of the N_2O. Still, the production is proportional to the amount of nitrous compounds in the soil. If we add fertilizers in the spring, these will thus cause formation of GHG, while if we administer the same amount of N-containing fertilizers in the green crops, we will see very little effect. At the same time administration of nutrients to green crops, approximately 10 cm high, will avoid leakage to the surrounding waters as well as reducing evaporation to the air.

It can be interesting to have a perspective on the energy development. The Swedish example is seen in Figure 5.1. Here the proportion of different types of primary energy is shown, where 40% losses in nuclear power plants is included with some 90–100 TWh. This is giving a total input of primary energy in the range 500 TWh/y compared to the total use around 400 TWh/y. If we would have no nuclear power we thus would increase the share of biomass, hydropower and wind power to be $(132 + 65 + 6)/400 =$ approximately 50%. This calculation is to show that statistical figures are always relative, and depends on what constraints are set.

5.1.1   Different biotopes

In Figure 5.2, we see a very typical view of a northern European landscape. There are small hills, small forests mixed with grass and farmland houses distributed in the countryside. Many areas at the coast or along rivers, have very good soil and humidity conditions, which make them fruitful. By administering fertilizers and using good seeds, large harvests can usually be achieved. A cold climate limits diseases, and where still these give problems the farmers use pesticides to keep parasites under reasonable control.

There are also many forests of Boreal type with a lot of pine and spruce in the north and more and more hard wood, broadleaf species towards the southern part of northern Europe, as seen in Figure 5.3. Here we also see very many lakes, actually thousands of lakes. This guarantees sufficient humidity for the crops, and thereby a reasonable annual production, although sometimes harsh winters gives a short growth season.

Along the coasts in Europe, we see a lot of grass-land as in Figure 5.4, where most trees were cut down to be used for building ships or houses, and the trees still have not been replaced by new ones. This is also typical in Scotland for example.

Another typical landscape view is seen in Figure 5.5. This is grassland where cattle or sheep normally can live and in the background *Phragmites australis* (Vass) is growing along the shores. We also see trees and bushes aside of the grassland and a wall of stone to keep the animals within the fences.

In Sweden, France, Finland, UK, Germany and many eastern European countries nuclear fission power replaced much of the oil imports during the 1970s and 1980s. In this way, the use of

Figure 5.2.   Typical Boreal landscape – Vasteras, Sweden.

Figure 5.3.   Southern Boreal landscape – Sweden.

Figure 5.4.   Typical landscape in coastal areas in Boreal regions.

Figure 5.5.   Typical landscape in arable areas in Boreal regions.

fossil fuels was reduced, although there has been a political discussion around nuclear power in many countries all the time. Nuclear power is not renewable, but in reality, no energy is renewable as the sun is also burning out in a long time perspective, as it is in reality a big nuclear fusion reactor. On the other hand – nuclear fuel is a limited resource. Today we only utilize a maximum 0.7% of the mined uranium – that is the percentage that is U^{235}, the rest is just discarded. In the future, it may be used as fuel in breeder reactors, or "generation 4 reactors".

The positive fact is that the nuclear fission power is not giving any significant greenhouse gas emissions, as they do not emit fossil CO_2. This has led to that almost no fossil fuel is used in Sweden for heating or electricity production. What the fossil fuels still are used for is for transportation, where biomass could be very suitable for production of ethanol, biogas and bio-diesel.

In Scandinavia, Finland, Russia and the Baltic States forests dominate. Of Sweden's total land area of 40.8 million ha 22.5 million ha is productive forests, 4.4 million ha wetlands, 0.9 million ha mountains, 3.5 million ha northern mountain areas with bushes or alpine forests, 3.4 million ha intensive and extensive agricultural land and 4.2 million ha natural reserves. The total forestry inventory is estimated to be 2 900 million m^3sk from which 39% pine, 42% spruce and 12% birch (and similar hard wood/broad leaf trees). The average inventory in productive forests is 131 m^3sk per ha. The average annual growth is 5.3 m^3sk per ha, or assuming an average density of 420 kg/m^3 5.3 × 0.42 = 2.2 tonnes DS/ha per year. This is the log wood. Branches, tops, stubs and similar adds another 0.8 tonne/ha per year at least. This gives a total production of 3 tonnes DS/ha per year. The split between different parts of trees is as follows at an average for pine (spruce) respectively: log 69% (59%), branches and tops 16% (27%), stubs and roots 15% (14%).

Since the 1920s, the wood inventory has increased by some 80% in Swedish forests and the growth is approximately 111 million m^3sk per year in Sweden in productive forests, and 117 million m^3sk including also the reserved areas. The possibility to improve growth in forests is a much debated area. Holgen and Bostedt (2004) discuss if we should plant more broad-leaved species instead of Spruce, and found that from an economic perspective this would be good. It could also be argued that many broadleaf trees increase the pH in the soil while spruce normally reduces pH. This can have positive effects for the growth of crops under the trees.

In Finland the situation is similar, although here much more peat is used aside of biomass. In both Sweden and Finland the new production of peat is extensive, in the range 15–25 TWh/y in each country. Finland also has a significant share of nuclear power and hydropower. In Norway hydropower is making up more than 100% of the electric power demand, while in Denmark fossil fuel, mostly natural gas and coal, is dominating for heat and power production, and oil for transportations. In Germany, fossil fuel from domestic brown coal is very important, but wind power and solar power has been expanding a lot. Today more than 10,000 MW is installed as solar power and 50,000 MW as wind power. A decision was made in 2011 to close all nuclear power plants within 10 years. This puts a strong push on using biomass much more efficiently and by larger volumes. In Holland, agricultural waste is very important, although natural gas is dominating as energy source. In the UK also natural gas is the dominating source for power production, although oil for transportation. In Poland and the Baltic states coal is very important, but the potential to use much more biomass is significant.

From this overall background, we will look more into the situation for producing biomass for energy utilization in northern Europe. In table 5.1 biomass production is shown in northern Europe including also Belarus and Russia, which are countries with similar climatic conditions and crops. The data are from the World Bank data base (2011) for the years 2008 and 2009. Energy use is for 2008 as the low economy 2009 caused a large reduction in energy use temporarily. In the first column, we have the cereal production in each country first doubled, as straw is equal amount as the grain measured. The heating value is assumed to be 5.4 MWh/tonne, which is reasonable. In the second column we have taken all agricultural and arable land minus the one used for cereal production and multiplied this with 10 t dry matter/ha × 5.4 MWh/tonne. In the third column we take the forestry area × 3 tonnes dry matter/ha × 5.4 MWh/tonne. The 3 top dry matter per hectare is calculated from an average 2.26 tonnes tree stem per ha in Sweden. This is for spruce approximately 59% (27% branches and top, 14% roots and stub) and for pine 69% stem (and 16% branches

Table 5.1.   Biomass production in Northern Europe in relation to total energy use.

2008/2009	Cereal including straw [TWh]	Other agro than cereal [TWh]	Forestry [TWh]	Energy use [TWh]	Prod-Use [TWh]
Austria	56	204	63	332	−10
Belgium	36	103	11	586	−436
Denmark	110	196	9	190	125
Esthonia	9	60	36	54	51
Finland	46	188	359	353	240
Germany	537	1207	179	3353	−1429
Netherlands	22	153	6	797	−617
Norway	10	86	163	297	−37
Ireland	22	275	12	150	159
Latvia	18	135	54	45	162
Lithuania	41	189	35	92	173
Poland	322	1108	151	979	602
Sweden	57	258	457	496	275
Switzerland	11	100	20	267	−136
UK	240	1128	47	2085	−669
Russian Federation	1027	16249	13107	6868	23515
Belarus	88	661	140	281	607

and top and 15% stub and roots). This corresponds to 75% as stems. The amount is probably conservative as it assumes Scandinavian conditions all over northern Europe. The column "energy use" from 2008 is using ktonne oil equivalents from the World-Bank data × 10 MWh/tonnes o.e. The last column adds all the biomass together as TWh/y and reduces the energy use as TWh/y. As can be seen the more populous countries like Belgium, Germany, Netherlands and UK have negative balances, but most of the others have a positive balance. For Russia the surplus is enormous, and if we take for the whole northern European region we have a surplus of 22,577 TWh/y! This corresponds to much more than the total energy use in EU27, which is 16,000 TWh/y, and a significant portion of the global energy use which is approximately 150,000 TWh/year. Russia obviously has a very important capacity within the energy sector also when the oil and gas has run out! The nomenclature in Table 5.1 is as follows: TWh/y of cereals including also the straw is the first column. In the second column, we have TWh/y of other agricultural crops than cereals. In the third column is TWh/y of forestry production including stems, bark, roots and tops. The fifth column shows the energy use for each country according to the official figures. In the last column we have the difference between the total production from all biomass compared to the official energy use. A positive figure means a surplus of available energy from biomass, while a negative figure means a deficiency. As can be seen Germany has a high deficiency while Russia has a very high surplus. It is interesting to note that 10 out of 17 countries have surpluses!

In Table 5.2, we also have collected the base data, which has been used for the calculations in Table 5.1. An interesting fact is that the sum of (forestry + agricultural area + arable area) for several countries is higher than the total land area registered for the country!

The cereal production differs a lot between different countries. This is seen in Table 5.3. To some extent, this depends on climatic issues, but probably also due to other conditions that can be affected by selecting better seeds, more efficient distribution of fertilizers/nutrients and perhaps introducing other crops than those dominating today.

Poland should have conditions to produce twice the amount compared to today, similar to Germany. Generally, it should be interesting to investigate special energy crops as seen in Chapter 1, or direct the selection of seeds towards the goal to enhance not only the grain production, but also the total biomass production. It should be possible to have another two tonnes dry

Table 5.2.   Areal distribution in Northern European countries.

2008/2009	Land area [km^2]	Forest [km^2]	Agriculture [km^2]	Arable land [km^2]	SUM [km^2]	Cereals [km^2]
Austria	82450	38870	31710	13740	84320	8380
Belgium	30280	6780	13730	8450	28960	3450
Denmark	42430	5440	26680	24000	56120	14977
Estonia	42390	22170	8030	5980	36180	3164
Finland	303900	221570	22960	22555	267085	11331
Germany	348630	110760	169220	119330	399310	69084
Netherlands	33760	3650	19293	10666	33609	2208
Norway	305470	100650	10243	8447	119340	3059
Ireland	68890	7390	42001	11010	60401	2935
Latvia	62190	33540	18250	11700	63490	5409
Lithuania	62670	21600	26721	18615	66936	11035
Poland	304220	93370	161540	125710	380620	85828
Sweden	410340	282030	30930	26260	339220	10321
Switzerland	40000	12400	15613	4083	32096	1530
UK	241930	28810	176840	60050	265700	31730
Russian Fed	16376870	8090900	2154940	1216490	11462330	417157
Belarus	202900	86300	89170	55160	230630	24180

Table 5.3.   Cereal production in northern European countries.

2008/2009	Cereal production [t/ha]						
Austria	6.1	Belgium	9.6	Denmark	6.8	Estonia	2.8
Germany	7.2	Netherlands	9.0	Norway	3.1	Ireland	6.8
Latvia	3.1	Lithuania	3.4	Poland	3.5	Sweden	5.1
Switzerland	6.6	UK	7.0	Russian Fed.	2.3	Belarus	3.4

matter/ha from straw for the cereals, which should be very interesting as raw material for biogas and ethanol production.

Concerning forestry the dominating trees in the northern boreal region is spruce, pine and birch. Further to the south, we have more broadleaf hardwood, although there is also a significant portion of pine and spruce as well. Birch is still common, but also beech (*Fagus silvatica*), aspen, ashtree, horse chestnut, maple, linden, oak, walnut and other trees. Hazel and juniper are also common. Larch is more common towards the east. The region consists primarily of Poland, Denmark, Germany, Netherlands, Belgium and UK/Ireland.

There is one region in-between the northern/eastern Taiga with very harsh conditions in wintertime and the milder broadleaf region in the south. This consists of southern Norway, southern Sweden, southern Finland, Estonia, Latvia, northern Lithuania, northern Belarus and central European Russia. The regions have a lot of mixed trees dominated by oak (*Querqus robur*), *Picea abies* (more to the north of the region where it is more moist) and *Pinus sylvestris* (where it is drier).

Concerning agricultural crops, the dominating species are wheat and other cereals, potato, maize, sugar beet and different vegetables. In the Netherlands a lot of vegetables are produced for export, like tomato, cucumbers, paprika and others. Carrots, onion, beetroot are other important crops. With respect to fruits apples, plums and berries like wine berries, grapes, raspberries and strawberries are common, but in Scandinavia and Russia, also wild berries like blue berries and lingo berries are of significant interest. Some wild trees and bushes like hazelnut also have a significant interest from the food perspective for both humans and animals.

5.2   BIOENERGY IN SOUTHERN EUROPE

We will now make a similar overview over southern Europe as has been done for northern Europe. What we can notice is that the climate varies quite a lot also here between the mountain areas in the Alps to the coastal areas along the Mediterranean. Especially in southern Spain, southern Italy and southern Greece the climate is quite different compared to even northern Italy or northern France. In the northern areas, the crops are quite similar to the southern parts of northern Europe. We have many trees with broad leaves in the forests, and in the high altitudes in the Alps we see trees with needles like spruce. Kazmierczyk *et al.* (2007) made a thorough investigation of the biomass production as well as consumption in South East Europe and Eastern Europe, Caucasus and Central Asia within a UNEP project. For the more western parts of southern Europe, Pascual and Saúl (2008) made an overview of Biomass Resources in the EU Chrisgas project, where the focus was on estimation of what available resources there are for gasification for production of fuels for vehicles, primarily as DME (dimethyl ether). The plants were grouped into the categories as indicated in Table 5.4.

This gives us a picture over what type of crops are growing in southern Europe and especially the rice, vineyards, olives and many of the fruits like oranges are much more common in Southern Europe than even the southern part of Northern Europe. Many crops are the same in northern and southern Europe, although there are also quite a few that are wild in southern Europe but only cultivated in gardens in northern Europe. Examples of herbs and habitats are given in Houdret (2000) for example. Here also conditions for cultivation of these are addressed.

The analysis in this project was primarily to identify resources for a number of gasifiers. This means that only waste of different kind were investigated: Herbaceous crop residues, woody crop residues and forestry residues. The results from the author's analysis are shown in Table 5.5 for 12 countries.

If we recalculate these figures into TWh/y the total will be for Agriculture $99 \times 10^6 \times 5.4$ MWh/tonne $\times 10^{-6} = 535$ TWh/year potential and 265 TWh/y available today. For forestry, the corresponding figures are 1430 and 840 TWh/year. Pascual and Saúl (2008) made a calculation for biomass in 12 countries and came to the conclusion that 205 Modt/y could be used to produce DME (dimethylether). This would then give 66 Mt DME (3.1 kg biomass dry matter gives 1 kg DME) which corresponds to 45 Mt diesel which is equivalent to 45 Mtoe (0.68 kg

Table 5.4.   Different type of crop categories relevant to southern Europeé. The numbers are different categories according to the standard presented in Pascual and Saúl (2008).

Agricultural	
12 Non-irrigated arable land	Rain-fed1
13 Permanently irrigated land	Irrigated2
14 Rice fields	Rice
15 Vineyards	Vineyard
16 Fruit trees and berry plantations	Orchards
17 Olive groves	Olive
19 Annual crops associated with permanent crops	Crop mixture
Forestry	
22 Agro-forestry areas	Dehesas
23 Broadleaved forest	Broadleaves
24 Coniferous forest	Conifers
25 Mixed forest	Mixture
29 Transitional woodland-shrub	Shrubs

diesel corresponds to 1 kg DME), or 17% of the final energy consumption in the transport sector in the 12 analyzed countries (265 Mtoe in 2007 in 12 countries; 70% of the 377 Mtoe in EU27).

If we just do as earlier for northern Europe and calculate the estimated production of biomass as cereals, other agriculture crops than cereals and total production in forests and compare the energy use in each country to the total biomass production, we get the results as seen in Table 5.6. We then can see that there is a net surplus in all countries except in Italy in the southern European countries.

The basis for this Table 5.6 is seen in Table 5.7. The source of the figures is as previously the World Bank database (2011). We here see the total land area, the area covered with forests, the agricultural area and the arable area, that is other type of useful area for biomass production than the agricultural and forestry areas. We also see the figures for the three categories compared to the official figures for the total land area and can once again see that these are not the same. The next columns show first the agricultural area except cereals and thereafter the area for cereals. In the last column, we have the cereal production in tonnes per year.

Table 5.5. Potential and available agricultural and forestry wastes in 12 countries in Europe (million oven dry tonnes/year).

Country	Agriculture		Forestry	
	Potential	Available	Potential	Available
Portugal	2.23	0.64	2.38	1.70
Greece	2.90	1.09	7.42	4.82
Italy	7.49	4.01	25.44	16.94
Spain	10.55	4.77	26.22	12.84
France	13.43	7.75	78.39	49.81
Denmark	0.74	0.37	12.99	8.17
Poland	4.11	2.05	28.09	10.36
Austria	5.26	2.63	6.25	3.79
Norway	8.6	4.33	1.85	0.83
Germany	9.54	4.27	67.34	42.56
Finland	13.59	6.94	3.08	1.12
Sweden	20.68	10.43	5.93	3.11
Total	99.12	49.28	265.38	156.05

Table 5.6. Biomass in southern Europe. In column 5, we have the production of biomass in TWh/y minus the total use of energy in TWh/y.

Country	Cereal including straw [TWh]	Other agro than cereal [TWh]	Forestry [TWh]	Energy use [TWh]	Prod-Use [TWh]
Portugal	14	228	56	242	57
Spain	258	1890	294	1388	1054
France	757	2096	258	2665	447
Italy	234	907	148	1760	−472
Greece	57	305	63	304	121
Production of biomass – Use of energy =					1207

Table 5.7.   Areal distribution in southern European countries.

2008/2009	Land area [km^2]	Forest area [km^2]	Agricultural area [km^2]	Arable land [km^2]	SUM [km^2]	Agri exclusive cereal [km^2]	Cereals [km^2]	Cereal prod [tonne]
Portugal	91 470	34 560	34 600	10 500	79 660	41 460	3640	1 309 600
Spain	499 110	181 730	279 000	125 000	585 730	343 567	60 433	23 903 987
France	547 660	159 540	2 924 206	182 603	634 564	381 142	93 882	70 075 131
Italy	294 140	91 490	133 960	71 320	296 770	164 904	40 376	21 624 474
Greece	128 900	39 030	46 250	21 000	106 280	55 502	11 748	5 238 150

Figure 5.6.   Mediterranean type of landscape (photo E. Dahlquist).

In Figure 5.6, we see a typical view of a central and southern European landscape. There are a lot of low mountains or hills covered by forests. In-between the hills we have farmland and villages.

In Southern Europe, we also have subtropical forests like in coastal areas. An example of this is Mediterranean region and the Azores in the Atlantic sea. An example of this biotope is seen in Figure 5.7. Here the productivity in both forests and farmland area is very high due to the humidity caused by the surrounding sea.

One typical crop is banana as seen in Figure 5.8, in this case from the Azores, but also growing in the Mediterranean areas.

There are also many different types of crops in southern Europe that do not exist further to the north. Some of these are seen in the foreground of Figure 5.9 showing a typical Mediterranean subtropical or tempered landscape.

Typical herbs in southern Europe are trees like *Pinus pinea* looking like an umbrella, the Greek fir, *Abies cephalonica*, which is a pyramidal tree with dark green foliage. This latter tree is found at elevations above approximately 600 m elevation in Greece and was once considered being the

Figure 5.7.   Subtropical landscape – the Azores (photo E. Dahlquist).

Figure 5.8.   Banana trees in the Azores (photo E. Dahlquist).

Figure 5.9.   Mediteranian subtropic or temperate landscape (photo E. Dahlquist).

best timber for building ships. *Castanea sativa* (Miller) or kastania in Greek is a tree with edible fruits, and are found especially in the mountain areas of northern Greece. *Quercus coccifera* or kermes is also relatively common in southern Europe. It is a good source for tannin, which is used as a dyeing substance. The insect *Coccus ilicis*, which gives a red color dye, is often found under these trees. *Juglans regia* or walnut is another tree with edible nuts. It is native from south-eastern Europe over India to China. *Ficus carica*, has edible fruits while the almond tree gives nuts. *Pistacia terebinthus* or the turpentine tree produces turpentine and while *Pistacia vera* produces pistage nuts, a very tasty nut. This tree is found also further to the east and is commonly cultivated in e.g. Iran. The most important oil tree in southern Europe is the olive tree or *Olea europea*. This gives a green oil that is used extensively in the southern European cuisine. It is supposed to be very healthy. *Nerium oleander* on the other hand is very common along the roads to stabilize the soil from erosion, but it's milky juice is poisonous.

There are also many well-known herbs growing naturally in southern Europe. On example is wild celery or *Apium graveolens*. This is the origin of the popular vegetable celery we have in our gardens. fennel (*Foeniculum vulgare*), dill (*Anethum graveolens*), French lavender (*Lavandula stoechas*), rosemary (*Rosmarinus officinalis*), sage (*Salvia officinalis*), wild thyme (*Thymus serpyllum*), oregano (*Origanum heracleoticum*), mint (*Mentha viridis*) are other popular vegetables or spices found wild in Greece and other south European countries (Niebuhr, 1970).

In the Alps, the climate is harsher or more similar to the climate in northern Europe, due to higher elevations. This causes more snow wintertime normally, but during the last decades, warmer climate is giving higher average temperatures in many mountainous areas. Glaciers are withdrawing due to this. This is also the case in the Andes in South America, where among others a roughly 6000 meter high mountain close to El Alto previously had downhill skiing. Today the lift is closed, as there is too little snow!

Still, we see higher mountains in the Alps compared to north and south of the Alps and thus see much higher difference in climate between areas quite close to each other, but at different

Figure 5.10.   Alpine landscape from Anecee in France (photo E. Dahlquist).

elevations. Example of this landscape is seen in Figure 5.10 and 5.11 from the French Alps close to Annecy.

Typical trees in the Alps are spruce, fir, white pine, larch, beech, sycamore maple and common juniper and more bush like species include bearberry and rock buckthorn (Picquot, 2003). At the lower elevations, *Xerophilus* spruce forests are dominating on the north slopes, followed by juniper heath and even higher bilberry heath. From around 1500 m and upward herbs are dominating. On the south slopes beech and fir forests are dominating at lower elevations followed by spruce forest a bit higher up and above that we find alder scrub and alpenrose heath. As a comparison, we can see that the landscape in southern Africa is quite similar to the one at high elevations in mountainous areas as seen in the Figure 5.12 from Cape of Good Hope.

Here we have more bushes than trees but also crops adapted to harsher environment with strong winds carrying in salt and fog covering the sun. Not very far from the Cape of Good Hope is the South African wine district, like the one close to Stellenbosch in Figure 5.13.

The climate is quite similar to the ones in the French and Italian wine districts here. The climate in the areas is mild enough for the wine to survive the winters but humid enough to give good conditions for growth and mature of the grapes. As a comparison the bush land and relatively dry areas in the more northern parts of South Africa and Namibia, Botswana and other countries further to the north like in Kruger Park in South Africa (Fig. 5.14) has similarities to the dry areas in the Mediterranean regions in Spain, Italy and Greece.

The crops in these drier forest and bush land areas also have special trees and bushes that can withstand not only the dry season, but also the interference by animals like the giraffes as seen in Figure 5.15. Although the bushes have strong, long needles, the giraffes can eat them without damaging their mouths. Here we also see many different types of acacia (Fig. 5.16).

These trees have adapted to many different biotopes in the subtropical regions of the world, and now have become important for production of wood for pulp and paper industry, aside of the eucalyptus. Eucalyptus or gum tree on the other hand is now grown in almost all parts of the world except in the northern, Boreal regions. It originates from Australia but has been brought

Figure 5.11.   Alpine landscape at Annecy in France (photo E. Dahlquist).

Figure 5.12.   Bush land at Cape of Good Hope, South Africa (photo E. Dahlquist).

Figure 5.13. Wine yards at Stellenbosh, South Africa (photo E. Dahlquist).

Figure 5.14. Bush land in the Kruger Park, South Africa (photo E. Dahlquist).

Figure 5.15.   Bush land in Kruger Park, South Africa (photo E. Dahlquist).

Figure 5.16.   *Acacia nigrens*, Kruger Park, South Africa (photo E. Dahlquist).

to all parts of the world due to their fast growth. There are more than 500 eucalyptus species and some can be more than 130 meters high. Aside of using eucalyptus as a source of wood for the pulp and paper industry it is also used for its volatile (essential) oil. Some species are frost hardy like *Eucalyptus gunnii*, which can tolerate even $-15°C$. One example of this is seen at a

Figure 5.17.   Eucalyptus, La Paz, Bolivia (photo E. Dahlquist).

Table 5.8.   Cereal production in southern European countries 2008/2009.

Country	Cereal tonne/ha
Portugal	3.6
Spain	4.0
France	7.5
Italy	5.4
Greece	4.5

slope in La Paz Bolivia, at almost 4 000 m a.s.l. This tree is now the most common tree for forest industries in Brazil, Indonesia and other countries in the temperate (Mediterranean), sub tropical and tropic regions, due to its fast growth rate. When trees in the north may take 80 or 100 years before harvesting, the eucalyptus may be ready after seven years, or even faster.

In Table 5.8, we have calculated the cereal production per hectare from the total production of cereals in tonnes/year in relation to the total land area used for cereal production. The source of the figures is as previously the World Bank data base (2011).

Here we can see that the cereal production is similar in southern Europe to northern Europe even though the climate principally should be better. The hot temperature during the summer is probably causing this combined with a shortage of water for irrigation. Thus France, with best availability of water, has the highest average production of cereals of the five.

Figure 5.18.   Tropical crops in Ghana (photo E. Dahlquist). Papaya originates from central and South America, but is now common in all tropical areas.

5.3   BIOMASS IN THE TROPICS

The focus of this chapter has been on biomass in Europe. In Europe, we have all the biotopes accept for real tropical areas, although some areas are almost tropical. As a complement, this section 6.3 will cover at least some aspects of tropical climate and biotopes as well. Cassava (*Manihot esculenta*) is an important crop in many tropical countries. Tweneboah (2000) states that cassava is the staple food for more than 200 million people in sub-Saharan Africa. The reason is that it can grow in marginal land areas where the soil is poor. It also has high calorific value and is relatively easy to store. It is also resistant to drought. Another important crop is yam, which is very important in West Africa, and especially in Nigeria, the most populous country in Africa. For yam, there is a high demand on enough moisture during the full growth period, but especially during the 14–20 weeks of growth. 1000–1500 mm rain per year and a dry season of 2–4 months gives optimum yield conditions. Still 20–25 million tonnes/year is produced, which is 90% of the world production. Sweet potato (*Ipomoea batatas*) and aoids (cocoyams) are other important food crops in sub-Saharan Africa, even if they are not as important as staple food as cassava and yam. Cocoyam (*Xantosoma* sp.) is normally cultivated in the tropical forest areas where is used to give shade to other crops. Sweet potato on the other hand is cultivated more in the savannah areas closer to the Sahara region, where it often grown on relatively poor sandy soils. Due to this, the productivity is relatively low. Information about both the crops and how to cultivate them is given in Opeke (1997). In Figure 5.18, we see a typical view of crops cultivated in gardens in Accra, Ghana. Also exclusive trees like jacaranda, hickory and others are species of importance for furniture and similar products.

Banana and plantain (musa) are other important crops. Plantain is a staple food crop in the central African low land forests. It is especially popular in Cameroon, Gabon, Republic of Congo

Figure 5.19.   Tropical trees in Ghana (photo E. Dahlquist).

Figure 5.20.   Palm trees in Ghana (photo E. Dahlquist).

and Zaire, but is also cultivated in West Africa. The most important cereals in tropical Africa are maize (*Zea mays*, corn), rice (*Oryza* spp.) and sorghum and millets. In the savannah areas in northern Ghana also Guinea corn is important and is a major staple food. Many vegetables are also cultivated in the tropics; condtions for growing are given in Sinnadurai (1992).

Of economic interest are also cacao, coffee and other crops produced for export. There are also many types of fruits like mango, kiwi and other. A typical view of rain forest is seen in Figure 5.19 while the view in Figure 5.20 show palm trees at the coast of Ghana, which is another type of trees common in this climate. Coconut is one type of palm tree of commercial interest, although oil palm is the most important. In Dupriez and De Leener (1988), Twenneboah (2000) and Longman (2002) production of tropical trees and crops is discussed from different perspectives.

Figure 5.21.   Jungle in Ghana (photo E. Dahlquist).

In Figure 5.21, we see another view from a rain forest. The trees can be very high and then the "solution" to handle this may be "support structures" as seen at the bottom of these trees. Throughout the years, many suitable solutions to different problems have evolved, and with so many different species as in rainforests, the solutions are very different. Here we have the opposite to monocultures!

5.4   QUESTIONS FOR DISCUSSIONS

- Why is biomass almost not considered as an energy resource in most European countries, although we have vast resources as seen in this chapter?
- What would a sustainable society look like where biomass is one part of the sustainable society, but combined with other renewable energy resources?
- How can we handle a sustainable biomass production long term?
- How would a global warming scenario affect the biomass production?
- How can we solve the problem of too little water for irrigation in Southern Europe today?
- How can the warmer climate affect the biotopes and the crops and animals in these?

REFERENCES

Bassam el N.: *Energy plant species – their use and impact on environment and development*. James&James, UK, 1998.
Dahlquist E., Thorin, E. & Yan, J.: Alternative pathways to a fossil-fuel free energy system in the Mälardalen region of Sweden. *Int. J. Energy Res.* (June 2007).
Dupriez, H. & De Leener, P.: *Agriculture in African rural communities – crops and soil*. Macmillan Publishers. 1988.

EC: Eurostat: Statistical pocketbook 2010, EU Commission, Brussels, Belgium, 2010, http://ec.europa.eu/energy/publications/statistics/statistics_en.htm (accessed July 2012).

Ernfors, M., von Arnold, K., Stendahl, J., Olsson, M. & Klemedtsson, L.: Nitrous oxide emis8sions from drained organic. *Biogeochemistry* 89: (2008), pp. 29–41.

Holgen, P. & Bostedt, G.: Should planting of broad-leaved species be encouraged at the expense of spruce? An economic approach to a current southern Swedish forestry issue. *J. Forest Econ.* 10 (2004), pp. 123–134.

Houdret, J.: The ultimate book of herbs & herb gardening. Hermes House, 2000.

Kazmierczyk, P., Tsutsumi, R. & Watson, D. (eds): Sustainable consumption and production in South East Europe and Eastern Europe, Caucasus and Central Asia. Joint UNEP-EEA report on the opportunities and lessons learned EEA Report No 3/2007, 2007, http://www.eea.europa.eu/publications/eea_report_2007_3/Sustainable-consumption-and-production-in-South-East-Europe-and-Eastern-Europe–Caucasus-and-Central-Asia (accessed July 2012).

Lambers, H., Chapin, F.S. & Pons, T.L.: *Plant physiological ecology*. Springer Verlag, 1998.

Larsson, M. & Granstedt, A.: Sustainable governance of the agriculture and the Baltic Sea – agricultural reforms, food production and curbed eutrophication. *J. Ecol. Econ.* 2009.

Longman, K.A.: *Preparing to plant tropical trees*. Blakton Hall, 2002.

Niebuhr, A.D.: *Herbs of Greece*. Library of (US) Congress card number 76-115791, 1970.

Opeke, L.K.: *Tropical tree crops*. Spectrum Books Limited. Ibadan. ISBN 978-246-152-0. 1997.

Pascual, E. & Saúl, L.: Biomass Resources – Chrisgas project. CEDER-CIEMAT, Spain, 2008.

Paz, A., Starfelt, F., Dahlquist, E., Thorin, E. & Yan, J.: How to achieve a fossil fuel free Malardalen region. *Conference Proceedings of 3rd IGEC-2007*, 18–20 June 2007, Västerås, Sweden, 2007.

Picquot Florence and Aedelsa Atelier (eds): *The French Alps*. Guides Gallimard, 2003.

Salisbury, F.B. & Ross, C.W.: Plant physiology. 4th edn, Wadsworth, and internal data from SLU, Dept of Plant Physiology, 1992.

Sinnadurai S.: *Vegetable cultivation*. Asempa Publisher, Accra, India, 1992.

Twenneboah, C.K.: *Modern agriculture in the tropics – food crops*. Co-wood Publishers, 2000.

Weidow, B.: The basics of crop farming (Växtodlingens grunder). LTs förlag, Stockholm, Sweden, 1998 (in Swedish).

World Bank: World Development Indicators & Global Development Finance. The World Bank, Washington, DC, 28 July, 2011.

CHAPTER 6

Organic waste as a biomass resource

Eva Thorin, Thorsten Ahrens, Elias Hakalehto & Ari Jääskeläinen

6.1 INTRODUCTION

The waste hierarchy promotes reuse and recycling of waste material but substantial amounts of wastes cannot be reused or recycled. However, the energy content of this waste can be utilized. Waste is constantly produced in agriculture, municipalities and industries and part of that thus provides a secure and a local source of energy. Wastes and biomass fuels are usually viewed as sustainable energy sources (Kothari *et al.*, 2010). Waste can consist of many different materials. Considering using waste as a biomass resource, the organic fractions containing material of biological origin is of interest, see the examples in Table 6.1.

Packed food waste from grocery shops is one important source of organic waste in industrialized countries. In Finland, for example, grocery shops' food wastage due to overdated products totaled 5.4×10^7 kg in 2007. This corresponds, on average, to 4.15×10^3 kg per M€. This includes products that have exceeded the "use-by date" such as fresh meat, mince, fresh fish, fresh cheese, uncooked ready-made food and other food easily contaminated microbiologically. It is prohibited to sell or even give away these after the use-by date. "Best before" marking is used for other groceries, such as bread, grain products, other dry products, frozen food and canned food. It is possible to decrease their wastage by selling them at a discount. Moreover, contaminated groceries and groceries with damaged packages are thrown away. In addition, frozen food that has melted or for which the cold chain has otherwise been broken, cause wastage of groceries (Ekoleima Ay, 2008).

In Table 6.1 possible conversion processes for examples of organic wastes are also given. Anaerobic digestion and fermentation are microbiological conversion methods while combustion, pyrolysis and gasification are thermochemical conversion methods. The product of the conversion processes of organic waste can be used as fuel in further energy conversion. Ethanol from fermentation and up-graded gas from anaerobic digestion can be used as a fuel in transportation. The gas from anaerobic digestion can also be used directly in an oven, gas engine or gas turbine for heat and power production. The gas from gasification and pyrolysis can be used in a similar way but can also be further reformed for production of other types of transportation fuels for example dimethylether (DME) or Fischer–Tropsch fuel.

6.2 PRE-TREATMENT

To utilize the organic waste as an energy resource, pre-treatment of a waste flow is required. The aim of the treatment will be to sort out the organic fractions and by that increase the energy content but also to get a more homogeneous, dense and/or less contaminated material. The latter reasons can simplify handling and use of the material as an energy source and also the handling of by-products and emissions from the conversion processes.

Pre-treated municipal solid waste with the purpose to prepare a fuel for thermal conversion processes is often called Refuse Derived Fuel (RDF). This fuel normally includes the plastic, paper, wood, textile and rubber parts of the solid waste. In addition, other terms are in use,

Table 6.1.   Examples of organic wastes and possible conversion processes (Thorin *et al.*, 2011).

Waste	Conversion processes	Comment
Sewage sludge	Anaerobic digestion to biogas Combustion Pyrolysis Gasification	
Grease trap sludge	Anaerobic digestion to biogas	
Manure	Anaerobic digestion to biogas Gasification	
Biowaste	Anaerobic digestion to biogas	
Municipal solid waste (MSW)	Anaerobic digestion to biogas Combustion Pyrolysis Gasification	
Food wastes	Anaerobic digestion to biogas	From grocery shops, restaurants and industry
Waste cooking oil	Combustion Pyrolysis	
Refuse derived fuel (RDF)	Anaerobic digestion to biogas Combustion Pyrolysis Gasification	
Lignocellulosic wastes	Fermentation to ethanol Combustion Gasification	For example straw, wood wastes, sugar cane bagasse, corn stover
Sugarcane molasses	Fermentation to ethanol	
Industrial waste water	Anaerobic digestion to biogas Fermentation to ethanol	From distilleries, food processing From breweries, sugar mills, food processing, tanneries, paper and pulp production
Packaging waste	Combustion	
Product specific industrial waste	Combustion	
Bulky waste	Combustion	Furniture, etc.
Clinic waste	Combustion	
Hazardous waste	Combustion Gasification	

such as Recovered Fuel (REF), Packaging Derived Fuel (PDF), Paper and Plastic Fraction (PPF), Processed Engineered Fuel (PEF), Solid Recovered Fuel and Specified Recovered Fuel (SRF). SRF is produced according to the requirements of a quality label. The CEN/TC 343 Working Group 2 "Fuel Specifications and Classes" which work with EU quality criteria for SRF mentions the following as important factors for the quality criteria:

- Economic aspect: net caloric value (the higher, the more regular fuel can be substituted)
- Technology aspect: Cl content (chloride leads to corrosion in the installations)
- Environmental aspect: Hg (+Cd) content.

The pre-treatment can include separation, drying, size reduction followed by formation of pellets, bales or briquettes. One possible drying method is to use biological drying where the drying is done by air convection and heat provided by exothermic decomposition of the readily decomposable waste fraction. Typical water losses of 25% of the waste weight have been reported. Collecting different fractions of the waste separately, for example biowaste from households, is

Figure 6.1.   Jätekukko Oy's mixed waste pre-treatment line. Photo: Jätekukko Oy.

also a possible pre-treatment strategy (Cherubini *et al.*, 2009; European Commission, 2006; Gendebien *et al.*, 2003; McDougall *et al.*, 2008; Rada, 2009).

6.2.1   *Examples of pre-treatment*

Kitchen biowaste is defined as easily degraded wastes of plant or animal origin. In the city of Kuopio and in the city of Pieksämäki, Finland, the collection of kitchen biowaste was started in June 2003 using two different methods. The regional waste management company arranged the collection. The collection of kitchen biowaste was started in housing estates with more than 10 households. The regional waste management company provided biowaste bins for every housing estate. In the city of Pieksämäki the waste management company provided bins with air holes (240 liters, made by Plastic Omnium) and furthermore every household was given one bucket, "ventilated" Combi BioBag holder system, and 150 compostable bags (bioMat® BioBags). In the city of Kuopio, the waste management company provided bins without air holes (240 liters, made by Plastic Omnium) with a compostable sack inside. In Kuopio, the households were given some compostable bags but the bucket was not given. The company studied whether the different biowaste collection methods have an influence on the quality (pH-value, moisture and microorganisms) of the biowastes and compared these two collection methods in practice. In addition to the measurements also cleanliness, odor and quantity of biowastes was assessed. The moisture and pH-values of the biowastes were at the same level in both collection methods. The mean of the pH-values was 5.57 in Pieksämäki and 5.55 in Kuopio. The mean of the moisture content was 66.5% in both Pieksämäki and Kuopio. In the city of Kuopio, people used newspaper and other papers to the packing of biowastes more than in Pieksämäki. This may have an effect on the moisture of biowastes as well as on the pH-values. According to the assessments, biowaste bins were cleaner, drier and more odorless in Pieksämäki than in Kuopio and the kitchen biowaste collection method, which was used in Pieksämäki, was better than the one used in Kuopio (Jätekukko Oy, 2004).

In Sweden, systems for the collection of food waste were used in more than half of the 290 municipalities in 2011. The experiences from these municipalities show that important factors for successful introduction of collection of source-separated food waste includes planning, adequate human resources, information and monitoring and control, i.e. factors that are not related to a specific design of the collection system. Different collection systems have their pros and cons but most important is that the systems are well adapted to the current municipality (Avfall Sverige, 2011; Biogas Öst, 2011).

An example of a RDF plant is the one run by Jätekukko Oy in Kuopio, Finland since 2009, see Figure 6.1.

The current fixed crushing line consists of a Lindner Jupiter 3200 fixed crusher, magnetic separator, a Doppstadt SM – 718 drum screen and a windscreen, all operated with electricity. All incoming mixed municipal waste is pre-sorted with an excavator, after which the waste is fed into the crusher. The drum screen has a diameter of 1.8 m, and a screen net with 40 mm holes. The permeate, with particle size is of 40–200 mm, is transferred to the wind screen and the retentate, with particle size of 0–40 mm, drops to a silo under the drum. The retentate fraction contains e.g. biowaste, sand and glass. The windscreen sorts the waste into a lighter fraction, consisting of e.g. paper, carton board, plastic and wood. The heavier fraction from the windscreen contains mostly food waste such as potatoes, carrots, oranges, apples and aluminum cans. The latter, containing a lot of biowaste, is mixed with the retentate from the drum screen and composted. The permeate drops to a silo, from which it is transferred with a wheel loader to a storage silo. Jätekukko Oy uses the standard SFS 5875 and Technical Specification CEN/TS 15359 for their quality ratings of the produced RDF. Representative samples from mixed, construction and energy waste are analyzed in the laboratory. Based on the results the classification according to SFS 5875 and CEN/TS 15359 are defined. Mass balances for the mixed and crushed construction waste have been studied by weighing different mass quantities produced in the crushing equipment. Hence, the company has been able to draw up product declarations and grades for the recycled fuels that will facilitate the marketing of the fuel. Based on the results the contaminant concentrations of the RDF produced from waste products are known. The results can be compared to the quality requirements set by combustion of RDF (Jätekukko Oy, 2009; Laitinen, 2011; Juusola, 2011).

6.3 BIOGAS PRODUCTION

6.3.1 *Basics of the biogas process*

The production of biogas is a natural process, occurring for example in swamps, soils or cattle paunches. Biogas plants are technical solutions for simulation of the natural degradation processes. In principle, most kinds of organic materials are suitable as substrate for the biogas process. Examples of substrates are (Ahrens and Weiland, 2007):

- liquid and solid wastes from livestock husbandries (like manure);
- ensiled energy crops;
- harvesting residues;
- residues from private and industrial waste material market.

According to latest developments in Europe, different and more complex waste fractions (like household wastes) also become substrates of interest for biogas applications. Hand-in-hand with waste utilization comes the demand for development of suitable process applications in terms of effectiveness and reliability. A focus on these aspects will be given in section 6.3.2. The organic load of a fermenter is catabolized by anaerobic bacteria (hydrolysis, acidogenesis, acetogenesis and methanogenesis) over carbonic acids as intermediate products to methane within a retention time of 20–30 days, see Figure 6.2

Methane concentration in biogas mixtures varies between 55 and 65% depending on substrate and fermentation conditions (Ahrens and Weiland, 2007). The biogas can be combusted in energy conversion systems for producing electricity and heat; depending on local conditions a gas upgrading to natural gas quality can be done, followed by utilization in natural gas grids or as vehicle fuel (Ahrens and Weiland, 2007).

General important aspects for implementing biogas technology are:

- electrical and thermal efficiency of the process;
- economic aspects;
- technical aspects concerning the fermentation process;
- process requirements (gas quality/gas production);
- substrate aspects.

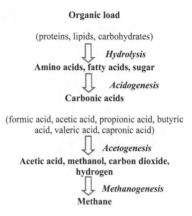

Figure 6.2.   Microbiological pathway for biogas production (Lens *et al.*, 2004).

Table 6.2.   Comparison of theoretical and practical methane yields. FM = Fresh mass, oDM = organic dry matter.

System and Substrate	Methane yield from wet digestion batch test		Methane yield from dry digestion full scale plant	
	$(Nm^3/10^3$ kg FM)	$(Nm^3/10^3$ kg oDM)	$(Nm^3/10^3$ kg FM)	$(Nm^3/10^3$ kg oDM)
Plug flow digester 1 Energy crops	157	482	126	388
Plug flow digester 2 Biowaste	99	352	69	243
Tower digester 1 Energy crops	142	459	108	349
Tower digester 2 Household waste	77	291	65	246
Garage digester Biowaste	49	374	33	256

Considering waste streams, several additional aspects become a matter of interest, for example (Ahrens and Weiland, 2007):

• waste pre-treatment;
• digestate quality (e.g. heavy metal content) and its suitability to be used as fertilizer;
• process technology (e.g. disturbing material or accumulation effects);
• suitable plant locations (e.g. waste infrastructure, digestate utilization, etc.).

6.3.2   *Technical background for waste-to-biogas utilization strategies*

According to substrate characteristics, waste as biogas substrate often requires different digestion strategies in comparison to energy crops. A matter of discussion is the so-called dry digestion, which might have a huge potential for waste-to-biogas utilization strategies. Reasons for this are the significantly lower demand of water for moisturizing the substrate and furthermore (as a parallel effect) the reduced amount of digestate to be treated afterwards. One basic question is the effectiveness of such applications in terms of methane productivity. This has been investigated by comparing examples of full-scale plant applications with parallel batch tests using the same substrate mixtures. Table 6.2 shows the results of this comparison.

Figure 6.3.   Batch test equipment in heating cabinet.

The conclusion is that dry fermentation applications are able to reach high methane yields. Nevertheless, there is still potential for optimization e.g. in terms of separation of disturbing material, substrate intake into the fermenter and inoculation of fresh substrate being added to the fermenter.

6.3.3   *Results from waste digestion*

Information about substrates is strongly demanded by biogas plant operators and future investors. Targets of research work are therefore examining reachable degradation rates and methane yields in mono-fermentation processes and co-fermentation processes of mixtures of different waste fractions being representative for different regions in Europe. This chapter will present comparison of process-relevant parameters of different single waste fractions and waste fraction mixtures being assumed in laboratory scale batch tests and semi-continuous tests. For appraisal of the used substrates, the following parameters have been acquired:

- organic dry matter (oDM);
- content of volatile fatty acids;
- degradation of organic load in the fermentation process;
- dry matter (DM) of substrate;
- ortho-phosphate (PO_4-P);
- total nitrogen content (Kjeldahl-N);
- ammonium nitrogen content (NH_4-N).

The assessment of the continuous fermentation process is done with degradation rate of liquid phase (concerning oDM, DM) (%):

- degradation rate (amount of formed methane related to the results of the individual batch tests);
- methane productivity (m^3 methane/m^3 fermenter volume);
- methane yield in (m^3 methane/kg oDM);
- reached biogas quality.

Figure 6.3 shows the batch tests with a volume of 3.5 liters being operated according to German VDI 4630 (VDI 2008) and Table 6.3 shows the process-relevant results from lab scale batch test analysis.

Table 6.3.   Results from batch tests of different suitable biogas substrates.

Region/city	EWC code	[Nm^3(CH$_4$)]/ [Mg(fresh mass)]	[Nm^3(CH$_4$)]/ [Mg(oDM)]	Methane content [Vol-%]
Kuopio household waste average	20 03 01	69.4	286.7	59.6
Lapinlahti waste water sludge	19 08 05	16.6–27.6	312.1–575.3	56.8–63.4
Kuopio household waste 1	20 03 01	67.1–127.6	210.5–394.1	56.4–61.7
Kuopio biowaste	20 01 08	67.0–71.6	215.9–230.7	67.0–68.8
Kuopio manure mix	20 02 01, 19 12 07, 02 01 06	46.3–46.6	210.4–211.8	65.0–65.4
Kuopio household waste 2	20 03 01	37.4–55.8	117.2–174.8	68.1–76.3
Klaipeda, waste water sludge	19 08 05	28.3–33.3	176.8–207.6	60.4–64.2
Siauliu, dump waste	20 03 01	15.1	98.3	67.9
Taurage, dump waste	20 03 01	13.4–24.1	90.6–162.6	51.3–58.8
Plunge, dump waste	20 03 01	11.7	57.8	53.0
Klaipeda, dump waste	20 03 01	6.7–9.9	39.1–57.4	52.0–53.0
Kretinge, cow manure	02 01 06	33.0	225.8	57.6
Taurage, dump waste	20 03 01	13.4–24.1	90.6–162.6	51.3–58.8
Plunge, dump waste	20 03 01	11.7	57.8	53.0
Lithuanian mix	02 01 06, 02 07 02, 19 08 05, 19 08 01, 20 01 25	83.6–86.7	453.7–470.7	64.0–67.7
Poland <80 mm	20 03 01	27.4–49.0	117.6–209.7	49.2–54.9
Poland <60 mm	20 03 01	22.2–32.7	119.1–174.9	55.7–62.6
Poland 20–80 mm 1	20 03 01	19.2–29.8	91.1–141.2	41.7–52.3
Västerås, biowaste	20 01 08	146.3–147.2	562.2–565.7	65.0
Västerås, household waste	20 03 01	66.1–84.6	221.1–283.2	60.6–72.1
Paper pulp 1	0303	3.0–3.4	16.5–18.4	38.0–40.0
Brewers' grains 1	0207	688.4–99.3	420.3–472.3	61.9–67.2
Grease traps waste 1	190809	305.6–323.3	643.2–680.3	68.4–70.0
Edible fat 1	200125	652.0–837.9	652.5–838.6	63.8–71.7
Kitchen & canteen waste 1	200108	197.8–241.3	468.5–571.5	53.9–67.9
Estonian mix 1	02 07, 19 08 09, 20 01 25, 20 01 08	253.0–266.6	605.1–637.6	74.7–74.8

In batch experiments, the essential retention time for the highest possible degradation rate of the examined substrate is acquired. With semi-continuous experiments, the whole process circle of a biogas plant is simulated at different retention times and with different substrate mixtures (mono- and co-fermentation). For collecting information about long-term behavior (e.g. with respect to accumulation of materials or inhibition effects) continuous tests are required. Figure 6.4 shows an exemplary laboratory test digester with a volume of 12 liters being operated at mesophilic conditions and Table 6.4 shows results from such tests being operated at stable conditions without any inhibition occurring for a continuous period of minimum three months.

With combining the test results presented in Tables 6.2 and 6.3, a basic up scaling to full-scale applications becomes possible. This allows the development of regional implementation strategies, which will be shown in section 6.3.4.

1.   Stirring device

2.   Water seal

3.   Double walled heating coat

4.   One of four water supply ports

5.   Sampling connection

6.   Gasbag with devices

Figure 6.4.   Self-constructed semi-continuous biogas digester.

Table 6.4.   Results from continuous tests of different suitable biogas substrate mixtures.

	Estonia	Finland	Lithuania	Sweden
Substrate	Brewers' grains, edible fat, biowaste, grease trap waste	Municipal household waste	Cow manure, sewage sludge, screenings, waste from spirit distillation, palm oil	biodegradable kitchen & canteen waste
Pre-sorting	–	✓	–	–
Testing period [d]	86	120	127	106
Range of organic load [g oDM/liter digester volume and day]	0.7–2.1	2.1–4.9	1.26–3.5	1.4–3.5
Recirculation of digestate	–	✓	–	–
Average final oDM of digestate [% of FM]	1.11	7.03	5.45	4.07
Average stable methane production [NL/h]	0.9875	0.5026	0.5725	1.2096
Average respective buffer capacity (maximum VOA/TAC)	0.12	0.13	0.25	0.19
Number of respective trace gases (>70% probability)	14	39	16	24
Average respective degree of degradation [%]	98.2	78.5	71.7	90.9
Average respective residence time [d]	200	80	100	74

6.3.4   *Example for a local implementation strategy*

When thinking about applying renewable energy strategies into regions, it is a matter of high importance to consider several regional aspects beforehand. The development of biogas applications in the last decades in Germany gave clear impressions about risk of problems, if basic aspects regarding biogas and its implementation are not being considered and regionally friendly adopted. These aspects are not only of a technical nature; furthermore, there are highly important social and economic backgrounds to be considered. Following a defined structure regarding these issues reduces the risk for missing important aspects before the practical implementation procedure starts.

The data needed for developing a regional biogas production strategy can be grouped as shown below:

General data:

- Population, population density, degree of urbanization;
- Income structure;
- Climate, topographic signs;
- Structure of administration (city, municipality, region);
- Rate of industrialization;
- Regional Infrastructure (streets, rails, gas pipelines, grid, heat supply).

Energy data:

- Energy demand (heat, electricity, gas, fuel; calculated in per capita factor);
- Electricity, heat and fuel production;
- Energy prices;
- Energy imports;
- Energy grids (heat, electricity, gas);
- Energy market structure and organization (energy providing companies, state owned, non-corporate, quantity);
- Locations of electricity, heat and fuel production.

Agriculture data:

- Area distribution (agriculture, forests, cities, agricultural areas);
- Crop growing species, resulting fertilizer demands;
- Already existing farm-located biogas applications for determination of already existing hot spots for further biogas technology implementation.

Substrate aspect data:

- Waste arising (calculated in per capita factor);
- Waste fractions, suitable wastes for biogas utilization;
- Alternative biomass resources (such as energy crops).

Waste sector data:

- Sorting, Waste pick-up service, Collection points;
- Waste disposal;
- Landfill sites, composting sites, Waste to energy infrastructures (e.g. incineration, anaerobic digestion, etc.) for determination of already existing hot spots for further or new biogas technology implementation.

Before starting detailed planning of technical implementation, the local conditions should be collected in the described way and as detailed and with reference to as many actual data bases as possible.

Outputs from such pre-study activities should be considered with respect to the following implementation focuses and their outputs; results from these scenarios are the best possible biogas plant locations concerning the individual scenario topics:

- *Focus I on products and co-products utilization*: best suitable scenario according to existing structure (heat, electricity, gas) including lowest demand for new infrastructure (such as grids for example).
- *Focus II on implementation to agriculture infrastructure*: best suitable scenario according to farm locations, farm sizes and digestate utilization for fertilizer demands (with respect to individual crop breeding strategies). This should consider place of combined waste, energy crop and agricultural residue utilization.
- *Focus III on optimum transporting distance*: best suitable scenario according to preferably short substrate and product transporting distances.

The challenge is to find the best suitable regional combination of all three individual focuses and scenarios. Depending on the local conditions to be determined beforehand, it is always somehow an individual decision to define the best suitable scenario or the most relevant focus for the individual region. For example in a country like Germany with a high number (and in some regions also a high local density) of biogas plants, the backgrounds are completely different in comparison to developing countries in terms of renewable energy production. Therefore, it is always important to think in terms of individual regions and to avoid the transfer of existing strategies simply from one country or region to another.

After defining the best suitable combination of scenarios the regional substrate-related biogas potentials should be determined (e.g. as described in section 6.3.3); referring to the resulting experimental data and the defined utilization and implementation strategies according to the above mentioned scenarios, regional substitution factors in terms of energy resources and CO_2 emissions can easily be calculated.

6.4   COMBUSTION OF WASTE

Combustion is a mature and well-proven technology that has been used for waste treatment of many types of wastes such as municipal solid waste (MSW), refuse derived fuels (RDF), agriculture wastes, wood wastes, packaging waste, industrial waste, hazardous waste, and sludge from wastewater treatment. Still, there are needs for development making the process more efficient as an energy conversion process, including improved emission control, plant efficiency and ash handling.

6.4.1   *Technical background*

The energy content that can be converted to thermal energy is dependent on the type of waste fuels. Table 6.5 shows heating values for some organic waste fuels.

The waste combustion plant consists mainly of the boiler and the flue gas cleaning system. The waste is combusted in the boiler and heat is recovered by boiling water. The steam can then be used for power production in a steam turbine or for heat production. Heat can also be recovered from the steam leaving the steam turbine.

The design and operation conditions of a waste combustion plant are dependent on the type of waste used as fuel. Especially, high concentrations of corrosive substances, as for example chlorides, are of importance. This is the reason to the usually rather modest steam pressures and temperatures (commonly around 400°C and 40 bar) used in waste combustion plants (European Commission, 2006).

6.4.2   *Examples of combustion of waste*

Techniques used for combustion of waste are grate furnace, fluidized bed and rotary kiln. Today grate furnace seems to be the most commonly used technique at least for MSW. For example,

Table 6.5.   The range of lower heating values for some organic waste fuels (EUBIA, 2011; European Commission, 2006).

	MSW	Bulky waste (furniture, etc.)	RDF	Packaging waste	Straw	Sewage sludge
Lower heating value [MJ/kg]	6–15	10–17	11–26	17–25	14	0.5–2.5

Table 6.6.   Some examples of existing waste combustion plants (Buchhorn, 2010; Hinge, 2009; Johansson *et al.*, 2009; Weiler and Grotefeld, 2010).

Plant (year built)	Technique	Steam data [°C/bar]	Capacity thermal [MW]	Capacity waste [10³ kg/year]	Efficiency [%]
Borås Energi och miljö, Sweden (2004)	Fluidized bed	405/49	20	100000 (20–30% household waste, 70–80% industrial waste)	89
Renova, Gothenburg, Sweden (1994 and 2001)	Grate furnaces	400/40	45	500000 (50/50 household and industrial waste)	89
Gevudo, Nordrecht, Netherlands (1972, 1997 and 2010)	Grate furnaces	400/40	75	396000 (municipal solid waste)	?
Bernburg, Germany (2009)	Furnaces	410/42	?	552000 (RDF)	?
Vattenfall, Fynsverket, Denmark (2009)	Grate furnace	540/110	(35 MW$_{el}$ 86MW$_{heat}$ (including 11 MW flue gas condensing)	150000 (straw)	El. 33

90% of the installations in Europe for combustion of MSW had grate furnaces in 2006 (European Commission, 2006) and in 2010, 70% of the installations in Germany for MSW and RFD used this technique (Fendel and Firege, 2010).

Fluidized beds have a limitation in that the fuel particles cannot be too large. Another problem for some types of waste, for example straw, can be that the bed particles in the fluidized bed stick together due to low ash melting point. However, developments are going on to optimize the conditions in fluidized beds to make it possible to use these types of wastes also. For straw, grate furnaces are recommended since it has a low ash melting point and it can cause the bed particles in a fluidized bed to stick together (Chunjiang *et al.*, 2011; Hinge, 2009). In Table 6.6, some examples on combustion plants using waste as fuel are given.

6.4.3   *Development considerations*

The development of combustion of waste is from treatment of waste to a competitive energy conversion process. This requires limiting the costs and improving environmental performance. Important factors concerning the costs are pre-treatment costs and maintenance costs. A study comparing mixed MSW and RDF as fuel show that using RDF might be more competitive if the energy prices increase when pre-treatment cost is not included. A disadvantage with RDF plants can be that they are less flexible concerning what fuels can be used compared to plants designed for using mixed waste. On the other hand, less pre-treatment can cause higher maintenance costs. Generally, waste combustion plants have higher operation and maintenance costs compared to combustion plants using other types of fuels due to more advanced handling but not many studies have been done and knowledge about possible improvements is low (European Commission, 2006; Fendel and Firege, 2010; Johansson *et al.*, 2009).

Another important issue is the risk for overcapacity due to several competing waste-to-energy concepts, for example:

- the combustion of mixed municipal waste;
- co-combustion of waste with other fuels;
- production of RDF followed by combustion;
- the use of waste fractions in cement kilns.

This can be caused because there are different actors on the market driven by different aims with the public sector increasing the municipal waste combustion capacity to decrease the lack of waste treatment capacities and the private sector increasing the RDF combustion capacity and waste use in cement kilns with the aim to reduce energy costs and using the possibility to sell electricity to the grid. (Fendel and Firege, 2010).

6.5   EXAMPLES OF USE OF ORGANIC WASTE IN OTHER CONVERSION PROCESSES

6.5.1   *Ethanol and butanol from organic waste*

Another biological conversion process that can use waste as feedstock is fermentation to ethanol or butanol. Raw materials that can be used for fermentation are materials containing sugars or substances that can be converted into sugars. The raw materials can be grouped as directly fermentable sugary materials, starchy materials and lignocellulosic materials. Wastes of interest for fermentation are industrial waste material and waste waters, for example by-products of the sugar industry, wastes from the food industry, waste water from breweries and lignocellulosic wastes such as straw, wood wastes and waste from the pulp and paper industry. Also part of MSW could be used for ethanol production. There are studies pointing out that the long-term sustainability of ethanol production will ultimately depend on the use of lignocellulosic wastes.

Fermentation of sugary wastes to ethanol is a mature and well-proven technology but fermentation of lignocellulosic waste and fermentation to butanol is still at the research stage. The development areas include increasing the yield, process design, and increasing the energy efficiency including process integration (Adler *et al.*, 2007; Farrell *et al.*, 2006; Flavell, 2007; Kim and Dale, 2005a,b; Kszos *et al.*, 2001; Prasad, 2007). In addition, the selection of the production organism could be an option for improved processes. For example, in the ethanol production the yeasts could sometimes be replaced by such anaerobic bacteria as *Clostridium thermocellum* or *Zymomonas* spp. (Righelato, 1980). Their involvement could increase such biocatalyst parameters as thermal resistance, or ethanol tolerance, which then could have profound effects on the overall process economics.

Biofuels produced with conventional technology from easily convertible raw materials, such as direct fermentable sugary material, are often referred to as first generation biofuels, while biofuels produced in more advanced processes from lignocellulosic material are called second-generation biofuels (UN-Energy, 2007).

Table 6.7. Methods for pre-treatment of lignocellulosic wastes for ethanol production.

Method	Characteristics	Performance	Comments	References
Steam explosion	High temperature, high pressure steam	Cost-effective for hard wood Less effective for soft wood		Mosier *et al.* (2005) Silverstein (2004)
Ammonia fiber/freeze explosion	Prewetting with NH₃ followed by pressurization (>12 bar)	Less effective for wastes with high ligning content	Recovery of NH₃ possible	McMillan (1997) Silverstein (2004)
Acid pre-treatment	Several types of acids possible, sulfuric acid, hydrochloric acid, peracetic acid, nitric acid, phosoric acid Continues (loading 5–10 mass-%+ >160°C) or batch (loading 10–40 mass-% +<160°C)		Dilute acid mostly used Re-precipitation of lignin need to be prevented	Brink (1993) Chung *et al.* (2005) Dale and Moelhman (2000) Hussein *et al.* (2001) Karimi *et al.* (2006) Knauf and Moniruzzaman (2004) Kurakake *et al.* (2005) McMillan (1997) Tucker (2003) Teixeira *et al.* (1999) Wyman (1999)
Alkaline pre-treatment	Ambient temperature, low pressure, dilute NaOH		Production of irrecoverable salts Easy removal of lignin content	Börjesson (2006) McMillan (1997)
Biological pre-treatment	Special microorganisms		Simple process Long reaction time Expensive microorganisms	Wyman (1999)

For the substrates that are not directly fermentable, substrate pre-treatment for conversion into sugar monomers are necessary. Possible pre-treatment methods for ethanol production from lignocellulosic material are presented in Table 6.7.

Agriculture waste and food industry wastewater can also be fermented to butanol. The process is similar to the production of ethanol but with other microorganisms. Butanol can be used as stand-alone transportation fuel or blended with petrol or diesel. Examples of wastes that have been tested for butanol production are corn stover, switch grass, straw, and whey (Kumar and Gayen, 2011; Qureshi *et al.*, 2010; Napoli *et al.*, 2010; Wang and Chen, 2011). There are still questions to be solved to reach high yields of butanol and high production rates. The process has yet not been widely demonstrated in scaled up plants or showed clear economic feasibility. However, remarkable development efforts are directed to this field of improved "real" biotechnologies utilizing the biocatalysis. For example, the problem of some compounds that are formed during the degradation of lignocelluloses inhibiting the butanol producing microorganisms, could be avoided by pretreatment.

The cost of butanol downstream processing is very sensitive to butanol concentration in the broth (Zhu and Yang, 2010). In order to increase the concentration in the production broth, in turn, it is crucial to increase the cell concentration of the production organism. This goal could be achieved by immobilized cells as biocatalysts. By this strategy, the reactor productivity has been increased to 6.5–15.8 g/L/h, instead of 0.5 g/L/h achieved by the traditional batch fermentation system. If immobilized cells or enzyme systems are used, also improved biocatalyst half-lives are obtained in biochemical engineering (Dunnill, 1980).

In practical biotechnology, one remarkable advantage is the possibility of combining various raw materials into same processes. Therefore, such a waste treatment plant exploiting several bio-catalytic processes could effectively exploit different biomass sources in the same reactor without causing overwhelming problems for the entire process run. This allows the combination of dry waste and sewage, or industrial wastes with the municipal residues, as well as seasonal variations. This ideal situation could be achieved by proper management of the entire process, selection of microorganisms and fermentation conditions, and process control and adjustment. The microbiological method could be used in sequence with other techniques such as a pretreatment for combustion where the overall environmental performance as well as the energy gain could then improve.

6.5.2   Hydrothermal carbonization of organic waste fractions

The hydrothermal carbonization was put forward in 1913 by the scientist Friedrich Bergius who explained the elementary process. Experiments involving the hydrothermal carbonization (HTC) started at the beginning of the 20th century as an attempt to understand the natural process of formation of coal. In the last years, the process of HTC has been rediscovered due to its likelihood of having a positive impact on future problems of energy use and agriculture. This process is very similar to the natural process of production of coal, but instead of thousands of years, it is done within a couple of hours. It can be said, that in a very simple way, the process consists in the separation of the water in the biomass with a parallel increasing of carbon concentration (Morrondo-Martin, 2011).

6.5.2.1   HTC reactions

In a very simplified way, the HTC process takes place in four steps. Firstly, the activation energy must be given to start the process and its reactions. This energy is reduced by a catalysator and given in the form of temperature (180–200°C) and pressure (up to 20 bar). After the reaction has started, the temperature reaches the expected stable value. The reactions continue as long as energy is given, since it is an endothermic process. At a constant temperature and pressure level, the water is separated from the biomass and the time of reaction determines the level of carbonization. The fourth step is the cooling phase.

Depending on the wanted products, the process of HTC can take between 4 to 24 hours. Sub-products besides bio coal are substances that remain in the water phase such as furfural, organic acids and aldehydes, and gas products such as CO_2, CH_4, etc. (Sevilla and Fuertes, 2009a).

There are three types of this process, which give different products, depending on the reaction time. The first option is to make an incomplete carbonization, which gives a product rich in nutrients that could be used, for example, to improve soil qualities. The second type is the complete carbonization where the product is carbon slurry that can be dried to bio-coal and used as an energy source. The third possibility is the so named "short carbonization". It gives a liquid, very reactive intermediate and hydrogen, which can perhaps be used for the production of fuel.

6.5.2.2   Substrates

A large amount of substrates seem to be suitable for the process of HTC (such as food industry wastes, waste from wood and paper industry, garbage coming from biotopes and parks, sports facilities, agricultural by-products, etc.). To the present, the potential of the HTC is still somehow uncertain since aspects like total energy balance, effects of the HTC coal in the soil and the life cycle assessment of the whole process are still unanswered.

Concerning the properties of the hydro char, analysis and laboratory results show that the carbon content increases from approximately 40–44% in the substrates to approximately 64–66%

Figure 6.5.   Original substrate sample.

Figure 6.6.   Substrate after dewatering.

in the bio char (Sevilla and Fuertes, 2009b). At the same time, there is a reduction in the oxygen and hydrogen contents. Furthermore, an increase of the temperature gives a reduced content of oxygen and hydrogen.

6.5.2.3   *HTC of a selected biowaste substrate*
As an example of the procedure, the HTC of pulp and paper waste sludge will be described. Figure 6.5 shows the raw material in original shape and Figure 6.6 presents the substrate after dewatering and preparation for HTC.

After preparation the substrate has been carbonized in a lab batch autoclave with a volume of approximately 300 mL. Figure 6.7 shows the most relevant process parameters within the experiment (two carbonizations were performed, firstly at 200°C autoclave temperature and secondly at 220°C autoclave temperature) (Krüger, 2011).

In Figures 6.8 and 6.9, a comparison of different substrate and product values is given to describe the change of certain parameters during the HTC process. Figure 6.10 provides a comparison of the heating values to lignite and coal-heating values (see corresponding marks). Additionally to this Figure 6.11 gives a visual impression about the produced biochar after dewatering and drying in comparison to the dried and dewatered substrate and the carbonization diagram in Figure 6.12 gives information about the quality of the HTC product out of paper waste material.

Finally, flaring tests were performed with the bio-char and the results gave a promising impression about the suitability concerning energy production (Fig. 6.13).

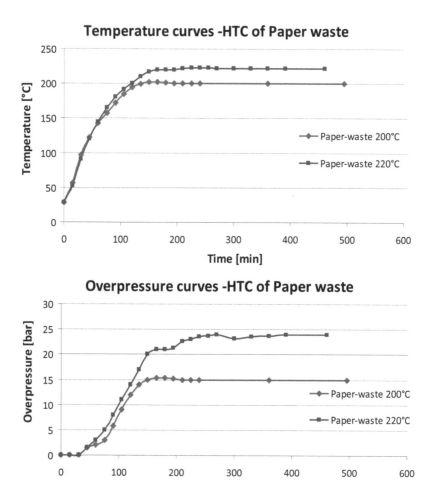

Figure 6.7. Experimental parameters for HTC batch process (Krüger, 2011).

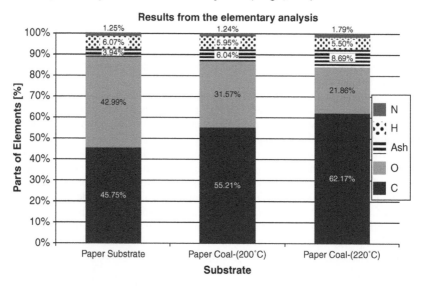

Figure 6.8. Elementary analysis of substrate and bio coals (Krüger, 2011).

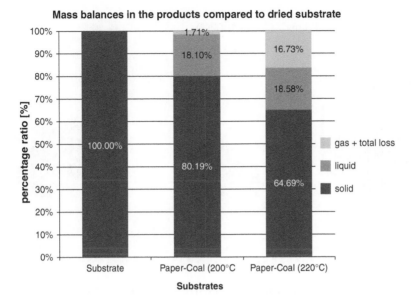

Figure 6.9.   Results of mass balances, substrate and products (Krüger, 2011).

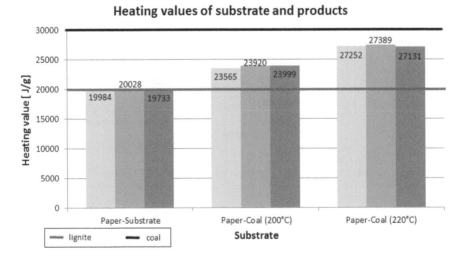

Figure 6.10.   Results of three heating value analysis, substrate and products (Krüger, 2011).

With respect to technological implementation of HTC, also the by-products have a strong matter of interest.

An important by-product from the HTC process is the process water. Several utilizations are imaginable, such as biogas production in high load fixed bed digesters or separation into individually utilizable fractions *via* continuous distillation (so-called fractioned rectification). Concerning development of such utilization strategies knowledge about the process water composition is absolutely required. Figure 6.14 shows quantitative GCMS analysis results of process waters from HTC batch processes being operated for 1, 3, 6 and 21 hours with the same substrate. The values were standardized to the highest value being measured for each single compound.

Figure 6.11.   Visual comparison of substrate with two HTC products (Krüger, 2011).

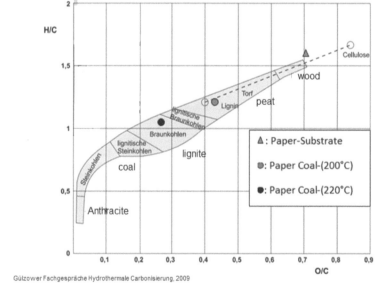

Figure 6.12.   Carbonization diagram of substrate and products (Belusa *et al.*, 2009, modified).

The presented results clearly show that organic waste fractions are suitable substrates for the HTC production, mainly in terms of producing an energy carrier. Other effects like soil improvement due to the forming of Terra Preta have not been taken into consideration yet; nevertheless all imaginable strategies should be taken into account and the best option, even besides to energy utilization strategies, should be chosen for full-scale implementations. Aspects of the future work on the matter of HTC from organic waste fractions include the identification of applicable technologies for either pilot or full scale applications under consideration of life cycle analysis and economic feasibility.

Figure 6.13.   Glowing bio-coal under flaring test conditions.

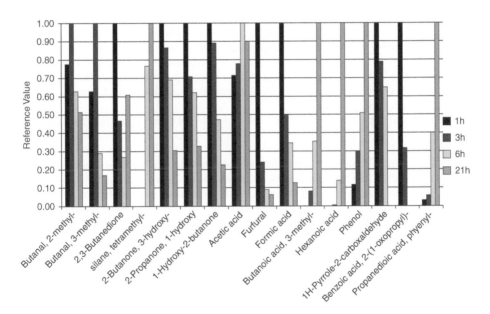

Figure 6.14.   GCMS analysis of process waters from different HTC batch processes (Pfingstmann, 2011; Müller, *person. commun.* 2011).

6.5.3   *Pyrolysis and gasification of organic waste*

Pyrolysis and gasification have been used for waste treatment but the technology is still in the development stage. Pyrolysis is the direct thermal decomposition, or degassing, of an organic matrix in the absence of oxygen producing an array of solid, liquid and gas products. The composition, characteristics and yield of the different pyrolysis products are dependent on the feedstock used and process parameters such as temperature, residence time and heating rate. High temperature, low heating rate and long gas residence time give a high yield of gas products. (Bridgwater and Peacocke, 2000; Cheremisinoff and Rezaiyan, 2005; Di Blasi, 2007; European Comission, 2006; Yaman, 2003).

All pyrolysis products can be used for energy purposes. The liquid and gas products can be used as vehicle fuel or combusted for production of power and/or heat production. The solid product, char, can be used as a solid fuel. An advantages of using pyrolysis instead of direct combustion are reduction of the flue-gas amounts and with that the requirements of flue-gas cleaning. The same techniques as used for combustion can also be used for pyrolysis. MSW has been used as feedstock in pyrolysis plants and RDF, waste cocking oil, sludge, rapeseed cake, crop residues and forestry residues are examples of other organic wastes that have been investigated for pyrolysis (Cheremisinoff and Rezaiyan, 2005; European Comission, 2006; Hossain et al., 2011; Karayildirim et al., 2005; Puy et al., 2011; Singhabhandhu and Tezuk, 2010; Smets et al., 2011).

Gasification is also a thermo-chemical process but compared to pyrolysis the aim is to convert as much as possible of the feedstock to gas. The process is also operated at higher temperatures than pyrolysis (up to 1800°C for conventional gasification and up to 20,000°C for plasma gasification compared to 250 to 800°C for pyrolysis). The gas mainly consists of CO_2, CO, CH_4, H_2, H_2O and small amounts of C_2-hydrocarbons. (European Comission, 2006; Gomez et al., 2009; Ptasinski, 2008; Valleskog et al., 2008).

The techniques mainly used for gasification are fixed beds, fluidized beds and entrained flow gasifiers (also called entrained suspension bed gasifiers). Fixed bed gasifiers have a simple design and give a low calorific gas with high tar content. The fluidized bed gasifier gives the possibility to a more uniform temperature distribution but the ash content of the fuel can cause slagging of the bed in the same way as for fluidized combustion of some feedstock's. The entrained flow gasifiers are operated at high temperatures and need feedstock divided to small particle sizes. Another technique is the so-called plasma gasification technique where an electrically generated plasma arc torch is used (Alimuddin et al., 2010; European Comission, 2006; Gomez et al., 2009; Li et al., 2004, McKendry, 2002).

Several types of organic waste have been tested in different types of gasifiers, for example:

- MSW, RDF, salmon waste, hazelnut shells and wood waste in fixed bed gasifiers (Arena, 2011; Dogru et al., 2002; Rowland et al., 2009; Sheth and Babu, 2009).
- MSW, RDF, forest residues, sawdust, straw, black liquor from pulp production and olive oil residues in fluidized bed gasifiers (Alimuddin et al., 2010; Arena, 2011; Göteborg Energi, 2012; Li et al., 2004; Naqvi et al., 2010; Palonen et al., 2006).
- liquid, pasty and granulated rice husk, black liquor from pulp production, and sawdust in entrained flow bed gasifiers (European Comission, 2006; Chemrec, 2012; Zhou et al., 2009).
- pelletized RDF from carpet and textile waste and MSW in plasma gasifiers (Clark and Rogoff, 2010; Lemmens et al., 2007; Willis et al., 2010).

Pyrolysis and gasification can be combined with combustion of the produced gas for heat and/or power production. An example of such a plant is a plant being built in Lahti, Finland where 250,000 tonnes of waste from businesses and households will be used to generate 90 MW of heat and 50 MW of power per year. It is also possible to further process the gas to vehicle fuels as for example dimethylether (DME), synthetic natural gas (SNG) and Fischer–Tropsch (FT) fuel. (Arena, 2011; Chemrec, 2012; European Comission, 2006; European Investment Bank, 2011; Ptasinski, 2008; Valleskog et al., 2008).

For pyrolysis the influence of process parameters on the gas and by-product quality and plant efficiency are of interest for further development. While gasification for gas cleaning, tar reduction, process operation conditions and increasing plant energy efficiency are development areas. (Arena, 2011, Cheremisinoff and Rezaiyan, 2005; European Comission, 2006; Kirkels and Verbong, 2011; Kumar et al., 2009; Malkow; 2004; Puy et al., 2011; Smets et al., 2011).

6.6 QUESTIONS FOR DISCUSSION

- What conversion processes are of interest when using organic waste as an energy resource?
- Why is it of interest to pre-treat organic waste before using it as an energy resource?
- How can organic waste be pre-treated?
- Identify the limitations and challenges from experimental biogas potential determination
- Develop a biogas implementation strategy for a self-selected region according to the aspects mentioned in Section 6.3.4
- Identify microbiological challenges regarding the different species of the biogas symbiosis
- Identify reasons for biogas outcome gap in comparison of dry and wet digestion (Table 6.2)
- What technique is most common for combustion of waste?
- What are typical characteristics for a combustion plant using waste as an energy source?
- What organic wastes can be used for ethanol production?
- How can lignocellulosic material be pre-treated to enhance ethanol production?
- Mention some benefits of using the immobilized biocatalysts and organism selection in the biotechnological processes.
- Why is productivity so important in a bioprocess that it often overshadows even the actual yield of fuels or chemicals?
- What happens when organic waste is pyrolysis?
- What products are formed in gasification of organic waste?
- How can organic waste be gasified?
- What development areas have been identified for pyrolysis and gasification processes?
- Define the chemical pathway for carbon enrichment during the process of HTC. Use the glucose molecule as biomass reference.
- Identify the pathways for raw oil refining procedures; are there steps and procedures that can be transferred to HTC water upgrading?
- In which ways could microbes influence the sustainability of a waste utilization process?
- What is the correct perspective in evaluating the environmental impacts of biomass waste treatment?

ACKNOWLEDGEMENT

Part of the work presented in this chapter has been done within the project Regional Mobilizing of Sustainable Waste-to-Energy Production (REMOWE) included in the Baltic Sea Region program and funded by the European Union (European Regional Development Fund). All partners and contributors to the project are also acknowledged.

REFERENCES

Adler, P.R., Grosso, S.J.D. & Parton, W.J.: Life-cycle assessment of net greenhouse-gas flux for bioenergy cropping systems. *Ecol. Appl.* 17:3 (2007), pp. 675–91.
Ahrens, T. & Weiland, P.: Biomethane for future mobility. *Landbauforschung Völkenrode* 1:57 (2007), pp. 71–79.
Alimuddin, Z, Alauddin, B.Z., Lahijani, P., Mohammadi, M. & Mohamed, A.R.: Gasification of lignocellulosic biomass in fluidized beds for renewable energy development: a review. *Renew. Sust. Energy Rev.* 14 (2010), pp. 2852–2862.
Arena, U.: Process and technological aspects of municipal solid waste gasification. A review. *Waste Manage.* 2011, doi:10.1016/j.wasman.2011.09.025.
Avfall Sverige: Guide för införande av system för insamling av källsorterat matavfall (Manual for implementing of collection of sorted food waste), Avfall Sverige, 2011 (in Swedish).
Belusa, T., Funke, A., Behrendt, F. & Ziegler, F.: Hydrothermale Karbonisierung und energetische Nutzung von Biomasse – Möglichkeiten und Grenzen. Gülzower Fachgespräche, Band 33, Hydrothermale Carbonisierung, 2009, pp. 42–54 (modified).

Biogas Öst: Matavfall blir biogas. Fem goda exempel (Food waste become biogas. Five good examples). Biogas Öst, 2011 (in Swedish).

Börjesson, P.: Energibalans för bioetanol — En kunskapsöversikt (Energy balance of bioethanol — A review). Report No. 59, Department of Environmental and Energy Systems Studies, Lund University, 2006 (in Swedish).

Bridgwater, A.V. & Peacocke, G.V.C.: Fast pyrolysis processes for biomass. *Renew. Sust. Energy Rev.* 4 (2000), pp. 1–73.

Brink, D.L.: Method of treating biomass material. US Patent 5221357, 1993.

Buchhorn, B.: Optimisation and extension of the WTE Plant Dordrecht, Netherlands. *ISWA World Congress*, 15–18 November 2010, Hamburg, Germany, 2010.

Chemrec, Internet site: www.chemrec.se (acessed January 2012).

Cheremisinoff, N.P. & Rezaiyan, J.: *Gasification technologies — A primer for engineers and scientists*. CRC Press, 2005.

Cherubini, F., Bargigli, S. & Ulgiati, S.: Life cycle assessment (LCA) of waste management strategies: landfilling, sorting plant and incineration. *Energy* 34 (2009), pp. 2116–2123.

Chung, Y.C., Bakalinsky, A. & Penner, M.H.: Enzymatic saccharification and fermentation of xylose-optimized dilute acid-treated lignocellulosics. *Appl. Biochem. Biotechnol.* 124 (2005), pp. 947–961.

Chunjiang, Y., Jianguang, Q., Hu, N., Mengxiang, F. & Zhongyang, L.: Experimental research on agglomeration in straw-fired fluidized beds. *Appl. Energy* 88 (2011), pp. 4534–4543.

Clark, B.J. & Rogoff, M.J.: Economic feasibility of a plasma arc gasification plant, City of Marion, Iowa. Paper no. NAWTEC18-3502, *18th Annual North American Waste-to-Energy Conference*, 11–13 May 2010, Orlando, FL, 2010.

Dale, M.C. & Moelhman, M.: Enzymatic simultaneous saccharification and fermentation (SSF) of biomass to ethanol in a pilot 130 l multistage continuous reactor separator. *9th Biennial Bioenergy Conference*, 15–19 October 2000, Buffalo, New York, 2000.

de Baar, H.J.W. & Stoll, M.H.C.: Storage of carbon dioxide in the oceans. In: P.A. Okken, R.J. Swart & S. Zwerver (eds): *Climate and energy — The feasibility of controlling CO_2 emissions*. Kluwer Academic Publishers. Dordrecht. The Netherlands. 1989, pp. 143–177.

Di Blasi, C. Modeling chemical and physical processes of wood and biomass pyrolysis. *Prog. Energy Combust. Sci.* 34 (2007), pp. 47–90.

Dogru, M., Howarth, C.R., Aka, G., Keskinler, B. & Malik, A.A.: Gasification of hazelnut shells in a downdraft gasifier. *Energy* 27 (2002), pp. 415–427.

Dunnill, P.: Immobilized cell and enzyme technology. In: S. Brenner, B.S. Hartley & P.J. Rodgers (eds): New horizons in industrial microbiology. *Phil. Trans. R. Soc. Lond.* B 290: 1980, pp. 409–420.

Ekoleima, Ay: Selvitys kaupan entisiä elintarvikkeita koskevien säädösvaihtoehtojen taloudellisista ja ympäristövaikutuksista, (Report on economical and environmental effects of statute alternatives concerning grocery shops' overdated food), Finnish Ministry of Social Affairs and Health, Helsinki, Finland, 2008 (in Finnish).

EUBIA: European Biomass Industry Association. www.eubia.org (acessed November 2011).

European Commission: Integrated pollution prevention and control. Reference document on best available techniques for the waste incineration, BREF (08.2006), EU Commission, Brussels, Belgium, 2006.

European Investment Bank. Energy-from-waste plant keeps Finnish city warm. http://www.eib.org (acessed March 2011).

Farrell, A.E., Plevin, R.J., Turner, B.T., Jones, A.D., O'Hare, M. & Kammen, D.M.: Ethanol can contribute to energy and environmental goals. *Science* 311 (2006), pp. 506–508.

Fendel, A. & Firege, H.: Competition of different methods for recovering energy from waste leading to overcapacities. *ISWA World Congress*, 15–18 November, 2010, Hamburg, Germany, 2010.

Flavell, R.: Biotechnology options for bioenergy crops: prospects for 2nd generation feedstock technologies. *USDA Global Conference on Biofuels*, Minneapolis, 2007.

Gendebien, A., Leavens, A., Blackmore, K., Godley, A., Lewin, K., Whiting, K.J., Davis, R., Giegrich, J., Fehrenbach, H., Gromke, U., del Bufalo, N. & Hogg, D.: Refuse derived fuel, current practice and perspectives. b4-3040/2000/306517/mar/e3, Final Report, Commission of EC Directorate General Environment, WRc, Swindon, 2003.

Gomez, E., Amutha Rania, D., Cheeseman, C.R., Deegan, D., Wise, M. & Boccaccini, A.R.: Thermal plasma technology for the treatment of wastes: a critical review. *J. Hazard. Mat.* 161 (2009), pp. 614–626.

Göteborg Energi: Gothenburg biomass gasification project. GoBiGas, Goteborg, Sweden, www.goteborgenergi.se (acessed January 2012).

Hinge J: Elaboration of a platform for increasing straw combustion in Sweden, based on Danish experiences. Värmeforsk project no. E06-641, 2009.

Hossain, M.K., Strezov, V., Chan, K.Y., Ziolkowski, A. & Nelson, P.F.: Influence of pyrolysis temperature on production and nutrient properties of wastewater sludge biochar. *J. Environ. Manage.* 93 (2011), pp. 223–228.

Hussein, M.Z.B., Rahman, M.B.B.A., Yahaya, A.H.J., Hin, T.Y.Y. & Ahmad, N.: Oil palm trunk as a raw material for activated carbon production. *J. Porous Mat.* 8 (2001), pp. 327–334.

Jätekukko Oy: Kotitalouksissa syntyvän biojätteen keräilymenetelmän vaikutus biojätteen ominaisuuksiin, (Effect of collection method to the properties of household biowaste). Jätekukko Oy, 2004 (in Finnish).

Jätekukko Oy: Jätekukko Oy:n vuosikertomus, (Jätekukko Oy Annual Report), 2009.

Johansson, A., Niklasson, F., Johnsson, A., Fredäng, J. & Wettergren, H.: Drift och underhåll av avfalls-förbränningsanläggningar – en jämförelse av två tekniker och strategier, (Operation and maintenance of Waste Combustion plant – development of a method for economical comparison of different techniques and strategies), WR-09, Waste Refinery, 2009 (in Swedish).

Juusola, M.: Kuopion jätekeskuksen jäteperäisen kierrätyspolttoaineen laatu- ja hyödyntämisselvitys, (Quality and Recovery Report on Waste-Derived Recovered Fuel at Kuopio Waste Center). Savonia University of Applied Sciences, 2011 (in Finnish).

Karayildirim, T., Yanik, J., Yuksel, M. & Bockhorn, H.: Characterisation of products from pyrolysis of waste sludges. *Fuel* 85 (2005), pp. 1498–1508.

Karimi, K., Emtiazi, G. & Taherzadeh, M.J.: Ethanol production from dilute-acid pretreated rice straw by simultaneous saccharification and fermentation with *Mucor indicus*, *Rhizopus oryzae*, and *Saccharomyces cerevisiae*. *Enzyme Microb. Tech.* 40 (2006), pp. 138–144.

Kim, S. & Dale, B.E.: Environmental aspects of ethanol derived from no-tilled corn grain: nonrenewable energy consumption and greenhouse gas emissions. *Biomass Bioenergy* 28 (2005a), pp. 475–489.

Kim, S. & Dale, B.E.: Life cycle assessment of various cropping systems utilized for producing biofuels: bio-ethanol and biodiesel. *Biomass and Bioenergy* 29 (2005b), pp. 426–439.

Kirkels, A.F. & Verbong, G.P.J.: Biomass gasification: still promising? A 30-year global overview. *Renew. Sust. Energy Rev.* 15 (2011), pp. 471–481.

Knauf, M. & Moniruzzaman, M.: Lignocellulosic biomass processing: a perspective. *Int. Sugar J.* 106 (2004), pp. 147–150.

Kothari, R., Tyagi, V.V. & Pathak, A.: Waste-to-energy: a way from renewable energy sources to sustainable development. *Renew. Sust. Energy Rev.* 14:9 (2010), pp. 3164–3170.

Krüger, C.: *Ermittlung von Betriebsdaten und Optimierungsparameter für den HTC-Prozess*. Bachelor Thesis, Ostfalia University of Applied Sciences, Braunschweig, Wolfenbüttel, Germany, 2011.

Kszos, L.A., McLaughlin, S.B. & Walsh, M.: Bioenergy from switchgrass: reducing production costs by improving yield and optimizing crop management. Oak Ridge National Laboratory, Oak Ridge, USA, 2001.

Kumar, A., Jones, D.D. & Hanna, M.A.: Thermochemical biomass gasification: a review of the current status of the technology. *Energies* 2 (2009), pp. 556–581.

Kumar, M. & Gayen, K.: Developments in biobutanol production: new insights. *Appl. Energy* 88: 6 (2011), pp. 1999–2012.

Kurakake, M., Ouchi, K., Kisaka, W. & Komaki, T.: Production of L-arabinose and xylose from corn hull and bagasse. *J. Appl. Glycosci.* 52 (2005), pp. 281–285.

Laitinen, M.: Kuopion jätekeskus, selostus katoksesta sekä murskaus — ja seulontalaitteista (Kuopio waste management centre, briefing about the shelter and crushing and sieving devices), Jätekukko Oy, 2011 (in Finnish).

Lemmens, B., Elslander, H., Vanderreydt, I., Peys, K. & Diels, L.: Assessment of plasma gasification of high caloric waste streams. *Waste Manage.* 27 (2007), pp. 1562–1569.

Lens, P., Hamelers, B., Hoitink, H. & Bidlingmaier, W.: Resource recovery and reuse in organic solid waste management. IWA Publishing, 2004, pp. 395–410.

Li, X.T., Grace, J.R., Lim, C.J., Watkinson, A.P., Chen, H.P. & Kim, J.R.: Biomass gasification in a circulating fluidized bed. *Biomass Bioenergy* 26 (2004), pp. 171–193.

Malkow, T.: Novel and innovative pyrolysis and gasification technologies for energy efficient and environmentally sound MSW disposal. *Waste Manage.* 24 (2004), pp. 53–79.

McDougall, F.R., White, P.R., Franke, M. & Hindle, P.: *Integrated solid waste management: a life cycle inventory*. John Wiley & Sons Ltd, 2008.

McKendry, P.: Energy production from biomass (Part 3): Gasification technologies. *Bioresour. Technol.* 83 (2002), pp. 55–63.

McMillan, J.D.: Bio-ethanol production: status and prospects. *Renew. Energy* 10 (1997), pp. 295–302.

Morrondo-Martin, A.-M.: *HPLC analysis method for water residues from hydrothermal carbonization.* MSc Thesis, Ostfalia University of Applied Sciences Braunschweig, Wolfenbüttel, Germany, 2011.

Mosier, N., Wyman, C., Dale, B., Elander, R., Holtzapple, Y.Y.L.M. & Ladisch, M.: Features of promising technologies for pretreatment of lignocellulosic biomass. *Bioresour. Technol.* 96 (2005), pp. 673–686.

Napoli, F., Olivieri, G., Russo, M.E., Marzocchella, A & Salatino, P.: Butanol production by *Clostridium acetobutylicum* in a continuous packed bed reactor. *J. Ind. Microbiol. Biotechnol.* 37 (2010), pp. 603–608.

Naqvi, M., Yan, J. & Dahlquist, E.: Black liquor gasification integrated in pulp and paper mills: a critical review. *Bioresour. Technol.* 101 (2010), pp. 8001–8015.

Palonen, J., Anttikoski, T. & Eriksson, T.: The Foster Wheeler gasification technology for biofuels: refuse-derived fuel (RDF) power generation. *Power-Gen Europe*, 30 May–1 June 2006, Kölnmesse, Cologne, Germany, 2006.

Pfingstmann, J.: *Optimierung einer kontinuierlich betriebenen HTC-Anlage inklusive der Ermittlung system-relevanter Betriebsparameter.* Bachelor Thesis, Ostfalia University of Applied Sciences, Braunschweig, Wolfenbüttel, Germany, 2011.

Prasad, S., Singh, A. & Joshi, H.C.: Ethanol as an alternative fuel from agricultural, industrial and urban residues. *Resour. Conserv. Recy.* 50 (2007), pp. 1–39.

Ptasinski, K.J.: Thermodynamic efficiency of biomass gasification and biofuels conversion. *Biofuels Bioprod. Biorefin.* 2 (2008), pp. 239–253.

Puy, N., Murillo, R., Navarro, M.V., López, J.M., Rieradevall, J., Fowler, G., Aranguren, I., García, T., Bartrolí, J. & Mastral, A.M.: Valorisation of forestry waste by pyrolysis in an auger reactor. *Waste Manage.* 31:6 (2011), pp. 1339–1349.

Qureshi, N., Saha, B.C., Hector, R.E., Dien, B., Hughes, S., Liu, S., Iten, L., Bowman, M.J., Sarath, G. & Cotta, M.A.: Production of butanol (a biofuel) from agricultural residues: Part II – Use of corn stover and switchgrass hydrolysates. *Biomass Bioenergy* 34 (2010), pp. 566–571.

Rada, E.C., Ragazzi, M. & Panaitescu, V.: MSW bio-drying: an alternative way for energy recovery optimization and landfilling minimization. *U.P.B. Science Bulletin* Series D 71: 4 (2009), pp. 113–120.

Righelato, R.C.: Anaerobic fermentation: alcohol production. In: S. Brenner, B.S. Hartley & P.J. Rodgers (eds): New horizons in industrial microbiology. *Phil. Trans. R. Soc. Lond.* B 290 (1980), pp. 303–312.

Rowland, S., Bower, C.K., Patil, K.N. & Dewitt, C.A.M.: Updraft gasification of salmon processing waste. *J. Food Sci.ce* 74 (2009), pp. E426–E431.

Sevilla, M. & Fuertes, A.B.: *Chemical and structural properties of carbonaceous products obtained by hydrothermal carbonization of saccharides.* Oviedo: Wiley-VCH Verlag GmbH & Co., 2009a.

Sevilla, M. & Fuertes, A.B.: *The production of carbon materials by hydrothermal carbonization of cellulose.* Oviedo: Elsevier Ltd., 2009b.

Sheth, P.N. & Babu, B.V.: Experimental studies on producer gas generation from wood waste in a downdraft biomass gasifier. *Bioresour. Technol.* 100 (2009), pp. 3127–3133.

Silverstein, R.A.: *A comparison of chemical pretreatment methods for converting cotton stalks to ethanol.* MSc Thesis, Biological and Agricultural Engineering, North Carolina State University, 2004.

Singhabhandhu, A. & Tezuka, T.: Prospective framework for collection and exploitation of waste cooking oil as feedstock for energy conversion. *Energy* 35 (2010), pp. 1839–1847.

Smets, K., Adriaensens, P., Reggers, G., Schreurs, S., Carleer, R. & Yperman, J.: Flash pyrolysis of rapeseed cake: influence of temperature on the yield and the characteristics of the pyrolysis liquid. *J. Anal. Appl. Pyrol.* 90 (2011), pp. 118–125.

Teixeira, L.C., Linden, J.C. & Schroeder, H.A.: Optimizing peracetic acid pretreatment conditions for improved simultaneous saccharification and co-fermentation (SSCF) of sugar cane bagasse to ethanol fuel. *Renew. Energy* 16 (1999), pp. 1070–1073.

Thorin, E., Daianova, L., Lindmark, J., Nordlander, E., Song, H., Jääskeläinen, A., Malo, L., den Boer, E., den Boer, J., Szpadt, R., Belous, O, Kaus, T. & Käger, M.: State of the art in the waste to energy area – Technology and systems. REMOWE Report No O4.1.1, 2011.

Tucker, M.P., Kim, K.H., Newman, M.M., & Nguyen, Q.A.: Effects of temperature and moisture on dilute-acid steam explosion pretreatment of corn stover and cellulase enzyme digestibility. *Appl. Biochem. Biotechnol.* 105 (2003), pp.165–178.

UN-Energy: Sustainable bioenergy: a framework for decision makers. UN-Energy, 2007.

Valleskog, M., Marbe, Å. & Colmsjö, L.: System- och marknadsstudie för biometan (SNG) från biobränslen (System and market study for bio methane from bio fuels). Report SGC 185, 2008 (in Swedish).

VDI-Gesellschaft Energietechnik: *Fermentation of organic materials — Characterisation of the substrate, sampling, collection of material data, fermentation tests*. VDI-Handbuch Energietechnik, 2008.

Wang, L. & Chen, H.: Increased fermentability of enzymatically hydrolyzed steam-exploded corn stover for butanol production by removal of fermentation inhibitors. *Process Biochem.* 46:2 (2011), pp. 604–607.

Weiler, C. & Grotefeld, V.: Licensation, design and erection of the RDF-CHP plant in Bernburg and enhancement of MWIP. *ISWA World Congress*, 15–18 November 2010, Hamburg, Germany, 2010.

Willis, K.P., Osada, S. & Willerton, K.L.: Plasma gasification: lessons learned at Eco-Valley WTE Facility. Paper no. NAWTEC18-3515, *18th Annual North American Waste-to-Energy Conference*, 11–13 May 2010, Orlando, FL, 2010.

Wyman, C.E.: Biomass ethanol: technical progress, opportunities, and commercial challenges. *Annu. Rev. Energy Environ.* 24 (1999), pp. 189–226.

Yaman, S.: Pyrolysis of biomass to produce fuels and chemical feedstocks. *Energy Convers. Manage.* 45 (2003), pp. 651–671.

Zhou, J., Chen, Q., Zhao, H., Cao, X., Mei, Q., Luo, Z. & Cen, K.: Biomass–oxygen gasification in a high-temperature entrained-flow gasifier. *Biotechnol. Adv.* 27 (2009), pp. 606–611.

Zhu, J.-H. & Yang, F.: Biological process for butanol production. In: J. Cheng (ed): *Biomass to renewable energy processes*. CRC Press, Taylor & Francis Group, Boca Raton. FL, 2010, pp. 271–336.

Part II
Systems utilizing biomass – system optimization

CHAPTER 7

System aspects of biomass use in complex applications: biorefineries for production of heat, electric power and chemicals

Erik Dahlquist & Jochen Bundschuh

7.1 TRADITIONAL USE OF WOOD

If we look back in history, biomass was first used as fuel for fire most probably. Thereafter people started to use it to build shelters, as food and even for the manufacture of tools and production of artifacts.

Since then, wood especially has been a very important material for manufacturing of houses, ships etc., and during the last few hundred years also to produce different type of chemicals as well as for the reduction of metal oxides to elementary metals.

During the last 100 years the focus has still been on manufacturing of furniture, building houses and especially for manufacturing of paper and paper products. Today the total production of paper and paper products is in the range of more than 400 million tonnes/year. Voith's CEO Hans Peter Sollinger (2011) predicts 500 million tonnes of paper to be produced 2015! In Sweden and Germany, huge amounts of biomass and organic wastes are used to produce district heat and electric power in thermal power plants. Also in many countries pellets are much used in houses for heating purposes.

In pulp mills, wood chips are digested with mostly sodium bi-sulfide (NaHS) or sulfite (Na_2SO_3). The first is mostly in Kraft pulp processes while the latter is mostly in CTMP plants, Chemo Thermo Mechanic pulp, but also in sulfite processes.

During the last decade, we have seen a continuously increased interest to produce more textile fiber from wood to replace synthetic fibers from oil as well as replace cotton. Together with dissolving pulp, products like ethanol, lignosulfonates, vanilla and others chemicals are also produced. Several existing batch digesters are converted into this type of production and from being a primarily fiber producer the plants becomes more of biorefineries (Rødsrud et al., 2012).

At Borregard in Sarpsborg, Norway, this has been the fact for quite a few years already, just like at Domsjö, Örnsköldsvik in Sweden.

7.2 USE OF WASTE AND WOOD FOR CHEMICALS

In Kuopio, Finland, the plan is to make a biorefinery using organic wastes as the feedstock. In China, Tiangang group in Nanyang has built a demonstration biorefinery using wheat, corn and jatropha so far, but also with a smaller plant operating with straw as the feedstock (Chapter 9, this book). The reason is that a decision was made some years ago in China to not use what can be used as food for energy purposes. Now ethanol is produced from the cereals, biogas from the residues from the ethanol manufacturing and production of different chemicals from the ethanol. In 2012, a decision was made by Henan province to produce 500,000 m^3 bioethanol/y from straw by 2015, and 3 million by 2020. Also 1000 million m^3 biogas will then be produced from the waste liquor. The aim is to use the bio-ethanol for E85, and the biogas to feed into the NG-pipeline net.

The biogas is used in the natural gas net to feed households with heating gas. For that purpose it isn't necessary to refine the gas, and the CO_2 can be left with the methane.

On the other hand, we have gone from only burning wood chips to producing heat by using it in cogeneration to also produce electric power, and now for cooling and chemicals production.

The major chemicals investigated are methane, DME and methanol (Chemrec, 2012). Sometimes also ammonia production is investigated together with the gasification plants. Residual gas then can also be combusted in a conventional boiler. This new trend is often called "polygeneration" (Naqvi, 2012). From this type of plant we have come very close to the biorefineries at the pulp mills and in the future we will most probably see many more integrated factories where production of chemicals go hand in hand with production of heat, power, cooling and different type of paper products. Integration of organic wastes from farmland, building materials and forestry will be seen and new processes combining high temperature gasification, combustion, biogas production through fermentation as well as torrefaction and pyrolysis. What has long been developed for coal will most probably be focused more on use for biomass as well.

This new trend will create new business opportunities for many companies and make countries independent of fossil fuels. We can foresee a fight between oil companies who want to keep their powerful position and companies converting and growing forests and agricultural crops, and what has been seen as an uninteresting waste will turn into a very profitable raw material source (Jakobsson, 2012). This will change the political balances between different regions, but also between companies and countries. Countries who do not support this transfer will most probably be losers, while those going for it will be the winners. Of course, biomass will not be the only renewable energy source. Wind power is already expanding dramatically and solar power is coming on very strong as well. These three legs will most probably be the dominating energy sources, while wood and cellulosic materials also will be the major source for production of polymers, chemicals and building materials (Dahlquist, 2011).

The fight between renewables and fossil fuels will not go quickly but most probably will last for the next 50–100 years. Still there is no doubt, which one will win. As the fossil fuels become more difficult to extract, the production cost and also the price will increase, so in the long-term the renewables will become more profitable. If we include also the global warming aspect, we just can wish that the renewables break through quickly and thereby reduce the use of fossil fuels and thereby reduce the risk for too intense global warming. In many countries, the politicians do not see the global warming as a problem, although the researchers are very worried, and thus we hope the market forces can act in a positive way. Still, the introduction of renewables needs economic incentives to drive the implementation on a large scale, and thereby give a reduced production cost.

For biomass and wind power we principally have already reached this point in many countries, and solar power has been increasing in volume to some 60,000 million US\$/y in 2009 as Germany has provided very attractive feed-in tariffs – 0.4 €/kWh delivered to the grid (Kazmerski, 2011)! Due to this, the price was falling to half during 2011, and is now around 2000 €/kW$_{el}$ installed. This can be compared to the new nuclear power plant in Finland costing some 3000 €/kW$_{el}$ in 2012.

7.3   USE OF HERBS FOR MEDICAL AND OTHER APPLICATIONS

Since many thousands of years back, humans have utilized herbs for medical purposes. Some of these herbs have been found to be very active in different ways. In the book *Green medicines – the search for plants that heal* (1965) Margaret Kreig goes through a number of interesting stories about how pharmacologists have found many different active substances by travelling to indigenous people and collecting herbs for testing in their laboratories. Examples of important medical substances found in this way are kinin for malaria, digitalis for heart diseases, curare to counteract allergic chock, acetyl salicylic acid, the most common substance for use against inflammations and fever, penicillium against bacteria etc. In the book *American Medicinal Plants* (1974), Charles Millspaugh has made a review of plants indigenous to and naturalized in the United States, which are used in medicine. Here we can see anything from wild strawberry (*Fragaria* spp.) that could be good towards diarrhea and dysentery according to old traditions, to helianthus (sunflower), which is now cultivated and used for food, but where the medical effects are still not well understood or investigated.

In old traditions all over the world, herbs and crops have been important parts of the culture. The Zulu, for instance, group medicines into black, red and white (Lipp, 1996). The black medicine can give energy to cause the disease while white medicine can counteract this effect. The red medicine is something in-between. In this case, the herbs are used to influence our bodies but mostly the soul. In South America, there are many different crops to use, but only a few hundreds have been used and medicine men found them to be active. Usually the herbal medicine is used in combination with different rites to enhance the effect. Sometimes it is probably these rites that help the body to fight down diseases, but that is also important. In "western medicine" it has been found that often "placebo medicine" is as active to cure people from diseases as the actual active substances, so it is nothing special for ingenious people.

It is often easier to start analyzing the effect of natural substances than to try to just develop any kind of substance for medical use. The chance that people have selected efficient herbs and crops throughout thousands of years is higher than if you try to randomly develop a new substance. New processes to extract active substances or modify them to get even stronger effects are research activities going on to utilize biomass in a more qualitative than quantitative way. Concerning microbial processes like biogas production, it has been noticed that combining different species like algae and household waste or green crops gives better yields than having only one type of substrate. In the same way is it better for us as living creatures to eat a little of many different types of food than to only concentrate on one type, although humans seem to be astonishingly adaptive to different types of food. Inuits can live on only meat and fish while many vegetarians can live on only vegetables.

REFERENCES

Chemrec: The fuels of the future combat the environmental challenges. HM King Karl XVI Gustaf broke ground for the BioDME plant. November 2009, http://www.chemrec.se/BioDME_ground_breaking_with_the_Swedish_King.aspx (accessed July 2012).
Dahlquist, E., Vassileva, I., Thorin, E. & Wallin, F.: How to save energy to reach a balance between production and consumption of heat, electricity and fuels for vehicles. Proceedings of the 6th International Green Energy Conference, 6–9 June 2011, Eshkeshir, Turkey, 2011.
Jakobsson, K.: *Petroleum production and exploration. Approaching the end of cheap oil with bottom-up modeling*. PhD Thesis, no 891, Uppsla University, Uppsala, Sweden, 2012.
Kazmerski, L.: Solar power today and in the future. Presentation at *World Renewable Energy Technology Conference*, 21–23 April 2011. New Delhi, India, 2011.
Kreig, M.B.: *Green medicine – the search for plants that heal*. Rand McNally, 1965.
Lipp, F.J.: *Les plantes et leur secrets*. Albin Michel, 1996.
Millspaugh, C.F.: *American medicinal plants*. Dover Publications, 1974.
Naqvi, M., Yan, J. & Dahlquist, E.: Energy conversion performance of black liquor gasification to hydrogen production using direct causticization with CO_2 capture. *Bioresour. Technol.*, BITE-D-11-04184, 2012.
Rødsrud, G., Lersch, M. & Sjöde, A.: History and future of world's most advanced bio refinery in operation. *Biomass Bioenergy* 46 (2012), pp. 46–59).
Voith's CEO Hans Peter Sollinger predicts 500 million ton paper to be produced 2015. Svensk Papperstidning, 2011.

CHAPTER 8

Biorefineries using wood for production of specialty cellulose fibers, lignosulfonates, vanillin, bioethanol and biogas – the Borregaard Sarpsborg example

Stefan Backa, Martin Andresen & Trond Rojahn

8.1 INTRODUCTION

Trees are seed-bearing plants that are subdivided into gymnosperms and angiosperms. Coniferous woods (softwoods) belong to the first category and hardwoods to the second group, and altogether 30,000 hardwoods and 520 softwoods are known (Sjostrom, 1993). Wood is mainly composed of cellulose, hemicelluloses and lignin. Simply put, cellulose forms a skeleton that is embedded in a matrix of the two other main components. In addition, minor amounts of extractives are also present in wood.

Cellulose is the most abundant renewable biopolymer in the world, and is a linear homopolymer composed of anhydroglucose units linked together by $(1 \rightarrow 4)$-glycosidic bonds (Sixta, 2006). Cellulose has a strong tendency to form intra- and inter-molecular hydrogen bonds, resulting in a strong semi-crystalline fibrous structure that is insoluble in most solvents. Hemicelluloses are branched polymers consisting of both pentose (C5) and hexose (C6) sugars. Due to their branched, amorphous structure, hemicelluloses are usually water soluble and, unlike cellulose, can easily be hydrolyzed by acids to their corresponding monomeric components. Lignin is the world's second most abundant biopolymer and is composed of randomly polymerized phenylpropane units. It has a heterogeneous, highly branched and relatively hydrophobic amorphous structure, resulting in poor water solubility (Sjostrom, 1993). Compounds extractable with organic solvents include terpenes, fats, waxes and low molecular weight phenols. The elementary composition of wood is approximately 49% carbon, 44% oxygen and 6% hydrogen (Sixta, 2006).

Wood is a versatile material. When leaving the forest or plantation, wood can be divided according to its use in fuel, sawn products, pulp wood and other industrial wood (Table 8.1). Sawn wood and veneer can be used as a construction material in buildings and pieces of furniture. Pulpwood is processed into particles and fibers and used in particle boards, fiber boards and paper. Wood can also be used as a fuel and finds its applications in many areas, from small bonfires to large CHP plants. Since wood can be employed in many areas, high quality wood may also be an expensive raw material for energy production. Low quality wood, such as scrap wood (used demolition wood) and sawdust, is better suited for bioenergy production.

Not only the wood itself but also the wood components are of different value. Differentiating the use of wood components is both sustainable and economically sound. By gentle chemical processing of wood a fraction of the lignin and hemicelluloses can be removed, resulting in cellulose fibers with relatively high hemicellulose content. These "crude" fibers are commonly used to make paper and board. Lignin and hemicelluloses can be nearly completely removed through more severe chemical treatments to obtain pure cellulose. These pure cellulose fibers are used as raw material for manmade fibers (e.g. viscose) and may also be converted to cellulose derivatives.

Today, the sulfate (or Kraft) process, in which a solution of sodium sulfide and sodium hydroxide is used for dissolving lignin from the wood fibers, is by far the most commonly used chemical pulping process (Sixta, 2006). While the fibers find their use in many types of paper products, the

Table 8.1.   The 2009 world production of round wood.
Data from FAOSTAT (FAO Statistics, 2011).

	Wood production [million m^3]
Wood fuel	1851
Sawlogs and veneer logs	782
Pulpwood, round and split	492
Other industrial roundwood	150

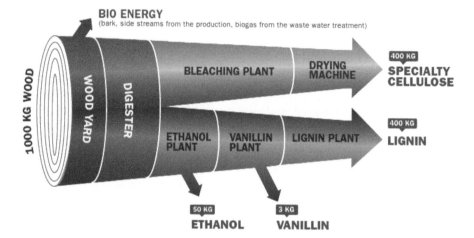

Figure 8.1.   Schematic illustration of how wood is converted to marketable products at Borregaard.

removed lignin (sulfate lignin) has limited commercial value, and is usually incinerated in order to recover energy and cooking chemicals. Some sulfate mills also isolate some of the extractives, which are fractionated into fatty and resin acid constituents. In addition, turpentine (mixture of monoterpenes) is produced.

In the less common sulfite chemical pulping process, the lignin in wood is dissolved from the fibers using an aqueous solution of either sulfite (SO_3^{2-}) or bisulfite (HSO_3^-) salts. The counter ion, often referred to as the cooking base, can be either calcium (Ca^{2+}), sodium (Na^+), magnesium (Mg^{2+}) or ammonium (NH_4^+). The sulfite process was initially developed around the acidic calcium bisulfite process, for which neither recovery of cooking chemicals nor energy production from lignin containing spent sulfite liquors (SSL) existed (Sixta, 2006). However, the lignin fraction resulting from the sulfite pulping process, so-called lignosulfonate, has several interesting properties (see section 8.2.4), making it an important raw material for a wide variety of useful products.

Today, many different biorefinery concepts exist, the vast majority of which have the aim to produce ethanol from biomass. Instead of listing and commenting on these current and future projects, we have chosen to present a biorefinery concept that has been running for several decades. This biorefinery, the Borregaard plant in Sarpsborg, Norway, is the most advanced biorefinery operating today, and as such is a good example of a wood-based biorefinery.

The Borregaard Sarpsborg mill operates a calcium-based sulfite pulping process, in which the high value wood components are extracted, while the less valuable components are used as fuel (Fig. 8.1). The extraction of cellulose fiber, lignin, vanillin and ethanol is very efficient, leaving only a small part of the wood to be converted to bio energy.

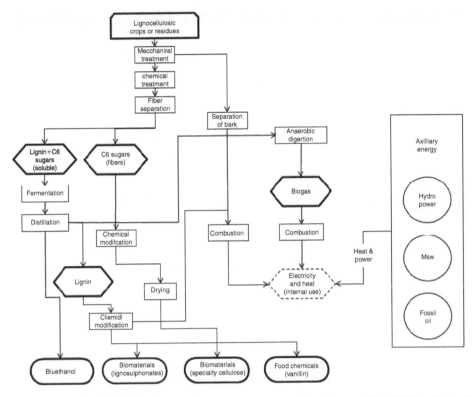

Figure 8.2.   The Borregaard Sarpsborg mill classified according to IEA Bioenergy, Task 42 "Biorefineries" (Cherubini *et al.*, 2009). MSW stands for municipal solid waste.

The effective use of the wood in the Borregaard mill fits in well with the definition of a biorefinery, which is a facility that integrates biomass conversion processes and equipment to produce fuels, power, materials and/or chemicals from biomass (Cherubini *et al.*, 2009). Other features of a biorefinery include an integrated use of the biomass components. The biorefinery must have low adverse environmental impacts and should be economically profitable. According to the classification of the International Energy Agency (IEA) Bioenergy Task 42, Borregaard can be considered as a lignin, biogas and C5 and C6 sugars biorefinery for bioethanol, bio vanillin, lignosulfonates, biomaterials and heat from lignocellulosic crops and residues (Fig. 8.2).

Borregaard pulp mill was founded 1889 and had already in 1892 a production capacity of 6000 tonnes per year of unbleached calcium sulfite pulp (Bergh and Lange, 1989). The mill was located near the river Glomma, which provided transportation of the feedstock to the mill and the products from the mill. A waterfall at the same location provided mechanical energy and later on, hydropower. The feedstock was small dimension timber of Norway spruce, rejected by the local sawmills. The cooking liquor contained calcium sulfite and sulfur dioxide and was made according to a new Austrian patent. In addition to the new technology, English capital and the local abundance of cheap labor made it possible to build the mill. Within the next ten years, both a paper mill and a bleach plant were in operation.

During the years to come, new products were continuously developed according to demands from the commercial markets and new mill departments were erected to produce a variety of materials and chemicals from wood (Fig. 8.3). However, in many cases the production ended due to competition from similar products produced from cheap fossil oil or produced in countries

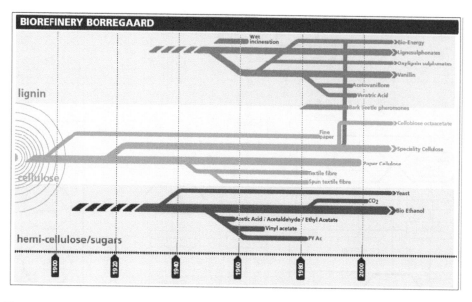

Figure 8.3.   The product portfolio of Borregaard during the last 120 years.

with lower production costs. SHE (Safety Health & Environment) aspects have also played an important role in when to terminate production.

Borregaard started up its bioethanol production in 1938, and in the 1960s and 1970s the ethanol was also refined to acetic acid and its derivatives. Borregaard has also produced carbon dioxide in this plant. In addition, Borregaard manufactured dioctyl phthalate by esterification of phthallic acid with 2-ethyl-hexanol, a branched alcohol made from ethanol. However, these products were soon outcompeted by fossil equivalents and production stopped, with only the ethanol production remaining today.

The production of lignosulfonates started in 1967. Already in 1936, Marathon Coporation in US had built a plant for producing lignosulfonate products including leather tanning agents and dispersants in Rothschild, USA. During the 1990s Borregaard acquired several lignosulfonate plants throughout the world and Borregaard's sales volumes increased more than 20 times from the modest start in 1967. Today Borregaard LignoTech, a Borregaard business unit, is the world's largest producer of lignosulfonates.

The vanillin plant was finalized in 1968 and at present, Borregaard is the only supplier of wood based vanillin in the world. Borregaard is a small player on the vanillin market as more than 90% of the world production of vanillin flavor, including ethyl vanillin, is produced from non-renewable raw materials. Also other lignin monomers, i.e. acetovanillone and veratric acid have been produced commercially in this plant.

8.2   THE BORREGAARD SARPSBORG BIOREFINERY OF TODAY

In 2011, Borregaard is still an efficient biorefinery, manufacturing a spectrum of marketable products such as biofuel, biochemicals and biomaterials. From 1000 kg dry spruce wood, our process yields approximately 400 kg specialty cellulose, 400 kg lignosulfonate, 70 kg solid bio-fuel, 50 kg bioethanol, 50 kg carbon dioxide and 3 kg vanillin. Well under 5% of the total input of wood ends up in the waste treatment plant, where a large portion is converted to biogas (Fig. 8.2).

The conversion of the old pulp mill to the biorefinery of today has been accomplished through specialization and continuous innovation, as illustrated in the product mixture manufactured at different time periods (Fig. 8.3). The innovative contribution is also seen in the development of value-added applications and products and in the ability to find and explore niches in the market place. In combination with an excellent mass balance, this has made Borregaard able to achieve a sustainable biorefinery business.

The Borregaard Sarpsborg mill still produces sulfite pulps in batch digesters using calcium as the cooking base. The calcium SSL has been transformed from a SHE problem to products of high quality. The mill has been enlarged several times to a design capacity of 160,000 tonnes of bleached market pulp per year (Bogetvedt and Hillstrom, 1996). In the following text, we will go into more detail according to the main features of the schematic presentation of the Borregaard biorefinery shown in Figure 8.2.

8.2.1   *Lignocellulosic crops and residues*

The Borregaard mill uses small dimension timber and wood chips of Norway spruce as raw material. The wood chips are delivered from sawmills while the timber logs come from the forests. The logs are debarked and chipped and are together with the sawmill chips fed to the mill digesters.

Cooking liquor is added to the digester and the chips are delignified according to given target specifications. The delignified chips are fibrillated at the end of the cooking process. The spent sulfite cooking liquor containing lignosulfonates and soluble sugars is separated from the fiber pulp by filtration, and is further processed to extract bioethanol, vanillin and lignosulfonate chemicals.

8.2.2   *Biomaterials, specialty celluloses*

The fibrous pulp is treated in a hot alkali extraction stage and further bleached with an ECF (Elementary Chlorine Free) sequence. This treatment removes hemicellulose, lignin and impurities (metals, bark residues and extractives) and adjusts cellulose chain length. The resulting pulp, known as specialty cellulose, is the raw material for production of cellulose ethers, cellulose acetate and nitrocellulose. Each pulp grade has different requirements, and the specialty celluloses are normally tailor-made for each individual customer. This requires a pulp production facility with a high degree of flexibility.

The main parameter of ether pulps is the intrinsic viscosity. High viscosity (long cellulose chains) or low viscosity pulps (short cellulose chains) are produced for cellulose ether customers. Cellulose ethers find their use in construction, paint, food and pharmaceutical applications.

Acetate pulps do not vary much with regard to cellulose chain length but require a high degree of purity (low hemicellulose content). Even small variations in purity can be crucial for the quality of the final cellulose acetate product. Cellulose acetate is used in filters, plastics, textiles and films (e.g. cellulose triacetate as a film layer in LCD screens).

8.2.3   *Bioethanol*

During the acid sulfite pulping process, the hemicelluloses and short chain celluloses are dissolved and hydrolyzed to monomeric sugars as mentioned in section 8.2.1. In the spent sulfite liquor from spruce, approximately 80% of the sugars are C6 sugars. The spent sulfite liquor is fermented by saccharomyces yeast to convert the C6 sugars to ethanol. The ethanol is distilled off and the desugared sulfite liquor is further processed to lignin chemicals.

Today Borregaard is the world's largest producer of second generation bioethanol produced from lignocellulosics, with a capacity of 20,000 m^3 per year. We produce both technical grades (A-grade 95.8% and absolute technical 99.9%) and pharmaceutical grades (rectified 96.2% and

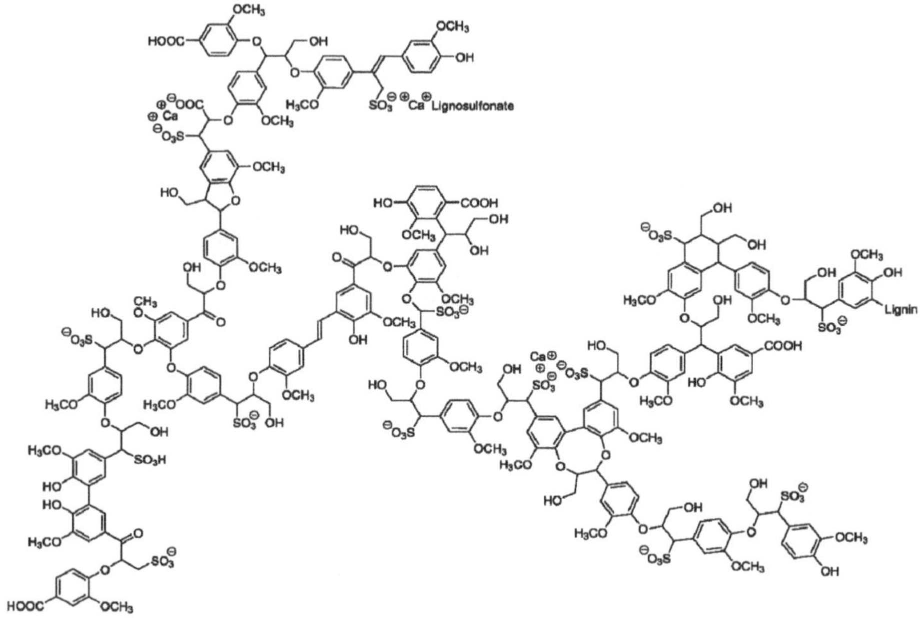

Figure 8.4.   Structural representation of spruce lignosulfonate.

absolute rectified 99.5%). All grades are certified as kosher/pareve and are utilized in a wide variety of applications, including biofuels, paints, solvents and pharmaceuticals.

8.2.4 *Biomaterials, lignosulfonates*

After fermentation and distillation, the remaining cooking liquor constitutes the raw material for production of different lignin chemicals. At this point, the main components in the liquor are sulfonated lignins, i.e. lignosulfonates, which are produced in the pulping process by hydrolysis and sulfonation of the wood lignin. Due to the introduction of the highly hydrophilic, charged sulfonate groups, the lignin is solubilized into the cooking liquor.

Lignosulfonates are highly complex aromatic polymers with multiple functionalities, including phenolic and aliphatic hydroxyl, methoxyl, carboxyl and sulfonate groups (Fig. 8.4). When leaving the fermentation plant, the cooking liquor is still a complex mixture, containing lignosulfonates of different molecular weights, but also certain amounts of residual (C5) sugar components, inorganic salts and insolubles.

In the lignin modification plant, this raw material is processed further by fractionation, purification and chemical modification (e.g. oxidation and desulfonation) depending on the end use of the product.

Due to their multifunctional nature, lignosulfonates have both dispersing, complexing and binding properties, making them useful in a wide variety of applications. Looking into the history of lignosulfonates, many current applications were also known 50 years ago. However, changes in the customers' technologies, introduction of new technologies, competition from other products etc. have changed the picture, and improvements and new lignosulfonate products have been the result of the company's ongoing R&D efforts. The interest for lignosulfonate products is also increasing since the products are produced from renewable resources, are environmentally friendly and sustainable. Today the most important applications include plasticizers in concrete, binder in animal feed pellets, extender in batteries, corrosion inhibitors in organic acids, dispersant in

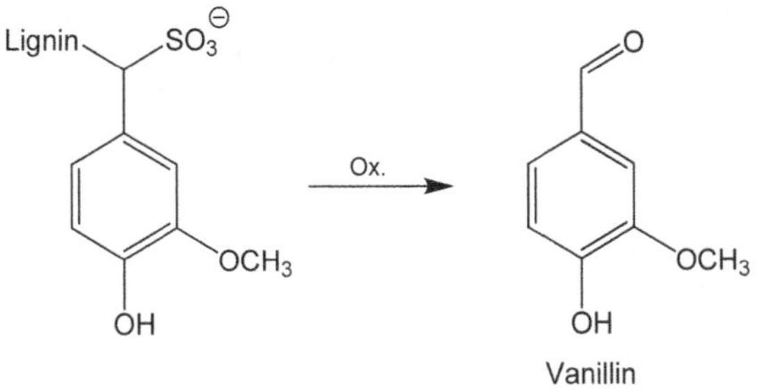

Figure 8.5.   The oxidation of lignosulfonate to vanillin (Bjørsvik and Minisci, 1999).

agrochemical formulations, and chelating agents for micronutrients (Gargulak and Lebo, 1999; Lebo *et al.*, 2000).

8.2.5   *Food/chemicals, vanillin*

The raw material for the vanillin production is part of the lignosulfonate fraction in the lye stream from the fermentation plant. This liquor is oxidized under alkaline conditions to produce vanillin as illustrated in Figure 8.5. After the vanillin has been extracted and purified, the rest of the oxidized liquor is returned to the lignin plant for further processing to produce special dispersing agents (Loe and Høgmoen, 2011).

Vanillin from wood is often described to have a creamier, rounder and more of a vanilla taste than vanillin produced from petrochemicals (i.e. guaiacol). It is likely that this is due to trace level of certain components, which are also found in vanilla from vanilla orchids.

The main application of vanillin is as vanilla flavor in foods, fragrances etc. The product is produced under strict quality requirements and is GMO-free as well as Halal and Kosher approved.

8.3   ENERGY

The highly efficient conversion of the wood into marketable products leaves little wood materials for energy production. The main internal sources for energy are bark removed before wood chipping, and effluents rich in organic content but not suited for the anaerobic treatment plant. These effluents come from the bleach plant and from the lignin plant. The effluents are evaporated before combustion and the steam produced is used in the various plants. The external energy is provided as hydropower (electricity) or as steam. The electrical power is produced in the hydropower plant on site, while additional steam is produced in 2 municipal waste combustion plants on site. The top load of steam, approximately 17% of total steam demand, is still produced from fossil fuels, however all fossil oil will be replaced by more environmentally friendly alternatives in the near future.

The biogas produced in the anaerobic effluent treatment plant is used internally in one of the mill's spray driers for lignins.

8.4   ENVIRONMENT

Waste treatment has also been greatly improved throughout the years. Gaseous and liquid discharges from the process are collected and led to high temperature (incineration) or to anaerobic treatment steps. The environmental performance of our products was assessed in a recent life cycle analysis (LCA, Table 8.2). The study was a cradle-to-gate analysis and confirmed a low

Table 8.2.   Environmental burdens from cradle to customer for Borregaard's products (Modahl Saur and Vold, 2010). Transport to customer (100 km) is included.

Environmental impact category	Unit	Cellulose [BDt]	Ethanol (96%) [m³]	Ethanol (99%) [m³]	Lignin (liquid) [BDt]	Lignin (powder) [BDt]	Vanillin [BDt]
Global warming potential	kg CO_2-eqv.	1160	324	666	666	1120	1090
Acidification potential	kg SO_2-eqv.	10.6	4.5	7.2	7.9	10.8	10.5
Eutrophication potential	kg PO_4^{3-}-eqv.	3.56	2.17	2.68	3.04	5.14	3.12
Photochemical ozone creation potential	kg C_2H_4-eqv.	0.77	0.29	0.49	0.5	0.78	0.75
Ozone depletion potential	kg CFC-11-eqv.	9.3×10^{-5}	2.6×10^{-5}	5.1×10^{-5}	4.9×10^{-5}	1.1×10^{-4}	8.9×10^{-5}
Cumulative energy demand	MJ LHV	32993	8718	18084	18216	31481	36490
Waste (solid)	kg	1386	408	793	701	1639	1330

BDt, Bone Dry tonne. Values on dry basis. For ethanol, the water content is subtracted.

environmental impact of our products (Modahl Saur and Vold, 2010). It was concluded that both the cellulose, bioethanol, vanillin, and lignin chemicals compared favorably to fossil-based equivalent products. As an example, vanillin from wood is associated with only 10% of the CO_2 emission of vanillin from petrochemicals.

It is also worth noticing that the remaining fossil fuels make a notable contribution in the figures in Table 8.2. Thus, these will be further improved when the use of fossil oil is eliminated.

8.5   THE FUTURE

The circumstances have changed greatly since the start of Borregaard. The once cheap labor and raw materials are now expensive relative to the other parts of the world. The financial situation has improved and so has also the local chemical and technical competence. This makes it possible to continue on the specialization and innovation path so fruitful in the past. The commercial goal is to always have more than 20% of the sales coming from products introduced to the market during the last 5 years.

The lignin and pulp production in the mill is continuously improved. This is done by optimizing the cooking and bleaching recipes, utilizing only the necessary amounts of chemicals and energy to obtain the target grades. Most of the investments in the mill the last 20 years have been of an environmental nature. Gaseous and liquid streams are collected and used in energy production. More efficient pulp washing equipment has been installed.

Another example of continuous improvements is a common control room for all the plants in the mill area (Kristiansen, 2010). The process from wood chips to marketable products is highly complicated and integrated, involving many different plants and departments. A common control room for all plants integrates the process control and makes it more efficient. It also reduces the need for manual labor.

Borregaard is also investigating the possibility to convert even more of the wood to marketable products. This will be interesting when the cost of products based on petrochemicals increases due to an increasing oil price. One example may be the acetovanillone mentioned above, which manufacturing was stopped due to cheaper alternatives made of fossil oil.

Borregaard needs to be competitive also in the future and to further increase the chemical and technical competence in the biorefinery area. Borregaard has therefore allocated more resources both for laboratory research and a pilot plant (Rødsrud et al., 2011). Borregaard strategy is to move in the direction of an "ideal process" depicted in Figure 8.6. This is in contrast to making fuel (e.g. ethanol) out of high cost wood raw materials.

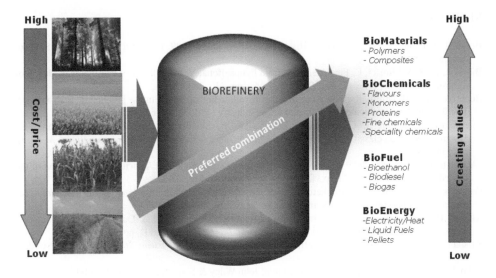

Figure 8.6.   From raw material to high value products *via* an ideal biorefinery process according to Borregaard.

As a result of the investment in R&D, Borregaard has developed a new biorefinery separation and pretreatment process, the BALI process, where the aim is to utilize low value biomass and convert it to various competitive products based on the hemicelluloses, the cellulose and the lignin in different plants (Rødsrud *et al.*, 2011). One of the great advantages with the BALI process is the flexibility in raw material. A new pilot plant will make it possible to test different raw materials like bagasse, straw, willow and spruce and to optimize the processing condition to get the optimum yield and quality of the different products.

Some of the new specialized products in Borregaard's portfolio will be based on micro-fibrillated cellulose (MFC). MFC consists of crystalline micro-fibrillated cellulose fibers obtained by mechanical disintegration of the pulp fiber cell wall. The diameter of the fibrils is in the nanometer scale, while the lengths can be several micrometers, resulting in very high aspect ratios and gel-like behavior in water (Klemm *et al.*, 2011). This combined with high strength and ductility, a high specific surface area and a surface that is amenable to chemical functionalization, expand the utilization of MFC far beyond that of conventional cellulose fibers. Suggested applications include reinforcement in nanocomposites; rheology modifier in foods, paint, cosmetics and pharmaceuticals; biodegradable films and barriers for packaging; stabilizing and emulsifying agents; and biomedical applications, such as wound dressings and bio-artificial and bioactive implants (Klemm *et al.*, 2011; Siró and Plackett, 2010). A pilot plant for dry MFC powder is currently under construction (Rødsrud, 2011).

8.6   CONCLUSION

Borregaard started out as a pulp mill nearly 120 years ago, converting spruce wood to paper products. The current Borregaard is a biorefinery, whose products find their use in niche products other than paper. The transformation has been possible through continuous innovation and specialization. For the future, more high value products are developed within our biorefinery concept, such as MFC.

Borregaard has demonstrated the commercial and technical viability of the biorefinery concept for many years. Current initiatives within "green chemicals", biorefining and second generation biofuels are the results of specialization and innovation and will create value-added products that match the cost of biomass and processing.

REFERENCES

Bergh, T. & Lange, E.: Foredlet virke. Historien om Borregaard 1889–1989, Ad Notam forlag AS, Oslo, Norway, 1989.

Bjørsvik, H.-R. & Minisci, F.: Fine chemicals from lignosulfonates. 1. Synthesis of vanillin by oxidation of lignosulfonates. *Org. Process Res. Dev.* 3:5 (1999), pp. 330–340.

Bogetvedt, K. & Hillstrom, R.: Upgrading of a aulfite mill to high alpha-cellulose production. In: Nashville, TN: Technical Association of the Pulp and Paper Industry (TAPPI), 1996, pp. 525–530.

Cherubini1, F., ungmeier, G., Wellisch, M., Willke, T., Skiadas, J., Van Ree, R. & de Jong, E.: Towards a common classification approach for bio-refinery systems. *Biofuels Bioprod. Bioref.* 3 (2009), pp. 534–546.

FAO Statistics, 2011. The 2009 world production of roundwood. Food and Agriculture Organization of the United Nations, Rome, Italy, http://faostat.fao.org (accessed July 2012).

Gargulak, J.D. & Lebo, S.E.: Commercial use of lignin-based materials. In: *Lignin: historical, biological and materials perspective*. ACS Symposium Series, Washington, DC, 1999.

Klemm, D., Kramer, F., Moritz, S., Lindström, T., Ankerfors, M., Gray, D. & Dorris, A.: Nanocelluloses: a new family of nature based materials. *Angew Chem. Int. Ed.* 50:24 (2011), pp. 5438–5466.

Kristiansen, T.: Experiences from biorefining operations. In: *Control Systems 2010*, Stockholm, Sweden, 2010.

Lebo, S.E., Gargulak, J.D. & McNally, T.J.: *Lignin*. John Wiley & Sons, Inc., New York, 2000.

Loe, Ø. & Høgmoen, H.: Vanillin from wood: a CO_2-friendly and sustainable bio-material. *Flavours Fragrances* 4 (2011), pp. 30–31.

Modahl Saur, I. & Vold, B.I.: The LCA analysis of cellulose, ethanol, lignin and vanillin from Borregaard. Sarpsborg Østfoldforskning, 2010, http://ostfoldforskning.no/publikasjon/the-2010-lca-of-cellulose-ethanol-lignin-and-vanillin-from-borregaard-sarpsborg-658.aspx. (accessed July 2012).

Rødsrud, G.: Teknologi og lønnsomhet for samproduksjon av lignocellulosebasert biodrivstoff og kjemikalier, Borregaard's pilotanlegg. *Nordisk Treforedlingssymposium*, Trondheim, Norway, 2011.

Rødsrud, G., Lersch, M. & Sjöde, A.: History and future of world's most advanced biorefinery in operation. *Biomass Bioenergy* 46 (2012), pp. 46–59.

Siró, I. & Plackett, D.: Microfibrillated cellulose and new nanocomposite materials: a review. *Cellulose* 17 (2010), pp. 459–494.

Sixta, H.: *Handbook of pulp*. Wiley-VCH Verlag, Weinheim, Germany, 2006.

Sjostrom, E.: *Wood chemistry: fundamentals and applications*. 2nd edn, Academic Press, 1993.

CHAPTER 9

Biorefineries using crops for production of ethanol, biogas and chemicals – a large-scale demonstration in Nanyang, Henan province, China of the bio-ethanol industry under Tianguan recycling economic mode

Du Feng-Guang & Feng Wensheng

9.1 INTRODUCTION

Henan Tianguan Enterprise Group Co., Ltd. (referred to here as "Tianguan Group") was founded in 1939. The company is China's oldest and most representative ethanol production company, one of 520 key national enterprises, one of the five designated fuel ethanol manufacturers, the only enterprise entered into the national recycling economic test point in the industry, and the country's main enterprise of new energy and high-tech industry base in Nanyang city.

Tianguan Group has two wholly-owned subsidiaries, eight subsidiaries and four equity subsidiaries. The enterprise has China's largest fuel ethanol production capability with an annual output of 500,000 tonnes and the largest gluten production line in the world with an annual output of 70,000 tonnes. At the same time, it has Asia's largest industrial biogas project with an annual output of 150 million cubic meters, and presently it is China's only enterprise, which owns a production line with three bio-energy products: fuel ethanol, industrial gas and bio-diesel. The sales income of Tianguan Group in 2010 is 5 billion Yuan.

Tianguan Group has constructed some high-level innovation platforms, including a state-level enterprise technical center and the joint center of national biomass fuel technologies. Meanwhile it is planning to build a state key laboratory on vehicle biomass fuel technology with a biomass fuel research and development platform, which represents China's highest level.

Besides, the national Fuel Ethanol Standardization Technical Committee Secretariat is located in Tianguan Group. The Commission is the drafting institution of China's fuel ethanol production and technical standards.

9.2 DOMESTIC AND INTERNATIONAL BACKGROUND AND CONDITIONS RELATED TO THIS CASE STUDY

Globally, the recycling economy is becoming a trend, and the trend has had a successful practice. From the aspect of enterprise, the most typical example of the recycling economy is the DuPont chemical company mode. This mode can be called internal recycling economy in the company, which is to organize the material cycle between the various processes in the plant. At the regional level, an eco-industrial zone can be formed through the inter-firm industrial metabolism and symbiotic relation. Denmark Kalundborg eco-Industrial zone is currently the most successful eco-industrial zone around the world, and is the most typical example in the industrial eco-system. In addition, it formed a dual recycling system mode – collection and disposal of packing waste in Germany, which is the typical mode of recycling economy operation.

In recent years, Chinese enterprises have been processing a large number of meaningful explorations and practices for the development of the recycling economy to strongly push China to take a new industrialization road and achieve sustainable economic and social development. The Green Island Ecological Agriculture Co., Ltd in Chao Zhou city, Guangdong province has formed a "Green Island mode". The company takes the tourism industry as a leading power to build a recycling agricultural production system, based on the concept of a recycling economy, which from its formation is a recycling agricultural production base taking an ecological agriculture tourism industry as carrier. The Inner Mongolia Ulan Cement Group Co., Ltd. has explored an Ulan power – cement symbiotic recycling economy mode (referred to as "Ulan-mode").

Huaibei Mining Group takes coal, rock salt and limestone resources as the basis and the "Coal–Salt Integration Project" as a symbol, to build several industry chains such as "coal → coke-forming" "coal → waste rock → building materials," "coal → electricity → calcium carbide → PVC" and to form a recycling economic structure of "small cycles in mine wells, big ones in mine areas".

Lubei Enterprise Group has formed a close symbiotic Lubei chemical-based eco-industrial ecosystem, through the organic communication and integration of three industrial chains: co-production by ammonium phosphate; sulfate cement by sulfuric acid; seawater, "a multi-purpose water" and saline cogeneration.

Compared with western countries, whether in terms of recycling economy theory studies or from practice, there is a big gap. The breakthrough on issues such as theory, means of realization and operation will determine the development speed of China's recycling economy, and is essential insofar as within the academic system there have been no systematic studies on the bio-fuel industry business cycle economic mode. The theorists often focus on the general analysis of recycling economy research and lack the quantitative research and systems analysis of the material cycle and the proliferation of the value, which is the most essential content to reveal the recycling economy. Enterprises are important carriers in industry and social and economic development so that to build recycling enterprises has become a core issue in establishing a resource-saving society. Therefore, it seems to be extremely important to explore and summarize the recycling economy mode of bio-fuel enterprises based on the concept of a recycling economy.

9.3   QUALITATIVE ANALYSIS OF THE CASE STUDY

9.3.1   *The scope of the case study*

As to Tianguan Group's fuel ethanol and its by-product industry, for the convenience of explanation, this case mainly takes the Tianguan Fuel Ethanol Company, the core company of Tianguan Group as an example to illustrate. Tianguan fuel ethanol company's existing production systems include: wheat flourmill → wheat gluten flour mill → ethanol plants → feed mill → CO_2 plant → water treatment plant, and besides there is a thermal power plant, which is to provide energy for the whole plant, and the plants, which are carrying out construction projects, including a carbon dioxide biodegradable plastic factory and large-scale biogas plants. The factory's general production process is shown in Figure 9.1.

9.3.2   *Description of the basic characteristics of the case study*

The fuel ethanol industry is a renewable biomass energy industry, in general, including fuel ethanol production, conversion, and comprehensive utilization etc. Wheat fuel ethanol by-products not only work as feed, but also as raw materials of biogas, organic fertilizer and other products. The industry characteristics of the fuel ethanol industry determine that it is a prerequisite for the development of the recycling economy. Out of the need for economic or environmental benefits, fuel ethanol enterprises and its related enterprises are symbiotic. Related enterprises are biogas, wheat bran and wheat gluten enterprises, DDG feed companies, organic fertilizer plants and other enterprises concentrated in a particular area. Together they are forming a symbiotic system, which has certain structure, and function in an industrial biotechnology community. The fuel ethanol enterprises are "major species" in the system. The bran and gluten flour mill is the upstream business, the CO_2 production plant is a symbiotic business, biogas and organic fertilizer plants and other enterprises are downstream enterprises of the fuel ethanol business, which can be called "secondary species" (Du and Hu, 2010).

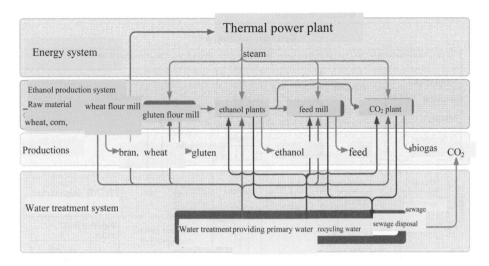

Figure 9.1. The factory's general production process.

The effective multi-level comprehensive utilization of the metabolic waste is one of the most typical features in the Tianguan mode. Tianguan Group has optimized the utilization of a variety of wastes through building an eco-industrial chain, which has significantly reduced environmental pollution. Take the resource reuse of fuel ethanol waste liquid as an example. The major pollutant generated during the production of fuel ethanol is waste liquid, which is an organic solution of high concentration. Tianguan Group uses lees to make DDG protein feed and biogas technology to turn the waste into treasure, and therefore has raised its utilization value, and meanwhile, it has reduced the COD effluent, which is of great significance to the protection of the regional river water quality.

Through waste recycling, the fuel ethanol production from wheat and the production rate of its relative industries will be raised. For example, the rate of lees waste utilization reaches 100%, and the rate of water cycle utilization reaches above 98%, and therefore the structural pollution of the regional alcohol industry will significantly improve.

9.3.3 *The recycling economic diagram and its analysis of this case*

Figure 9.2 Tianguan Group fuel ethanol recycling economic mode with wheat raw materials as its materials.

Similar to natural eco-systems, fuel ethanol eco-industrial systems are formed by producers, consumers, decomposers, and the external environment. Wheat, fuel ethanol and related companies and environmental comprehensive management are its basic composing units, which form a horizontal coupling and vertical closed flexible network through material exchange and energy flow.

The cornfield subsystem mainly provides basic raw materials for fuel ethanol enterprises, which is the "producer" of the fuel ethanol eco-industrial system. Wheat bran and wheat gluten flour mill and fuel ethanol enterprises take wheat as initial raw materials to produce wheat bran, wheat gluten flour and fuel ethanol, which are primary "consumers" of the fuel ethanol eco-industrial system, while other companies takes the "waste" generated during the fuel ethanol production process as raw material for production activities, to be called sub-prime "consumers". The environmental comprehensive management system carries on recovery, decomposition, reuse and recycling to the waste generated among each production link in the fuel ethanol industrial symbiotic system or during the production process, and in fact acts as "decomposers" in the fuel ethanol eco-industrial system.

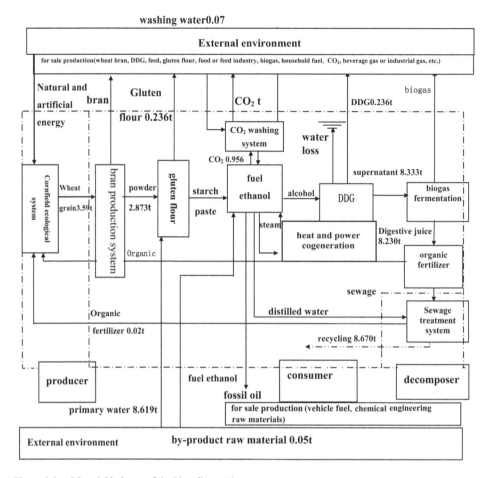

Figure 9.2.   Material balance of the biorefinery plant.

In the fuel ethanol symbiotic, the symbiotic unit has established an ecological link and formed eco-industrial chains through the logistics (raw materials, wheat, flour, starch, alcohol lees, etc.) exchange. It can be seen that the cornfield is taken as the beginning to form several eco-industrial links such as cornfield → wheat bran manufacturers → Gluten flour mill → fuel ethanol plant → DDG plant → biogas plant → organic fertilizer plant → wastewater treatment plant and fuel ethanol plant → CO_2 → biodegradable plastic factory, etc. Each link takes the environmental comprehensive management system as an end, and each eco-industrial chain intervenes and cross-couples through material, energy, information flow and sharing, to make the whole body a symbiotic eco-industrial network (Li, 2005).

9.4   QUANTITATIVE ANALYSIS OF THIS CASE STUDY

9.4.1   *Changes in four major indicator systems*

Tianguan Group's assets output indicators primarily show an upward trend, among which the land output rate increased from 16.8 million Yuan/ha in 2006 to 22.8 million Yuan/ha in 2009. The rate of increase is 35.7%. The energy output rate increased from 4.7 million Yuan/thousand tonnes standard coal in 2006 to 5.4 million Yuan /thousand tonnes in 2009. The rate of increase is 14.4%. The water output rate changed from 0.25 million Yuan/thousand tonnes in 2006 to 0.22 million Yuan/thousand tonnes in 2009, and to some extent, there has been a downward trend.

In Tianguan Group resource consumption indicators, the unit comprehensive energy consumption of main products shows a continuously downward trend, among which, the fuel ethanol comprehensive energy consumption was reduced from 0.58 tonnes standard coal/unit production in 2006 to 0.38 tonnes standard coal/unit production in 2009. The DDGS feed comprehensive energy consumption was reduced from 1.11 tonnes standard coal/unit production in 2006 to 0.4 tonnes standard coal/unit production in 2009, and the wheat gluten for comprehensive consumption declined from 0.57 tonnes standard coal/unit production in 2006 to 0.36 tonnes standard coal/unit production in 2009. The unit comprehensive water consumption of main productions shows a downward trend except wheat gluten flour, among which, the fuel ethanol reduced from 5.39 tonnes/unit production in 2006 to 3.74 tonnes standard coals/unit production in 2009. The DDGS feed comprehensive water consumption declined from 3.09 tonnes standard coal/unit production to 1.9 tonnes standard coal/unit production in 2009.

Among Tianguan Group resource comprehensive utilization indicators, there are three indicators such as the comprehensive utilization rate of industrial solid waste, sewage centralized treatment rate and harmless garbage treatment rate are always maintained at 100%. In addition, the industrial water recycling utilization rate and industrial wastewater regeneration rate and some other indicators maintain an increasing trend. Industrial water recycling rate rose from 92.8% in 2006 to 95.4% in 2009, while the industrial wastewater-recycling rate rose from 17% in 2006 to 35% in 2009.

Although Tianguan Group's four major indicators meet the need of the recycling economy, however, there is still a certain distance against the level of overseas developed countries. The main difference is as follows:

- The high-temperature liquefaction time of foreign production devices is short (104°C, 6 min or so) and there is a vacuum flash cooling, while the spray liquefaction temperature of Tianguan Group's production devices is lower (95–100°C), and there is no pre-liquefaction and post-liquefaction process. The further development trend abroad is adopting high-quality composite liquid enzymes to cancel the direct injection liquefaction steps, and good liquefaction results can be achieved even at 85°C. Therefore, the cost is significantly reduced whether in terms of equipment investment or consumption.
- The majority of foreign companies have canceled the saccharification process, and directly perform saccharification and fermentation at the same time, which can effectively solve the excessive growth of yeast caused by excessive nutrients, and meanwhile, the ethanol generated by the large amount of sugar consumption influenced the feedback suppression caused by yeast metabolism. However, Tianguan Group's devices all maintain an independent process of saccharification.
- Most of the foreign devices adopt continuous thick mash fermentation, and the yeast can be reused 2–3 times, significantly reducing production costs, while Tianguan group mostly used semi-continuous thick mash fermentation, instead of yeast recycling technology. The above are all factors that affect parts o f Tianguan Group indicators and are the main reasons for the gap between Tianguan Group and overseas developed countries.

9.5 ENERGY FLOW ANALYSIS

The energy that Tianguan Group mainly uses is coal, which equals to standard coal at 37 t/h, about 740.9 MJ per hour. The ultimate effective energy that can be used is about 419.28 MJ (the amount of materials 277.08 MJ, power 81 MJ and refrigeration work 54.2 MJ), which accounts for 56.6% of the total efficiency of the energy used by whole plant.

As to the company's heat losses, process emissions, boiler flue gas emissions and condensate heat loss are the major ones, but it is materials that carry the most heat. For the utilization of the steam, the production process, which uses most energy, is ethanol and feed production.

The specific energy flow analysis is shown in Figure 9.3.

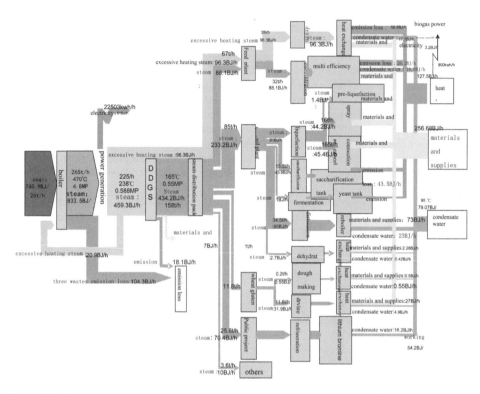

Figure 9.3.   Fuel ethanol production energy flow diagram. System general material flow analysis (2).

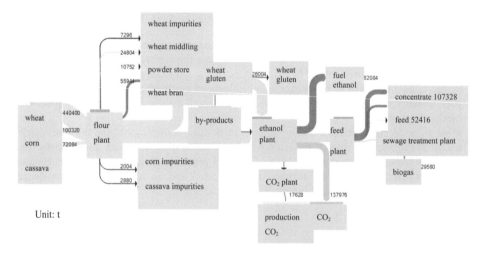

Figure 9.4.   General material flow diagram of fuel ethanol (the whole year 2007).

9.5.1   *The diagram of system general material flow*

The general material flow of the fuel ethanol in the whole year 2007 is as shown in Figure 9.4. This figure clearly shows the material flow and relative flow rate from raw material to products and by-product materials (the water added during the process is not shown in the figure).

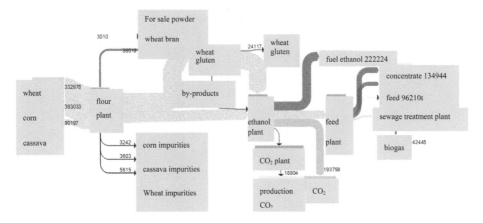

Figure 9.5.   General material flow diagrams in the zone (2009 the whole year).

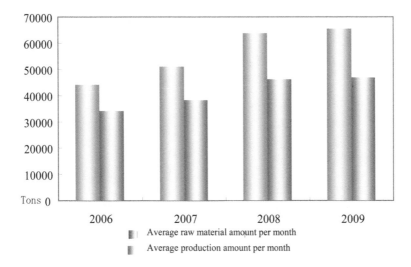

Figure 9.6.   Total raw material input and total amount of production in the plant from 2006–2009.

The general material flow of the fuel ethanol in the whole year 2009 is as shown in Figure 9.5. Compared with 2007, the production system structure in the zone remains the same, but the production scale has improved significantly.

9.6   GENERAL MATERIAL FLOW ANALYSIS

Figure 9.6 shows the total amount of raw material and total production output from 2006 to 2009. The input of raw material that the company uses for production (excluding coal energy consumption and non-coal process water consumption) is increasing every year. From 2006 to 2009, the rate reached 48%. The corresponding increasing rate of production amount was 37%, and the average material use efficiency was above 70%.

Figures 9.7 and 9.8 shows the system input and output in 2007 and 2009. The material utilization of the raw materials for the existing fuel ethanol system has already reached a relatively high level. Almost all of the useful substances have been put into use, and only the impurities in

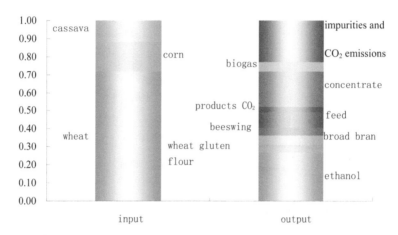

Figure 9.7.   Input and output structure diagram with respect to material flows (in 2007).

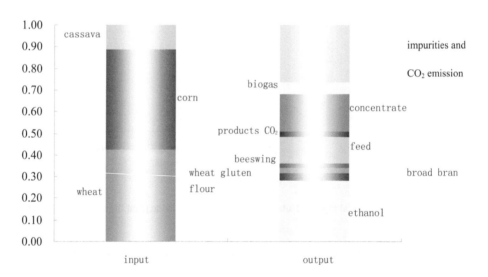

Figure 9.8.   Input and output structure diagram with respect to material flows (in 2009).

the raw materials and non-recovery carbon dioxide emissions occupy about 20–30% of the total amount. It is the increase of impurities and carbon dioxide emissions that causes the year by year reduction of the total material utilization (production amount/raw material consumption) in the years 2006, 2007, 2008 and 2009.

9.6.1   Analysis of systems group diversion

We take the year 2007 as an example to analyze two main components: starch and protein. First, we make a *starch flow system diagram and analysis*.

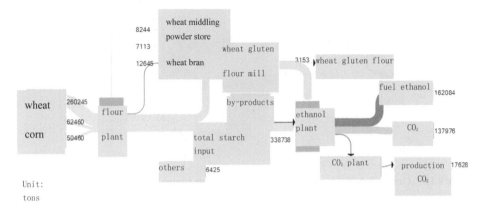

Figure 9.9.   Starch system flow diagram (material balance in tonnes the whole year 2007).

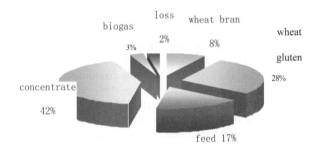

Figure 9.10.   Protein component material distribution structure.

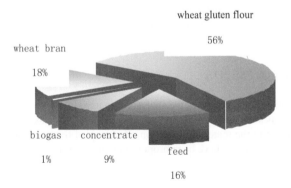

Figure 9.11.   Protein benefit structure.

Take the Figure 9.9, which shows the starch flow diagram of the year 2007 in the production system as an example. The total amount of starch in the raw material entering into the system is 373,000 tonnes, and the ethanol finally produced is 162,000 tonnes. Based on the ethanol plant's starch input, the starch alcohol yield is calculated to be 47.8%, but according to the alcohol fermentation reaction equation, the theoretical rate of starch liquor is 56.78%, based on the system starch utilization rate of 84%, so there is still room for improvement.

Analysis of the protein system shows that although the protein has been fully utilized, and the loss was only 2%, from Figure 9.10 and Figure 9.11, there is still room for improvement of

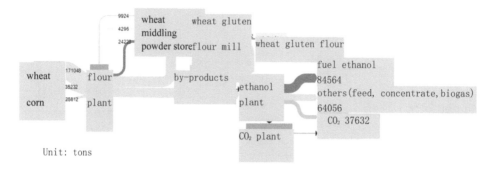

Figure 9.12.   Carbon system flow diagrams (material balance in tonnes whole year 2007).

the efficiency: the concentrate's value, which occupies the 42% of the total, only occupies 9% of the protein efficiency ratio, because the concentrate itself is not a high value-added product. Its price is low and the concentrate can be made into feed after being dried. Therefore, reducing the amount of concentrate emission and expanding feed production is one of the effective means to increase the value of the fuel ethanol production system.

The carbon flow system diagram was drawn according to the material flow condition and the carbon content in each material. The system carbon flow diagram in the year 2007 is as shown in Figure 9.12. The carbon contained in the wheat in the materials occupies 74% of total carbon, and in the production, the product bran, wheat gluten, ethanol etc. all have a certain fixed amount of carbon, respectively 10%, 5%, 36%. During the fermentation process, the carbon produced with ethanol, which has not been recovered and directly emitted, is 37,632 t/year, accounting for 16% of the total input.

The analysis shows that the carbon dioxide production plant still has great potential for improvements. In addition, in the existing carbon flow, it is not possible to count the carbon dioxide emitted by the energy consumption of coal, etc., (it is approximately 7 times that of the carbon dioxide emissions during the production process). There are still some imperfections.

9.7   SYSTEM IMPROVEMENTS

According to the statistics and results of the analysis of material flow and diversion, it can be seen that in the four aspects such as material, process, product and business chain, there is still some room for improvement in the material utilization efficiency of the existing fuel ethanol system, and therefore the improvements have been proposed as shown in Figure 9.13.

9.8   CONCLUSION

Tianguan Group targets the eco-enterprise development, reduction of material inputs, production process clean technology, changing waste into resources, and enlargement of the research investment. The goal is also to extend the industrial metabolism process, and optimize enterprise system architecture, turning the point block style development into a chain network. This will lead to building an industrial ecology network chain of intermediate products, waste and energy in each system and form a "multi-level recycling utilization mode". Finally, Tianguan Group will be built to be a state grade recycling economical advanced enterprise, which has obvious characteristics, advanced technologies, civilized management, friendly environment, harmonious structure and optimized system.

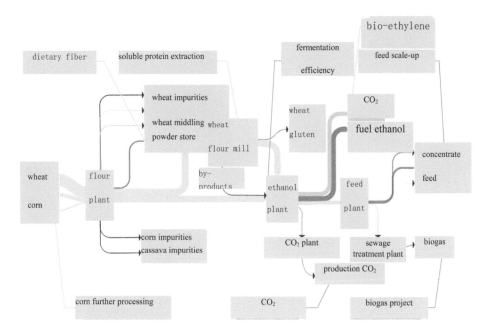

Figure 9.13.   System improvements.

During Tianguan's bio-ethanol production process, the main raw material is still food, and this does not match with "no competing with people for food, no competing with food for room" raised by the Chinese government. Besides, the current production cost of fuel ethanol is still high, which makes the ethanol produced not competitively priced compared with gasoline. Therefore, Tianguan Group is actively developing the technology research and industrialization demonstration to turn the cellulose-rich biomass (including stalks, wheat straw, rice straw, tree branches, etc.) into ethanol. In 2009, Tianguan Group completed the construction of a cellulosic ethanol industry model line with an annual output of 10,000 tonnes ethanol, which is currently in continuous operation, and its running condition is good. The breakthrough of cellulosic ethanol industrialization technology will bring profound influence to the bio-ethanol industry.

To speed up the pace of industrial waste utilization, to achieve a full utilization of fuel ethanol's industrial waste, to build a recycling industrial ecological chain network of intermediate products, waste and energy in each system, and to support the existing industrial base, several advanced bio-energies and chemical and industrial technologies are used in realizing the industrial breakthrough of key link technologies. The focus is to develop a biomass waste → ethanol → ethylene and downstream products industry chain and ethanol by-product carbon dioxide → biodegradable materials (or chemicals such as dim ethyl carbonate, etc.) and downstream industry chain, biodiesel byproduct glycerin → 1,3-propanediol → polyethylene terephthalate, etc.

REFERENCES

Du, F. & Hu, S.: The development and application of key technologies of the resource recycling in alcohol industrial zone (R). Studies and reports of fund projects, China National Technology Support Program (2006BAC02A17) Beijing, China, 2010.

Li, S.: *The studies of biomass fuel ethanol business under recycling economic mode* (D). PhD Thesis, Agricultural University, Beijing, China, 2005.

CHAPTER 10

Bioenergy polygeneration, carbon capture and storage related to the pulp and paper industry and power plants

Jinyue Yan, Muhammad Raza Naqvi & Erik Dahlquist

10.1 INTRODUCTION

Worldwide energy consumption is projected to expand by 50% between 2005 and 2030, which will cause depletion of known fossil fuel resources (EIA, 2007). Fossil fuels currently account for 85% of the world's total energy consumption, which has created issues such as global warming, fuel security and the depletion of non-renewable resources. The rapid depletion of fossil fuels has resulted in increased utilization of biomass as a bioenergy resource in future systems. Currently, bioenergy represents one of the most significant available opportunities. The UN's Food and Agriculture Organization (FAO) defines bioenergy as the energy derived from biomass-based fuels, where biomass is defined as 'material of biological origin excluding material embedded in geological formations and transformed to fossil' (FAO Views on Bioenergy, 2012). According to the US Department of Energy, bioenergy is the energy derived from any available renewable organic matter that includes energy crops, wastes and residues from the agriculture industry, waste from the forestry industry, aquatic plants, wastes from animals, and municipal wastes (US Department of Energy, 2012).

A number of initiatives are currently being developed that aim to reduce oil dependency and fossil-based greenhouse gas emissions. There is great promise in using renewable energy resources, including the increased production of bio-based fuels, as a possible way to solve environmental issues. Renewable fuels are likely to play an important role in the future as replacement of fossil fuels, due to increasingly strict regulations for greenhouse gas (GHG) emissions reduction. The European Union (EU) has set a target for biofuel use in the transportation sector of 5.75% by 2010 (EU Directive, 2003). According to a 2009 renewable directive (Fig. 10.1), the share of renewable sources should be 20% of energy supplied by 2020 and a share of 10% biofuels in the transportation sector (Swedish Energy Agency, 2009). Bioenergy currently provides 69 million tonnes of oil equivalent (Mtoe) in the EU, which is equivalent to approximately 4% of EU's total primary energy consumption. With the increasing consumption of fossil fuels, developing innovative bioenergy polygeneration technologies will help meet these challenging targets.

Bioenergy can help in reducing greenhouse emissions related to fossil-based fuels. Biomass is considered to be CO_2-neutral if it is sustainably managed, while it is also important to improve the efficiency of the current bioenergy systems (Thuijl et al., 2003; Möllersten and Yan, 2001). Theoretically, carbon emissions from the bioenergy production and consumption stages may be counterbalanced by photosynthesis during biomass growth. There are different technologies and various energy alternatives exist for bioenergy production, including biofuels for transport, combined heat and power generation, and upgraded biofuel such as pellets. Figure 10.2 shows various bioenergy polygeneration alternatives.

Since biomass is a limited resource, bioenergy production should be as efficient as possible and the integration with other industrial processes should be energy efficient; that is the production of high-quality products should require as little energy as possible. Biomass energy resources can be used sustainably if they are produced in a sustainable manner. Forest-based biomass plays an important role as a raw material for wood-based bioenergy products and as a renewable fuel.

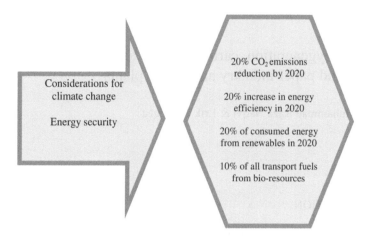

Figure 10.1.   20/20/20 Challenge by 2020 (Source: CEPI, 2012).

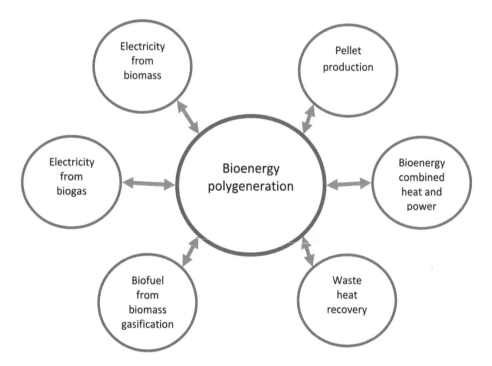

Figure 10.2.   Bioenergy polygeneration options.

The pulp and paper industry consumes a large proportion of the world's biomass and is the sixth-largest energy consumer industry in Europe (CEPI, 2008). The industry has made a significant contribution to the European economic cluster, generating an annual turnover of more than EUR 400 billion (Gebart, 2006). According to the Swedish forest industries, Sweden's pulp and paper industry is the third largest in Europe and represents 6% of the world's total pulp production (Swedish Forest Industries, 2010). The modern pulp industry has an established infrastructure for handling and processing biomass, which means it is possible to lay the foundation of biorefineries to co-produce different bioenergy products including electricity, chemicals or biofuels together with the pulp. There is potential to export electricity or biofuels by improving the energy systems of existing chemical pulp mills and integrating bioenergy systems.

In this chapter, we will focus on potential of bioenergy systems with carbon capture and storage integrated in the pulp industry. The system analysis of electricity, biofuel and pellet production together with CO_2 mitigation opportunities is presented. To give an overview, the chapter first describes the potential bioenergy systems at the pulp mills replacing conventional energy systems and CO_2 capture and storage (CCS) integration opportunities with the bioenergy systems. The chapter covers performance analysis and sustainability aspects such as technical, economical, and environmental aspects to determine if potential bioenergy systems at the pulp mills lead to a step towards sustainable energy systems especially in the transport sector. The chapter draws major conclusions in the light of discussed system performance and sustainability analysis of bioenergy systems with CCS.

10.2 BIOREFINERY SYSTEMS IN THE PULP INDUSTRY

A biorefinery system is defined as efficient utilization of biomass for economically optimized production of bioenergy products (Berntsson *et al.*, 2006). There is a vision that bioenergy production could lead to sustainable energy systems especially in the regions with large pulp industry by integrating biorefinery systems and optimized conversion of forest biomass to valuable energy products like biofuels or pellets. A number of research institutions in North America and Nordic countries are investigating different approaches and bioenergy production routes to develop technologies for successful implementation of biorefinery systems.

10.2.1 *Black liquor gasification (BLG) based biofuel production*

An important approach towards potential bioenergy systems at the pulp mills is black liquor gasification (BLG) based biorefineries to produce biofuels. Biofuel production using black liquor gasification could potentially decrease fossil fuel dependency and associated CO_2 emissions. Figure 10.3 shows different biofuel alternatives from black liquor gasification.

Black liquor (BL) is considered a major bioenergy resource e.g. about 40 TWh of black liquor energy is available in Sweden (Swedish Energy Agency, 2009). Black liquor is the residue of lignin and spent cooking chemicals obtained from the delignification unit in the Kraft pulping process. Table 10.1 shows typical black liquor composition. Conventionally, black liquor is fired in the recovery boiler to generate steam and electricity. In the gasification system, black liquor is gasified to produce the synthesis gas that mainly contains CO, CO_2, H_2 and CH_4. The synthesis

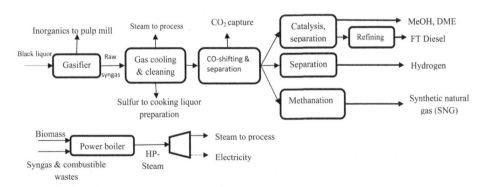

Figure 10.3.   Polygeneration of different biofuel alternatives in the pulp mill (Source: Naqvi *et al.*, 2010b).

Table 10.1.   Elemental composition black liquor solids (BLS) (Source: KAM report, 2003).

	Component		[%Mass]
Composition	Carbon	C	35.7
	Oxygen	O	35.8
	Hydrogen	H	3.7
	Sulfur	S	4.4
	Sodium	Na	19.0
	Potassium	K	1.1
	Chlorine	Cl	0.3
	Nitrogen	N	<0.1
Black liquor solids (BLS)	[%Mass]		80
Higher heating value (HHV)	[MJ/kg]		14.5
Lower heating value (LHV)	[MJ/kg]		12.3

gas is further processed to bioenergy products e.g. synthetic natural gas (SNG), methanol (MeOH), dimethylether (DME), hydrogen, FT-liquids etc.

However, there will be an energy deficit in terms of electricity and steam demand in the pulp mill with biofuel production that shall be compensated by using a biomass-fuelled power boiler. The synthesis gas obtained in the gasification process requires cleaning from particulates, tars and sulfur components before it can be converted to biofuels. There is no commercial biofuel production from black liquor gasification today but the development and interest in producing transport biofuels has been rising over the last couple of decades. The efficient production of transport biofuels can be achieved by integrating biorefinery systems at the pulp mills, co-production pulp together with transport biofuels.

10.2.2   *Black liquor gasification-based power generation*

Black liquor gasification combined cycle (BLGCC) is another alternative of bioenergy polygen-eration in the pulp industry achieving higher system energy efficiency with electricity generation than conventional recovery boiler technology (Larson *et al.*, 2003). The synthesis gas obtained in black liquor gasification has similar composition to the synthesis gas obtained from coal-based gasification and the electricity is generated from the gas turbine fired with the synthesis gas. The heat recovery steam generator (HRSG) is used to produce high-, medium- and low-pressure steam by cooling flue gas from the gas turbine.

Figure 10.4 presents a simplified design of a black liquor gasification combined cycle (BLGCC) system. The BLGCC configuration requires existing falling bark and extra biomass to be com-busted in the power boiler to generate more high-pressure steam since the high-pressure steam generated in the HRSG is not sufficient. High-pressure steam from the HRSG and the power boiler is used in a backpressure steam turbine to generate electricity. The amount of electricity is increased with pressurized high temperature black liquor gasification and a gas-turbine based cogeneration system.

10.3   BIOFUEL UPGRADING WITH PELLET PRODUCTION

Another salient approach is upgrading biomass to a higher energy density fuel like pellet pro-duction. There is a potential for pellet production in bioenergy systems integrated with the pulp mills that could be used to replace fossil fuels in the combined heat and power (CHP) plants. In Sweden, there has been a 300% increase in use of pellets from 1997 to 2005, mainly due to tax-driven changes in the Swedish energy system (Andersson, 2007). The low quality excess process

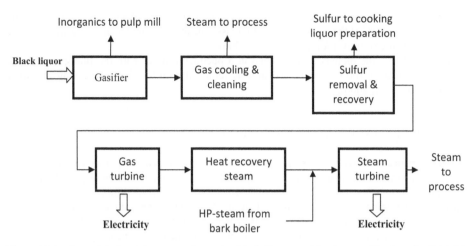

Figure 10.4. Potential electricity generation using black liquor gasification combined cycle (BLGCC) (Source: Naqvi *et al.*, 2010b).

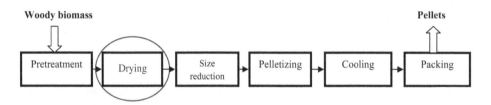

Figure 10.5. Simplified flow diagram of pellet production process.

heat from the pulp industry could be used for integrated pellet production that requires drying of the biomass. From the results of the Ecocyclic Pulp Mill (KAM) report, the chemical pulp mills have surplus of biomass in the form of bark i.e. about 70,700 tonnes of bark per year (KAM, 2003). It is worthwhile for the pulp mills to produce pellets due to continuous increasing demand for biofuel in the heating sector. The major steps in the pellet production process are discussed and is shown in Figure 10.5 (Zakrisson, 2002).

1. Pretreatment: the biomass is chopped and metals are removed.
2. Drying: the dry solid contents in the biomass are increased from 45% to 89%.
3. Size reduction: the size of particles is reduced using a hammer mill.
4. Pelletizing: the pellet press is used to press wood particles to shape pellets.
5. Cooling: the pellets are cooled using air.
6. Packing: the pellets are sacked or transport to storage facility.

The most important process step in pellet production is drying of the biomass, which requires most of the heat energy. A brief description of dryers that could be considered in the integrated pellet production and pulp mill is presented:

• *Steam dryer*: the medium pressure steam can be used to dry the biomass and the heat from removed moisture can be recovered as low-pressure steam. The steam drying technology is efficient but the steam obtained is relatively at low pressure and temperature in comparison with the steam fed.

- *Vacuum dryer*: the vacuum drying technology requires sub-atmospheric pressure to lower the boiling temperature of the moisture in the biomass. The moisture is evaporated using hot water at a temperature of 90°C (Andersson *et al.*, 2006).
- *Flue-gas dryer*: the biomass is circulated in a pneumatic dryer with the flue gas from the recovery boiler at 160°C and the moisture is removed with the flue gas leaving the dryer (Andersson *et al.*, 2006).

10.4   PERFORMANCE AND SUSTAINABILITY ANALYSIS

The performance analysis presents the objective to integrate bioenergy polygeneration systems with the pulp industry and to produce bioenergy products in relation with technological limitations that hinder sustainable commercialization of bioenergy polygeneration. Since a modern pulp mill has an excess of internal biomass, additional advantages can be achieved with electricity, biofuel or pellet production with enhanced energy efficiency as a step towards sustainable development. The integrated bioenergy systems offer significant improvements in the energy efficiency of the pulp mill as compared to the conventional recovery boiler system.

In this chapter, we have discussed performance of bioenergy systems integrated with the pulp industry and specific sustainability aspects to analyze, share and compare information based on technical, economic and environmental perspectives. The environmental performance of bioenergy polygeneration is presented together with potential of CO_2 reductions and CCS opportunities as a separate section.

10.4.1   *Performance of BLG-based biofuel production*

A number of black liquor gasification technologies are available to produce variety of biofuels e.g. dry black liquor gasification with direct causticization (DBLG) (Dahlquist and Jones, 2005), Chemrec black liquor gasification (CBLG) (Berglin *et al.*, 2002; Ekbom *et al.*, 2005), catalytic hydrothermal gasification (CHG) (Naqvi *et al.*, 2010a; Sricharoenchaikul, 2009), etc. The performance of pulp mill integrated bioenergy systems can be evaluated as the energy ratio of the system which is defined as the sum of mill process steam, net electricity import or export, and biofuel production divided by the sum of energy inputs i.e. black liquor, biomass to the power boiler, and biomass to the lime kiln. In recent studies, Naqvi *et al.* compared and evaluated the energy conversion performance of different black liquor gasification systems for various biofuel polygeneration alternatives based on system performance indicators (Naqvi *et al.*, 2010a, 2012a, 2012b, 2012c). Based on a reference pulp mill capacity of 1000 air dried tonnes per day (ADt) of pulp production, the study showed better performance results of synthetic natural gas (SNG) production from the dry black liquor gasification (DBLG) system as compared to other gasification systems due to highest energy ratio and biomass-to-biofuel conversion efficiency. The Chemrec gasification system to produce methanol and dimethyl ether (DME) showed competitive results with substantial biofuel production. Table 10.2 shows energy efficiencies of different biofuel options compared with the recovery boiler.

Ekbom *et al.* (2003) presented a study on black liquor gasification-based motor fuel production (BLGMF) to investigate the integration of biorefinery systems in terms of technical and economic feasibility. The energy withdrawn from black liquor to biofuel is compensated by additional biomass import. Based on 2000 ADt per day of pulp production, about 1183 tonnes per day (273 MW) of methanol and about 824 tonnes per day (275 MW) of DME could be produced with an additional biomass import (414 MW) resulting in 66% biomass-to-biofuel conversion efficiency (Ekbom *et al.*, 2003). The identical pulp mill could produce 188 tonnes per day of hydrogen fuel equivalent to 261 MW (Andersson and Harvey, 2006). There is a large potential of motor fuel production as a replacement for fossil fuels especially in countries with a large pulp industry e.g. Sweden and Finland. Black liquor production in the pulp industry located in Sweden, Europe, and world is substantial and is estimated about 47 TWh, 184 TWh, and 733 TWh

Table 10.2.   Process energy ratios of biofuel alternatives (Source: Naqvi *et al.*, 2010a, 2012a, 2012b, 2012c.

	DBLG O$_2$ (SNG)	DBLG air (SNG)	DBLG (H$_2$)	DBLG (MeOH)	CHG (SNG)	CBLG (DME)	CBLG (MeOH)
Inputs							
Black liquor [MW]	243.5	243.5	243.5	243.5	243.5	243.5	243.5
Bark to power boiler [MW]	77.2	55.1	90.7	80.8	117.4	79.9	70.8
Bark to lime kiln [MW]	–	–	–	–	29.7	29.7	29.7
Total inputs [MW]	320.7	298.6	334.2	324.3	390.5	353.1	344
Outputs							
Mill process steam [(MW]	125	125	125	125	78.4	125	125
Electricity[a] [MW]	−21.8	−14.7	−19.1	−8.4	1.1	−26.2	−22.2
Biofuel produced [MW]	162.2	63.8	141.4	55.7	240.2	131.9	143
Total outputs [MW]	265.4	174.1	247.3	172.3	319.7	230.7	245.8
Process energy ratio	*0.82*	*0.58*	*0.74*	*0.53*	*0.81*	*0.65*	*0.71*

[a]electricity import/export (−/+)

Table 10.3.   Potential motor fuel replacement from annual produced biofuel based on black liquor solids availability in 2010 (Source: Naqvi *et al.*, 2010a, 2012a, 2012b, 2012c).

		Sweden	Europe	World
Black liquor availability[a]	[TWh]	47	184	733
Motor fuel consumption[b]	[TWh]	84	3910	17633
DBLG O$_2$ (SNG)				
Fuel production	[TWh]	31.3	122.5	488.7
Motor fuel replacement	[%]	37.2	3.1	2.8
DBLG air (SNG)				
Fuel production	[TWh]	12.7	50.1	200.1
Motor fuel replacement	[%]	15.1	1.3	1.1
DBLG (H$_2$)				
Fuel production	[TWh]	27.4	107.1	427.3
Motor fuel replacement	[%]	32.6	2.7	2.4
DBLG (MeOH)				
Fuel production	[TWh]	10.8	43.7	174.1
Motor fuel replacement	[%]	12.9	1.1	0.9
CHG system				
Fuel production	TWh	46.9	181.7	725.7
Motor fuel replacement	%	55.8	4.6	4.1
CBLG (DME)				
Fuel production	TWh	26	101.4	402.4
Motor fuel replacement	%	30.9	2.6	2.3
CBLG (MeOH)				
Fuel production	TWh	28.4	113.2	451.9
Motor fuel replacement	%	33.8	2.9	2.6

[a]Based on the Food and Agriculture Organization (FAO) data base (FAO, 2010)
[b]Motor fuel consumption combines gasoline and diesel as transport fuel. Data taken from Earth Trends: Environmental Information (Earth Trends, 2010)

respectively during 2008 (FAO, 2010). Table 10.3 shows potential motor fuel replacement in Sweden, Europe and world based on black liquor production (Naqvi *et al.*, 2012c).

Despite positive performance and considerable biofuel production potential, some issues related to successful integration are vital. One of the key issues is increased load on the lime

Table 10.4.   Estimated internal rate of return (IRR) and payback time (Source: Ekbom et al., 2005).

		Methanol	DME
Additional investment cost[a]	[million euro]	174	190
Total incremental operation cost	[million euro/year]	66.3	66.1
Production capacity[b]	[million m³/year]	0.26	0.24
Cost of production[c]	[euro/liter]	0.26	0.27
Selling fuel price	[euro/liter]	0.51	0.62
Profit	[euro/liter]	0.25	0.35
Profit	[million euro/year]	66.5	82.3
Payback	[years]	2.6	2.9
Internal rate of return (IRR)	[%]	40	45

[a]The recovery boiler requires total investment = 171 million euro
[b]Production capacity equivalent to gasoline and diesel in m³
[c]Annuity factor (%) = 11.1.

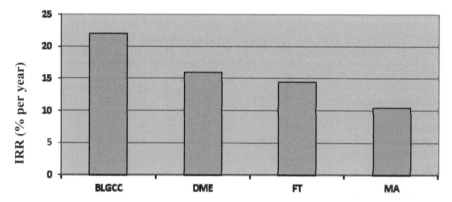

Figure 10.6.   Estimated internal rate of return (IRR) of different biorefinery configurations (Source: Consonni et al., 2009).

kiln for causticization using Chemrec black liquor gasification systems in the black liquor recovery cycle for chemical recovery resulting in extra biomass utilization (Consonni et al., 2009). In a conventional recovery system, sulfur is recycled as sodium sulfide whereas a large amount of sulfur is recycled as hydrogen sulfide in the Chemrec system that results in about 25% increased causticization load on the lime kiln. The increased biomass usage raises critical issues for future biomass supply as the limited biomass will result in less availability of biomass to other processes. The dry black liquor gasification (DBLG) system resolves this issue using direct causticization within the gasifier eliminating the energy-intensive lime kiln in the recovery cycle.

The electricity deficit is an important system consequence of biofuel polygeneration. The conventional recovery boiler system has a potential to export electricity after meeting the pulp mill electricity demand (Table 10.2). However, biofuel polygeneration systems require electricity to be imported from the grid or generated in biomass-based power plant.

The biofuel polygeneration systems using black liquor gasification indicate economic viability since the economics of the pulp mill would be less sensitive to pulp prices, due to the diversified overall economics of co-production of biofuels together with the pulp (Ekbom et al., 2005). The internal rate of return (IRR) considering investment and operating cost of black liquor gasification based methanol and DME production is presented in Table 10.4. Figure 10.6 shows internal rate of return (IRR) results from another study (Consonni et al., 2009) estimated for different biorefinery options e.g. Black liquor gasification combined cycle (BLGCC), dimethyl ether (DME) production, Fischer–Tropsch (FT) liquid, mixed-alcohols (MA).

From the economic perspectives, there are some limitations that must be resolved for sustain-able commercialization of transport biofuel polygeneration with black liquor gasification e.g. incremental biofuel production cost, biofuel distribution cost that is an area of concern especially for gaseous fuels (e.g. DME, compressed natural gas (CNG), H_2 gas) but less important for liquid fuels (e.g. methanol, bio-diesel), high cost of fuel-flexible vehicles, high cost of fuel cell vehicles (e.g. using hydrogen fuel cells). Moreover, the economic risks would be considerable in terms of poor operational reliability and absence of technical guarantees from technology suppliers.

10.4.2 *Performance of BLG-based electricity generation*

Black liquor gasification combined cycle (BLGCC) has a net electricity efficiency of 22%, which means twice the electricity output as compared to a conventional recovery boiler system (Näsholm and Westermark, 1997). The increased electricity generation is mainly due to efficient energy utilization of the gas turbine and a large amount of latent heat recovered from the synthesis gas. Maunsbach *et al.* (2001) studied the combined cycle with the integration of advanced gas turbines e.g. evaporative gas turbine (EvGT) (Jonsson and Yan, 2005), steam injected gas turbine (STIG) and externally fired gas turbine (EFGT) (Eidensten *et al.*, 1996; Wolf *et al.*, 2002; Yan *et al.*, 1995; Yan and Eidensten, 2000). The STIG cycle showed better energy efficiency, cost and load performance as compared to the combined cycle. For a pulp mill producing 1000 ADt per day of pulp, the STIG cycle has potential to double 576 kWh/ADt electricity surplus meeting the internal steam demand of the mill (Maunsbach *et al.*, 2001). However, the gas turbine without steam injection results in electricity surplus with reduced steam demand and the EvGT cycle has high net electricity efficiency.

Based on a reference mill producing 2700 tonnes dry solids of black liquor, a high efficiency of the BLGCC system can be achieved with a net electricity generation of about 115 MW as compared to about 65 MW of electricity generation in the conventional recovery boiler system (Larson *et al.*, 2003). The conventional recovery boiler system is required to purchase electricity from the grid whereas the BLGCC system has potential to export about 14 MW of electricity to the grid. To produce identical process steam in the BLGCC system relative to the recovery boiler system, additional biomass is required to be combusted in the bark boiler to generate steam to meet the steam demand of the reference pulp mill. The BLGCC system with Chemrec gasification results in higher lime-kiln load resulting in increased demand for additional fuel for the lime kiln.

For a base capacity of 2000 ADt per day of pulp production, the BLGCC system can export electricity to the grid generating about 87 MW of electricity as compared to about 45 MW of electricity from the recovery boiler system (Andersson and Harvey, 2006). The combined heat and power (CHP) plant configurations with the recovery boiler system and the black liquor gasification system were studied by Eriksson and Harvey (2004). The study showed that the integrated black liquor gasification CHP units had electrical efficiency between 60% and 70% as compared to the heat-only powerhouses requiring extra biomass import. Table 10.5 shows combined performance results from black liquor gasification-based biofuel and electricity polygeneration systems.

10.4.3 *Performance of pellet production system*

Andersson (2007) conducted a detailed study on the possibilities of integrating pellet production with the pulp mills. The study evaluated different options to integrate drying and pellet production with the Ecocyclic pulp mill (KAM) producing 630,000 air-dried tonnes of pulp per year (KAM, 2003). The excess heat from the pulp mill was used to dry the biomass that contributed to an efficient integrated pellet production and pulp mill. Three technologies for biomass drying were studied i.e. steam drying, flue gas drying and vacuum drying, based on the energy demand, CO_2 emissions and economics of pellet production. Table 10.6 shows results from drying technologies. The flue gas dryer using flue gas from the recovery boiler showed better performance than other dryer options (Andersson, 2007) and about 70,000 tonnes per year of potential pellet production based on the reference pulp mill capacity at a production cost of about 25 Euro per tonne of

Table 10.5.   Performance analysis of different bioenergy systems (Source: Naqvi et al., 2010c).

Parameter	BLGCC		BLG for biofuel production					
Reference	Larson et al. (2003)	Eriksson and Harvey (2004)	Ekbom et al. (2003)		Andersson and Harvey (2006)	Larson et al. (2006)		Naqvi et al. (2010a)
Product	Electricity	Electricity	MeOH	DME	H$_2$	DME	Mixed alcohol	CH$_4$
Pulp [ADT/day]	1600	2000	2000	2000	2000	1600	1600	1000
Black liquor solid [tDS/day]	2724	3420	3420	3420	3420	2724	2724	1700
BLS flow [MW]	350.7	487	487	487	487	350.7	350.7	243.5
Purchased biomass [MW]	27.1	21.3	129	125	123.5	77.4	89.2	106.5
Electricity [MW] Import/Export [−/+]	15.2	86.5	−45.9	−48.7	−56.7	−99.6	8.21	1.1
Fuel production [MW]	–	–	272	275	261	168	60	240.2

Energy content based on lower heating value (LHV) of fuels; black liquor (12.3 MJ/kg), bark (19.4 MJ/kg), DME (28.8 MJ/kg), methanol (21.1 MJ/kg), methane (50 MJ/kg), hydrogen (120 MJ/kg).

Table 10.6.   Pellet production using different drying technologies (Source: Andersson, 2007).

Type of dryer	Steam dryer		Flue gas dryer	Vacuum dryer
	26 bar	10 bar		
Pellet production [tonnes/year]	110,700	110,700	70,500	209,000
Alternate heat use	Electricity production	Electricity production	–	–
Heat cost[a] [million Euro/year]	1.2	0.63	0	0
Dryer cost[b] [million Euro]	3.0	5.0	2.0	28
Specific production cost [Euro/tonne]	32.0	31.4	24.6	38.9
Payback period [year]	4.6	4.8	3.4	7.4

[a]Based on lost electricity generation using steam dryers
[b]References: Wimmerstedt and Linde (1998), Andersson et al. (2006) and Eklund (2002).

pellets. In addition, the pellet production using vacuum drying also showed high potential but at high production cost and about 7 years of payback period.

10.5   BIOENERGY SYSTEMS AND CCS POTENTIAL

A bioenergy system in pulp mills with carbon capture and storage (CCS) is an interesting CO_2 mitigation technique, decreasing the amount of CO_2 emissions to the atmosphere and which can also be termed as "negative CO_2 emissions". The biomass use as an alternative to fossil fuels and capture of CO_2 leads to long-term removal of carbon from the biogenic cycle. It is important to note that currently the bioenergy systems integrated with CCS have not been fully commercialized. According to the Intergovernmental Panel on Climate Change special report on renewable energy

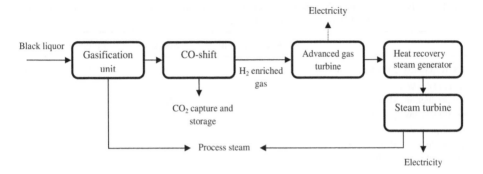

Figure 10.7.   Integration of CCS using BLGCC.

sources and climate change mitigation, the bioenergy systems with CCS (BECCS) could result in reduction of greenhouse gases from the atmosphere at economically feasible mitigation cost (IPCC, 2006).

The net CO_2 emissions in the pulp industry mainly come from different sources: (i) On-site CO_2 emissions from fossil fuels, (ii) CO_2 emissions from fossil fuel used for purchased electricity generation, (iii) CO_2 emissions from fossil fuels to extract, manufacture and transport feed-stocks. In addition, CO_2 emissions in large quantities come from biomass combustion. There are opportunities for CO_2 mitigation in the pulp industry and significant research has estimated the considerable potential to reduce CO_2 emissions from fossil fuel consumption and electricity in pulp mills based on the most efficient existing technologies.

10.5.1   *BLG systems with CCS*

Figure 10.7 shows an example of a bioenergy polygeneration system integrated with CCS using a black liquor gasification combined cycle (BLGCC). Larson *et al.* (2003, 2006) investigated environmental benefits of black liquor gasification technology for electricity production as a replacement for conventional recovery boilers integrated with a reference mill producing about 6 million lbs. per day of black liquor solids (BLS). The electricity generation could result in significant reductions in net CO_2 emissions than the recovery boiler case and the BLGCC systems in United States has potential to reduce up to 35 million tonnes net CO_2 emissions and 160,000 tonnes net SO_2 emissions and 100,000 tonnes net NO_x emissions within 25 years (Larson *et al.*, 2003). However, the net CO_2 emissions from black liquor gasification based biofuel production are larger than the recovery boiler system due to additional biomass consumption to compensate for the total energy deficit (Larson *et al.*, 2006).

The potential of CO_2 emissions reduction could be achieved if hydrogen is produced from black liquor gasification and using hydrogen as transport fuel replacing gasoline (Andersson and Harvey, 2006). About 59,000 tonnes per year of hydrogen could potentially be produced, reducing about 830,000 tonnes per year of CO_2 if integrated with a reference mill producing 630,000 air-dried tonnes per year of pulp. About 60% of Sweden's domestic gasoline could be replaced with hydrogen if all black liquor from market pulp mills in Sweden was converted to hydrogen, which could have resulted in a reduction of about 8% of CO_2 emissions in Sweden in 2006. The hydrogen production with CCS resulted in about 1 million tonnes per year of CO_2 reduction.

The potential and economic feasibility of CCS in the Swedish pulp and paper industry was studied by Möllersten *et al.* (2003a, 2003b, 2004), which showed a potential of about 6 million tonnes of CO_2 reductions per year, equivalent to about 10% of Swedish net CO_2 emissions from black liquor gasification integrated with pre-combustion CO_2 capture. The integration of black liquor gasification and CCS with large pulp and paper mills could result in capture of a large

Table 10.7.   Potential on-site CO_2 offset based on black liquor solids (BLS) availability in 2010 (Source: Naqvi *et al.*, 2010a, 2012a, 2012b, 2012c).

	Reference mill	Sweden	Europe	World
BLS availability[a], Mtonnes/year[b]	0.6	14.1	55.2	220
DBLG O$_2$ (SNG)				
Fuel production, Mtonnes/year	0.1	2.4	9.4	37.4
CO_2 on-site savings, Mtonnes/year	0.43	10.1	39.2	156.2
DBLG air (SNG)				
Fuel production, Mtonnes/year	0.04	0.96	3.68	14.7
CO_2 on-site savings, Mtonnes/year	0.18	4.3	16.6	66.1
DBLG (H$_2$)				
Fuel production, Mtonnes/year	0.03	0.71	2.76	11.2
CO_2 on-site savings, Mtonnes/year	0.54	12.7	49.7	198.1
CHG (SNG)				
Fuel production, Mtonnes/year	0.15	3.6	13.8	55
CO_2 on-site savings, Mtonnes/year	0.31	7.48	29.1	115.8

[a] Based on the Food and Agriculture Organization (FAO) database 2008 (FAO, 2010)
[b] Mtonnes/year $= 10^6$ tonnes/year.

amount of CO_2 from the synthesis gas concentrated with hydrogen after the water-gas shift reaction (Yan *et al.*, (2007). In Sweden, the fossil CO_2 emissions in the pulp and paper industry are not very large but CO_2 emissions from biomass use are substantial.

By using CO_2 capture technologies integrated with various biofuel production routes, a substantial amount of CO_2 can be captured. The theoretical annual on-site CO_2 reduction from various biofuel options using different black liquor gasification technologies based on black liquor availability in a reference mill case, Sweden, Europe and the world is reported in Table 10.7 (Naqvi *et al.*, 2010a, 2012a, 2012b, 2012c). The on-site CO_2 capture potential is important for such biofuel alternatives where CO_2 separation from the product gas is an integral process of biofuel production. Note that the scale up is used as a scenario to estimate theoretical CO_2 reductions if all recovery boilers were to be replaced with black liquor gasification systems and all available black liquor would be used to produce biofuel. The pulp mills with different energy balances will result in different on-site CO_2 reductions and CO_2 savings replacing fossil fuels.

10.6   CONCLUSIONS

The use of bioenergy is rapidly increasing to achieve climate change goals as well as to ensure energy security for sustainable development. The electricity and biofuels can be produced using black liquor gasification as an additional energy resource replacing a substantial amount of fossil fuels. The combined cycle of electricity generation using black liquor gasification has the potential to convert pulp mills from electricity importers to electricity exporters. The potential of biofuel production in the pulp industry is considerably large, and can replace significant amounts of motor fuel in the transport sector especially in countries with a large pulp and paper industry. The biofuel upgrading with pellet production can increase the energy efficiency of the pulp mill with a possibility of heat recovery from the flue gases from the recovery boiler in the conventional black liquor recovery cycle. From the economic perspective, bioenergy production such as electricity, biofuel and pellets in the pulp industry has the potential to diversify the overall economics with less dependence on pulp prices, integrating co-production of bioenergy products together with the pulp. The pulp industry is a major consumer of biomass that could potentially contribute to bioenergy generation with CO_2 capture and storage enabling negative CO_2 emissions. By using CO_2 capture technologies integrated with bioenergy production routes, a substantial amount of on-site CO_2 can be captured at the pulp mill.

REFERENCES

Andersson, E.: *Benefits of integrated upgrading of biofuels in bio refineries – systems analysis*. PhD Thesis, Chalmers University of Technology, Sweden, 2007.

Andersson, E. & Harvey, S.: System analysis of hydrogen production from gasified black liquor. *Energy* 31:15 (2006), pp. 3426–3434.

Andersson, E., Harvey, S. & Berntsson, T.: Energy efficient upgrading of biofuel integrated with a pulp mill. *Energy* 10:11 (2006), pp. 1384–1394.

Berglin, N., Lindblom, M. & Ekbom, T.: Efficient production of methanol from biomass via black liquor gasification. *TAPPI Engineering Conference*, San Diego, CA, 2002.

Berntsson, T., Axegård, P., Backlund, B., Samuelsson, Å., Berglin, N. & Lindgren, K.: Swedish pulp mill biorefineries — A vision of future possibilities. Swedish Energy Agency, 2006.

CEPI: Key statistics 2008. European Pulp and Paper Industry. Confederation of European Paper Industry, 2008.

CEPI: Confederation of European Paper, Bio-energy and the European pulp and paper industry – an impact assessment summary of a study conducted by McKinsey & Company and Pöyry Forest Industry Consulting, 2012.

Consonni, S., Katofsky, R. & Larson, E.: A gasification-based bio-refinery for the pulp and paper industry. *Chem. Eng. Res. Des.* 87 (2009), pp. 1293–1317.

Dahlquist, E. & Jones, A.: Presentation of a dry black liquor gasification process with direct caustization. *TAPPI J.* (2005), pp. 15–19.

Earth Trends: Environmental Information, 2010. Data is maintained by World Resources Institute, http://earthtrends.wri.org/searchable_db/index.php?theme=6/ (accessed July 2012).

Eidensten, L., Yan, J. & Svedberg, G.: Biomass externally fired gas turbine cogeneration. *J. Eng. Gas Turb. Power* 118 (1996), pp. 167–174.

Ekbom, T., Lindblom, M., Berglin, N. & Ahlvik, P.: Technical and commercial feasibility study of black liquor gasification with methanol/DME production as motor fuels for automotive uses-BLGMF. Report for contract no. 4.1030/Z/01-087/2001, European Commission, Altener program, Stockholm, Sweden, 2003.

Ekbom, T., Berglin, N. & Lögdberg, S.: Black liquor gasification with motor fuel production – BLGMF II: a techno-economic feasibility study on catalytic Fischer-Tropsch synthesis for synthetic diesel production in comparison with methanol and DME as transport fuels. Stockholm, Sweden, 2005.

Energy Information Administration (EIA): World energy projections. International Energy Annual (June–October 2007), Paris, France, 2007.

Eriksson, H. & Harvey, S.: Black liquor gasification — consequences for both industry and society. *Energy* 29 (2004), pp. 581–612.

EU: Directive 2003/30/EG, May 8 2003, Brusseles, Belgium, 2003.

FAO: FAO views on bioenergy. Food and Agriculture Organization of the United Nations, Rome, Italy, 2012, http://www.fao.org/bioenergy/47280/en/ (accessed July 2012).

FAOSTAT. Industrial roundwood production 2010. Food and Agriculture Organization of the United Nations, Rome, Italy, January 2010, http://faostat.fao.org/site/630/default.aspx (accessed July 2012).

Gebart, R.: The BLG2 Program: BLG–enabling technology for renewable transportation fuels. Application for financial support to MISTRA, Stockholm, Sweden, 2006.

IPCC-Intergovernmental Panel on Climate Change: IPCC guidelines for national greenhouse gas inventories. IPCC and IGES, Hayama, Japan, 2006, http://www.ipccinggip.iges.or.jp/public/2006gl/index.htm (accessed July 2012).

Jonsson, M. & Yan, J.: Humidified gas turbine – a review of proposed and implemented cycles. *Energy* 30 (2005), pp. 1013–1078.

KAM: Final Report. Eco-cyclic pulp mill. KAM project, Report no. A100, STFI, Stockholm, Sweden, 2003, www.stfi-packforsk.se (accessed July 2012).

Larson, E., Consonni, S. & Katofsky, R.: A cost-benefit assessment of biomass gasification. Power Generation in the Pulp and Paper Industry. Final report, Princeton University and Politecnico di Milano, 2003.

Larson, E., Consonni, S. & Katofsky, R.: A cost-benefit assessment of gasification-based biorefining in the Kraft pulp and paper industry. Final report, Volume 1. Princeton University and Politecnico di Milano, 2006.

Maunsbach, K., Isaksson, A., Yan, J., Svedberg, G. & Eidensten, L.: Integration of advanced gas turbines in pulp and paper mills for increased power generation. *J. Eng. Gas Turb. Power* 123:4 (2001).

Möllersten, K. & Yan J.: Economic evaluation of biomass-based energy systems with CO_2 capture and sequestration – The influence of the price of CO_2 emission quota. *World Resource Rev.* 13:4 (2001), pp. 509–525.

Möllersten, K., Yan, J. & Westermark, M.: Potential and cost-effectiveness of CO_2-reduction in the Swedish pulp and paper sector. *Energy* 28 (2003a), pp. 691–710.

Möllersten, K., Yan, J. & Moreira, J.: Potential market niches for biomass energy with CO_2 capture and storage. *Biomass Bioenergy* 25:3 (2003b), pp. 273–285.

Möllersten, K., Lin, G., Yan, J. & Obersteiner, M.: Efficient energy systems with CO_2 capture and storage from renewable biomass in pulp and paper mills. *Renew. Energy* 29 (2004), pp. 1583–1598.

Naqvi, M., Yan, J. & Fröling, M.: Bio-refinery system of DME or CH_4 production from black liquor gasification in pulp mills. *Bioresour. Technol.* 101 (2010a), pp. 937–944.

Naqvi, M., Yan, J. & Dahlquist, E.: Black liquor gasification integrated in pulp and paper mills: a critical review. *Bioresour. Technol.* 101 (2010b), pp. 8001–8015.

Naqvi, M., Yan, J. & Dahlquist, E.: Synthetic gas production from dry black liquor gasification process using direct causticization with CO_2 capture. *Appl. Energy* 97 (2012a), pp. 49–55.

Naqvi, M., Yan, J. & Dahlquist, E:. Bio-refinery system in a pulp mill for methanol production with comparison of pressurized black Liquor gasification and dry gasification using direct causticization. *Appl. Energy* 90 (2012b), pp. 24–31.

Naqvi, M., Yan, J. & Dahlquist, E.: Energy conversion performance of black liquor gasification to hydrogen production using direct causticization with CO_2 capture. *Bioresour. Technol.* 110 (2012c), pp. 637–644.

Näsholm, A. & Westermark, M.: Energy studies of different cogeneration systems for black liquor gasification. *Energy Convers. Manage.* 15:38 (1997), pp. 1655–1663.

Sricharoenchaikul, V.: Assessment of black liquor gasification in supercritical water. *Bioresour. Technol.* 100 (2009), pp. 638–643.

Swedish Energy Agency: Report on the annual Energy in Sweden. Swedish Energy Agency, Stockholm, Sweden, 2009, http://www.energimyndigheten.se/en/ (accessed July 2012).

Swedish Forest Industries: The report on the Swedish Forest Industries – Facts and figures 2010. Stockholm, Sweden, 2010, http://www.skogsindustrierna.se/web/Utgivningsar_2011.aspx (accessed July 2012).

Thuijl, E., Roos, C. & Beurskens, L.: An overview of biofuel technologies, markets and policies in Europe. ECN-C–03-008, 2003.

US Department of Energy: Views on renewable energy and biomass. 2012, http://www.energy.gov/energy sources/bioenergy.htm (accessed July 2012).

Wolf, J., Barone, F. & Yan, J.: Performance analysis of evaporative biomass air turbine cycle with gasification for topping combustion. *J. Eng. Gas Turb. Power* 124:4 (2002), pp. 757–761.

Yan, J., Eidensten, L. & Svedberg, G.: An investigation of heat recovery system in externally fired evaporative gas turbines. *ASME Paper* 95-GT-72, 1995.

Yan, J. & Eidensten, L.: Status and perspective of externally fired gas turbines. *J. Propul. Power* 16 (2000), pp. 572–576.

Yan, J., Dahlquist, E., Jin, H., Gao, L. & Tu, S.: Integration of large scale pulp and paper mills with CO_2 Mitigation Technologies. *The 3rd International Green Energy Conference*, Västerås, Sweden, 2007.

Zakrisson, M.: Internationell jämförelse av produktionskostnader vid pelletstillverkning. Department of Forest Management and Products, SLU, Uppsala, Sweden, 2002.

CHAPTER 11

Alternative fuels and green aviation

Emily Nelson

11.1   INTRODUCTION

For engineers and scientists who are new to the field of biofuel production for aviation, it can be bewildering to come to terms with the conflicting conclusions drawn in the literature or evaluate the claims found in a casual web search. It is a multidisciplinary field that spans chemical, petroleum, environmental, mechanical, aerospace, materials and industrial engineering. The field is also overlaid by a web of geographically varying public policies that profoundly impact economic feasibility and the pace and directions in technology development. It is a field that has grown rapidly in all of these areas, so that there is a vast, rich literature base to peruse. Finally, there are passionate and articulate advocates for every conceivable side. This can make the leap across the interdisciplinary aspects rather challenging. Some of the most embedded sources of confusion arise from the following:

- Terminology may be conflicting or otherwise difficult to understand. For example, the naming of the same set of hydrocarbon compounds in petroleum engineering differs from the conventions in organic chemistry.
- The standard units in which values are reported vary widely. Often this is because the convention had already been established in various disciplines and geographical regions.
- There may be no clear standards or protocols for measurements, which leads to difficulty in comparing values, e.g., the manner in which to quantify biomass yield from algae.
- Some studies do not clearly state their assumptions and identify their uncertainties, which are critical to evaluating its conclusions.
- In order to create assessments for global applications, the numbers used to describe the physics may be scaled up from laboratory results which grossly magnify the uncertainties present in the small-scale data (e.g., processing conditions to yield 500 milliliters (mL) of crude algal oil over a few days in a laboratory) to huge numbers (e.g., annual production of millions of barrels per hectare per year (MMbbl/ha/y) of biodiesel).

With regard to scaleup, a number of factors come into play: To be useful, the lab-scale experiment must meticulously keep track of all nutrients and additives, gas exchange and energy input. Sometimes the processes or equipment that are useful for lab experiments for mixing, separation or combustion do not translate up to industrial scale facilities. For example, centrifuging to remove excess water from algal broth is both simple and effective for small quantities of fuel, but it is too expensive for commercial production. Finally, the lab has the luxury of a controlled environment and may run its tests over only a few days, whereas commercial producers and refiners will be subject to the vagaries of seasonal and diurnal temperature swings, cloudy days, too much or too little rainfall, and the intermittent interest of predators or competitors.

The next generation of engineers, scientists and policymakers must find a way to bridge these incompatibilities in order to provide the most useful, comprehensive, evidence-based data on current technologies and new strategies to make those technologies commercially viable. The purpose of this monograph is to present basic concepts in the key areas and provide a foundation for critical thinking. Where available, uncertainties in the data are given. Such uncertainties

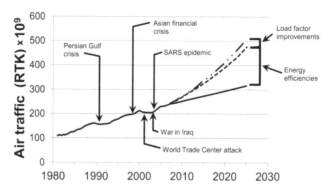

Figure 11.1.   Chronological patterns in air traffic, including passenger and cargo traffic, from 1960–2006, overlaid with relevant events. Air traffic is shown in terms of revenue tonnes per kilometer (RTK). Predicted growth in jet fuel demand is shown from 2006 through 2025 (solid line), which includes expected efficiency gains from technology development (dotted line) and logistics improvements (dashed dotted line). Adapted from Chèze *et al.* (2011), with permission from Elsevier, Inc.

can arise due to incomplete access to all the relevant data or processes, or they can be due to real statistical variation (e.g., variation in fuel sulfur content depending on whether the fuel was extracted from shale or more conventional petroleum reservoirs). Either way, it is useful to develop an understanding of how much variation can be expected in any given factor. Now, let us dig into some of the details.

Over the last several decades, air traffic has increased by about 5% annually. As shown in Figure 11.1, its momentum was slowed only briefly through 2006 by major events such as the attack on the World Trade Centre in 2001, wars, the SARS epidemic, and gloomy economic conditions. Since the global economic crisis in 2008–2009, the airline industry has recovered to pre-recession levels and is expected to grow at or above historical levels (Boeing, 2011). However, even when accounting for technology advancements (dashed line) and logistics optimization (dash-dotted line), jet fuel demand is expected to increase for the foreseeable future. In other studies that attempt to predict energy demand for the aviation industry during the coming decades, the details may vary somewhat, but the fundamental conclusion is that the demand for jet fuel will continue to increase. Better air traffic management could provide a potential fuel consumption savings up to 15% (Blakey *et al.*, 2011), such as separation distance control and speed optimization. With the right infrastructure on the ground, continuous descent approaches could improve fuel efficiency over the current stepped approaches. Although there is some ongoing work in alternative modes of propulsion, such as fuel cells (Novillo *et al.*, 2011; Renouard-Vallet, *et al.*, 2012), this technology will not be commercially viable for decades, at best. The major efforts at reducing fuel consumption in engine and airframe design include increased efficiency through new turbofan designs, e.g., high pressure-ratio cores and super high bypass-ratio fans, reduced drag airframes, and advanced materials in ceramic composites. The recently unveiled Boeing 787 delivers a 20% fuel reduction relative to similarly sized planes through the use of advanced materials, new engine design, and improved fuselage integration. Aircraft lifetimes are typically in the range of 20–30 years or more (Moavenzadeh *et al.*, 2011), so that it will be some time before there is a complete turnover to more efficient aircraft designs. To work with the fleet as it exists, some aircraft can be retrofit with winglets to improve aerodynamics, but this is only an aid, not a fix to the problem. In addition, airport supply chains and infrastructure are currently set up for petroleum-based fuels. The automobile industry is well-suited take advantage of all-electric and hybrid engines to reduce overall fuel consumption. However, the aviation sector has no practical alternative to the internal combustion engine in the coming decades, and in spite of technological and logistical improvements, jet fuel demand will continue to grow.

At the same time, the availability of suitable energy sources to meet this demand will depend on strategic and innovative actions. To date, economic growth in developed and developing nations has been predicated on the availability of cheap oil. According to the International Energy Agency (IEA), the global peak of conventional oil reserves was reached in 2006 (IEA, 2010). Historical data indicates that production in oil reservoirs have declined sharply after the peak is reached (Hirsch, 2006), which is also predicted in the IEA report. The sharp decline in output can be at least somewhat mitigated through development of new oil reservoirs, advanced techniques for recovering oil from unconventional sources such as oil shale, and increased reliance on coal, natural gas and alternative fuels. However, huge environmental impact could result without appropriate clean coal technologies, safe recovery methods particularly for natural gas, and sustainable practices in alternative fuel manufacture. There is a real sense of urgency, since there are consequences associated with waiting too long to put mitigation strategies into place (Vaughan *et al.*, 2009). The rapidly fluctuating price of oil puts pressure on the global economy that a stable source of energy could help to alleviate. One thing that all sides should agree on is that it is crucial to make informed public policy decisions that are based on high-quality, up-to-date science.

We are feverishly seeking a way to satisfy our thirst for energy in a way that respects our planet and the life that depends on it. This will require the creative development of new energy sources, and biofuels are widely considered to be the best, if not the only, solution. Aviation is responsible for ~10% of global transport energy consumption (Moavenzadeh *et al.*, 2011). Air transport dumps hundreds of millions of tonnes of greenhouse gases (GHG) into the atmosphere annually, currently accounting for ~3.5% of all GHG emissions. Legitimate concern over this contributor to climate change has led to a deluge of environmental and economic analyses that attempt to provide guidance as to a rational way forward at the global scale. The Life Cycle Assessment (LCA) has emerged as a relatively new framework that is used to quantitatively evaluate local and worldwide biofuel approaches in terms of their environmental and/or economic impact (see section 11.5).

Until the biorefining industry becomes more established or another serious global oil crisis is at hand, it is difficult to make a purely commercial case for biofuel manufacture. Public policy can make all the difference in providing incentives. Despite a rocky beginning in 2005, the European Union (EU) has been the leader in instituting carbon trading mechanisms in response to the Kyoto Protocol (Gourlay *et al.*, 2011). To account for air traffic's contribution to climate change, the European Union will include air travel in its comprehensive carbon trading system in January 2012 (Convery, 2009). All domestic and international flights that arrive or depart from the EU will be covered by the EU Emissions Trading System. Airlines will receive carbon credits through the use of biofuels; if their net output exceeds the specified targets, the airline must offset the overproduction through carbon trading.

Other countries have also developed goals for reducing dependence on petroleum-based fuels through the use of alternative fuels. China has set targets of reducing energy consumption by 16% and CO_2 emissions per unit of GDP by 17% in 2015 from its baseline in 2010 (Li *et al.*, 2011). The US Energy Independence and Security Act of 2007 specifies a ramping-up of biofuel production through 2012, which includes 36 billion gallons (136×10^9 liters (L)) of renewable fuels by 2022, which must include 21 billion gallons (79.5×10^9 L) of next-generation biofuels that are not derived from corn ethanol. In addition, life cycle GHG emissions must be reduced by at least 50% relative to 2005 output of petroleum-based transportation fuels (DOE, 2010). The US Air Force has targeted that half of its aircraft will use blends of conventional and alternative fuels by 2016 (Byron, 2011), and the US Navy plans to build the "Great Green Fleet" of carrier ships by 2016 powered entirely by non-fossil fuels (Karpovitch, 2011). The commitment to purchase large quantities of biofuel will help bridge the gap between a product that is viable in the laboratory or small-scale pilot plant and the establishment of large-scale biofuel production facilities that will significantly bring down the cost of manufacture.

Although biofuels are an attractive solution, whether developed from feedstocks such as vegetable oils, agricultural waste or the much-anticipated algae, there are many issues that hinder wide availability. Most studies agree that the cost of the feedstock is by far the largest contributor to the final biodiesel cost, although the numbers vary from about 60–95% (Balat, 2011a; Razon, 2009). Embedded within that issue are considerations of the economics and logistics of manufacturing, economies of scale (Knothe, 2010a), sustainable land development, lack of refining infrastructure, and market considerations, particularly with regard to the fluctuating price of oil.

In section 11.2, the international standards for jet fuel are outlined, along with a description of the most important physical characteristics for aviation fuel. Fuel composition and its corresponding effects on fuel properties are the subject of section 11.3. The following sections discuss alternative fuel feedstocks (section 11.4), biorefining techniques (section 11.5), and an introduction to Life Cycle Analysis for aviation fuel (section 11.6). For those without a solid background in chemistry and petroleum engineering, see Appendix A for a review of definitions and basic hydrocarbon chemistry.

11.2   AVIATION FUEL REQUIREMENTS

11.2.1   *Jet fuel specifications*

Aviation fuel must operate in extreme conditions, and so must live up to carefully developed standards, which have evolved over time (Edwards, 2007). If the temperature and pressure on the ground are 15 degrees Centigrade (°C) and 101.3 kiloPascals (kPa), then at a cruise altitude of 11,000 meters (m), the external temperature is about −56.5°C and the pressure is ∼22.6 kPa.

All aviation fuels are blends of various hydrocarbons. There are two basic types of jet fuel, which differ by the proportions of hydrocarbons present in the fuel. The carbon number is a measure used to indicate the number of carbon atoms present in a hydrocarbon molecule. For example, methane (CH_3) is assigned a carbon number of C1, and octane (C_8H_{18}) is C8. The most common type of jet fuel is a kerosene blend with carbon numbers from C8–C12, while the less typical naphtha/kerosene blend is a "wide-cut" fuel with a broader range of about C5–C12. By including the lighter hydrocarbons, the fuel's vapor pressure is reduced and it has better cold temperature properties. For civilian aircraft, the most common aviation fuels for powering jet and turboprop engines are:

- *Jet A*: a kerosene-grade fuel used throughout the US, designed to operate under the demanding conditions of flight. Its freeze temperature must be $\leq$−40°C
- *Jet A-1*: a kerosene-grade fuel widely available outside the US. It has a lower freeze point of $\leq$−47°C, and there are other minor differences relative to Jet A.
- *Jet B*: a naptha/kerosene blend, used primarily in cold climates such as northern Canada. Although it operates more effectively at lower temperatures, it is also more volatile, so it exhibits greater evaporation loss at high altitude. In addition, it is a greater fire hazard on the ground, and it makes a plane crash less survivable.

The most common military fuels are:

- *JP-4*: the military equivalent of Jet B with the addition of corrosion inhibitors and anti-icing additives. It used to be the primary fuel of the US Air Force, but it was phased out in the 1990s due to safety concerns. Although still in use by other air forces around the world, it is in limited production.
- *JP-5*: a high flash-point, wide-cut kerosene fuel used by the US Navy, primarily for aircraft carriers.
- *JP-8*: the military equivalent of Jet A-1 with the addition of corrosion inhibitor and anti-icing additives.

Other common additives include antioxidants to prevent gumming; antistatic agents to dissipate static electricity; metal deactivators to remediate the effects of trace materials in the fuel that affect thermal stability; and biocides to reduce the likelihood of microbial growth within the system, which could plug filters and produce corrosive metabolites (Raikos *et al.*, 2011). Many countries

Table 11.1.   Key specifications for aviation fuels.

Property	Jet A-1	Jet A	Jet B	JP-4	JP-5	JP-8
Density at 15°C [kg/m^3]	775–840	775–840	750–801	751–802	788–845	775–840
Viscosity at −20°C [mm^2/s]	≤8	≤8.0	–	–	≤8.5	≤8.0
Flash point [°C]	≥38	≥38	–	–	≥60	≥38
Freeze temperature [°C]	≤−47	≤−40	≤−51	≤−58	≤−46	≤−47
Distillation end point [°C]	≤300	≤300	–	≤270	≤300	≤300
Vapor pressure [kPa]	–	–	<21	14–21 at 37.8°C	–	–
Specific energy [MJ/kg]	≥42.8	≥42.8	≥42.8	≥42.8	≥42.6	≥42.8
Lubricity: wear scar diameter [mm]	<0.85	–	–	–	–	–
total acidity [mg KOH/g]	≤0.015	≤0.010	≤0.010	≤0.015	≤0.015	≤0.015
aromatics [%v/v]	≤25	≤25	≤25	≤25.0	≤25.0	≤25.0
sulfur [%m/m]	≤0.30	≤0.30	≤0.40	≤0.4	≤0.30	≤0.30
hydrogen [%m/m]	–	–	–	≤13.5	≤13.4	≤13.4
Specification	UK DEF STAN 91-91	ASTM D 1655	Canada CGSB-3-22	US MIL-DTL-5624U	US MIL-DTL-5624U	US MIL-DTL-83133E or UK DEF STAN 91-87

and organizations provide somewhat different sets of specifications for jet fuels, but some of the most broadly based provisions are given in Table 11.1.

In the governing standards identified in the bottom row of Table 11.1, there are many other specifications given that are not shown. Some of them set minimum or maximum values for various parameters; others simply require reporting. For example, the standards for Jet A-1 defined in UK DEF STAN 91-91 require that fuel manufacturers report the percentage of fuel by volume (%v/v) that has been hydrotreated. This has implications for synthetic fuels, since hydrotreatment is the method of choice used to process biologically derived oils (or "green crude") into aviation-grade fuel (see section 11.5). Each specification is associated with one or more standard measurement protocols, since the method of measurement can have an impact on the detected value. For a more detailed outline of specifications for other fuel types and the standards of other countries, see ExxonMobil (2005). For an excellent review of aviation fuel and testing methods, see Chevron (2006).

Between 1993 and 2011, kerosene-type jet fuel comprised 9–11% of the crude oil content in US refineries (EIA, 2011). The presence of aromatics, sulfur and other trace components are highly correlated to the geographical source of the extracted crude. In Figures 11.2 and 11.3, representative properties are shown, as identified from 56 samples of aviation fuel obtained from around the globe in the World Fuel Sampling Program (WFSP) (Hadaller and Johnson, 2006). Not all of these samples were known to have been used as jet fuel, and one sample failed its thermal stability testing. Furthermore, some of the compiled regional results (Fig. 11.2a) or fuel type results (Fig. 11.2b) were predicated on one or two samples and should not be considered statistically significant. The samples from South Africa were either partially or completely comprised of

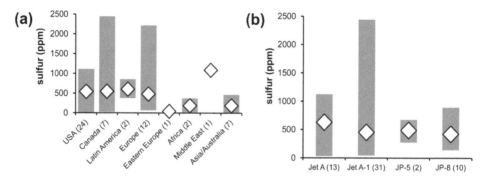

Figure 11.2.   Sulfur content in 56 samples of jet fuel from around the world. The number of samples in each category is shown in parentheses. Average values are marked with a diamond on a gray window representing minimum and maximum sample values. (a) Sulfur content by geographical region; and (b) Sulfur content by jet fuel type. Adapted from Hadaller and Johnson (2006).

synthetic Fischer-Tropsch fuel. Three of the samples from Asia/Australia were extracted from oil shale. In Figure 11.1, the average values are marked by a diamond on a gray window representing the envelope of minimum and maximum values. Although a few samples of Jet A-1 exhibited relatively high values of sulfur, most of the samples had sulfur content of about 500 parts per million (ppm) or less, which is significantly below the specified limit of 3000 ppm (Chevron, 2006). Sulfur can be corrosive to the engine and it is of concern in emissions, but extremely low levels of sulfur (below ~100 ppm) have also been correlated with increased engine wear.

The WFSP study presented results for a range of other fuel parameters as well, some of which are reproduced in Figure 11.3, and will be discussed in the following paragraphs. The units for the parameters are identified by the symbols for mass [M], length [L], time [T], temperature [K] and amperes [A].

Density (Figs. 11.3a and 11.4a) measured in ML^{-3}, here in kilograms per cubic meter [kg/m^3]: Fuel density represents the mass per unit volume of fuel. Density is a key parameter because increased fuel weight means that more energy must be supplied to move the loaded aircraft, but it is also correlated with other performance parameters discussed below, such as specific energy (heat of combustion).

Fuel injectors meter their output of fuel by its volume, not its density. Consequently, when the fuel is injected into the combustion chamber, the density of the fuel will govern the fuel/air ratio. Thus, density is directly related to the thrust through the injected fuel volume and fuel reaction properties (Fazal *et al.*, 2011).

Since the density of liquid fuel decreases linearly with temperature (Fig. 11.4a), the standards in Table 11.1 require that the measurement be taken at a standardized reference temperature of 15°C. All of the fuel sampled in the WFSP met the minimum requirements for aviation fuel; none of the samples exceeded the maximum value of 845 kg/m^2 (JP-5) and 840 kg/m^2 (Jet A, Jet A-1 and JP-8).

Kinematic viscosity at −20°C (Figs. 11.3b and 11.4b), measured in units of L^2t^{-1}, here square millimeters per second [mm^2/s]: Viscosity is a measure of fluid resistance to shear stress. Consequently, lower viscosity fluids deform upon application of shear forces more readily, and they pour more easily. In aircraft operation, this property is important in cold starts, reignition at altitude, lubrication and combustion quality. Higher viscosity causes larger pressure drops across the fuel lines, requiring the pumps to work harder to maintain a given flow rate. In addition, higher viscosity fuels have an impact on combustion quality: Liquid fuel enters the combustion chamber as an atomized spray. Higher viscosity fuels tend to cause larger droplets, and the spray pattern does not penetrate as deeply into the chamber. Incomplete combustion can result, with

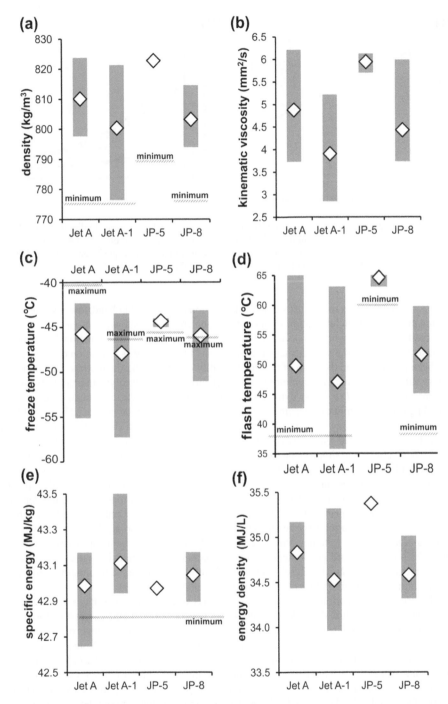

Figure 11.3.   Representative properties of aviation fuels Jet A, Jet A-1, JP-5 and JP-8 as found by the World Fuel Sampling Program. The properties are (a) density; (b) kinematic viscosity; (c) freeze temperature; (d) specific energy; (e) specific energy density; and (d) flash temperature. Maximum and minimum values per the specifications of Table 11.1 are identified by hashed lines. Adapted from Hadaller and Johnson (2006).

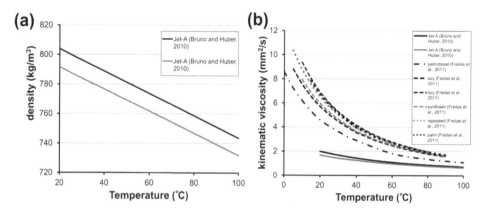

Figure 11.4.   Experimentally measured density and viscosity of jet fuel, petroleum diesel and biodiesel. (a) Density of two samples of Jet A; and (b) viscosity of two samples of Jet A, petrodiesel and several biodiesels derived from various vegetable oils. Data adapted from Bruno and Huber (2010) for jet fuel; and Freitas *et al.* (2011) for petrodiesel and biodiesel.

accompanying increases in exhaust smoke and emissions. On the other hand, low-viscosity fuels may not provide enough lubrication for moving engine components to work properly.

Viscosity decreases with increasing temperature, but the form of the functionality is not as simple as for density. Many mathematical descriptions for viscosity as a function of temperature have been proposed, and the form of that functionality differs among the models. To complicate matters further, there are two commonly used definitions for viscosity: kinematic viscosity v measured in units of $L^2 t^{-1}$ and dynamic (or absolute) viscosity (usually denoted as μ or η, depending on the discipline) measured in MLt^1. The viscosities can be converted back and forth through the density ρ, i.e., $v = \mu/\rho$. Strictly speaking, the dynamic viscosity is the factor that relates fluid response to shear stress, but the kinematic viscosity comes in handy since this converted parameter can simplify the form of the equations of fluid motion in some cases. From a physics standpoint, the dynamic viscosity of a typical Newtonian liquid should obey an Arrhenius-type expression in temperature T, i.e., an exponential, $\mu \propto \exp(-A/T)$. The constant A is specific to the type of liquid under consideration. Since density is inversely proportional to temperature, $\mu \propto 1/T$, we should expect that $\mu \propto T \cdot \exp(-A/T)$. However, over small temperature ranges, even a linear approximation may be sufficient. Near the freezing point, the viscosity behavior becomes more sensitive to temperature decrease, and the viscosity shoots up (Kerschbaum and Rinke, 2004). Freitas and co-workers evaluate several viscosity models against experimental measurements of diesel and biodiesel, shown in Figure 11.4b, that vary substantially over a range of temperature and the fuel's hydrocarbon content (Freitas *et al.*, 2011).

In Figure 11.4b, experimental data on two samples of Jet A (the two traces at the bottom) shows that the temperature dependence is not linear between 20 and 100°C, but it is not grossly nonlinear. However, viscosity does not behave linearly over the broad range of temperatures encountered during aircraft operations, particularly at low temperatures. Note that, in Figure 11.4b, the petrodiesel (third from bottom) and biodiesels (clustered above the petrodiesel curve) all exhibit increasing sensitivity to lower temperatures. In this case, the petrodiesel was a commercial product suitable for automotive use, and the biodiesel was comprised of pure methyl esters from a variety of vegetable oils. For jet fuels, the WFSP found that the temperature dependence became markedly nonlinear when the temperature approached −40°C and below (Hadaller and Johnson, 2006). For the purposes of fuel qualification, all of the fuel in the WFSP survey easily met the specification at the upper bound, 8 mm^2/s at −20°C. For other discussions on the temperature dependence of diesel-grade fuel viscosity, see Freitas *et al.* (2011), Pratas *et al.* (2011a,b), Yuan *et al.* (2003, 2005, 2009), and Hansen and Zhang (2003).

Freeze temperature (Fig. 11.3c) measured in units of K, here in °C: Since jet fuels are a blend of different hydrocarbon compounds, it freezes over a temperature range rather than at a single temperature, as for a pure liquid. This is due to the fact that, as the temperature is decreased, the heaviest hydrocarbons freeze into waxy crystals before the lighter components solidify. To create a systematic means of comparing the freezing properties of jet fuels, the term "freeze temperature" (not "the freezing temperature") is defined through the following procedure: the hydrocarbon fuel blend is cooled until wax crystals form. As the fuel is gradually warmed back up, the lowest temperature at which all of the wax crystals have melted is defined as the freeze temperature (also sometimes denoted as the "freeze point"). Consequently, the freeze temperature is well above the temperature at which the fuel completely solidifies. A related term is "cloud point", which is the temperature at which wax crystals first start to form as the temperature is lowered. Roughly 10°C below the cloud point, the freezing fuel reaches the "pour point", at which the wax in the fuel has built up sufficient solid structure to prevent pouring. The combination of viscosity and freezing point define the pumpability of a fuel, that is, the ease of pumping fluid through the fuel lines (Chevron, 2006).

Flash temperature or flash point (Fig. 11.3d) measured in units of K, here in °C: The flash point is the lowest temperature at which vaporized fuel above a flammable liquid will burn when exposed to an ignition source. Vapor burns only when the air/vapor mixture is in a certain range. Below the lower flammability limit, there is insufficient fuel in the mixture to combust. For kerosene-type jet fuel, the range is 0.6–4.7 volume percent (%v/v) vapor, while for wide-cut fuel, it is 1.3–8.0%v/v (Chevron, 2006). The upper flammability limit is a function of the local temperature and pressure. The flash point of wide-cut fuels like Jet B is not specified, but is below 0°C (Chevron, 2006).

Specific energy (heat of combustion) (Fig. 11.3e) measured in units of L^2T^{-2}, here in mega-Joules per kilogram [MJ/kg]: The specific energy can be used to compare fuels by the relative energy content that a kilogram of fuel could release through complete and perfect combustion. The WFSP found that all but two samples reached the minimum required energy content.

Energy density (Fig. 11.3f) measured in units of $ML^{-1}T^{-2}$, here in megaJoules per liter [MJ/L]: The fuel energy content per unit volume is measured by the energy density, which is the product of the specific energy and density. If these properties are measured by the units used for this work, that product must also be multiplied by the necessary factors to convert between liters and cubic meters (1 m^3 = 1000 L). When a system is limited by volume, as in a completely filled fuel tank, a fuel with higher energy density can release more net energy, which can be used to travel a longer distance.

Aromatics content measured as a nondimensional number, the percentage of aromatic hydrocarbons in terms of volume per total fuel volume (%v/v): Aromatics are unsaturated hydrocarbon molecules with one or more carbon rings. As the fuel combusts, carbonaceous particles form that become incandescent at the high temperatures and pressures in the combustor. The hot particles emit infrared radiation, which can heat up the surrounding walls and create hot spots. The result can be a loss in combustion efficiency, or even worse, a loss of structural integrity. Carbon that deposits on the wall can change the carefully designed flow pattern in the combustor by inhibiting the entrance of diluting air through combustor walls. If the particles are not completely consumed by the time they reach the turbine blades and the stator, they can damage these key engine components. Since aromatics tend to produce more of these carbonaceous particles, their content is limited to a maximum of 25%v/v in aviation fuels. On the other hand, the presence of aromatics can be critical for maintaining aircraft seals. Aromatics can cause seals and sealants to swell, developing a "set" to a particular swell level, which is a function of the aromatic content and exposure time. If the seals are subsequently exposed to jet fuel with either very low aromatics content or a sufficiently different mix of aromatics, the absorbed petroleum leaches out of the seal material, resulting in geometric shrinkage and possibly a leak (Hadaller and Johnson, 2006). An aging aircraft would be susceptible to this condition if it has used a wide-cut fuel like JP-4 for a long time and it is switched over to a narrower cut like JP-8 or one of the synthetics. Synthetic fuels tend to have fewer aromatics than petroleum fuels and may require supplementation to meet jet fuel requirements (Corporan *et al.*, 2011). DEFSTAN 91-91 requires a minimum of 8% aromatics

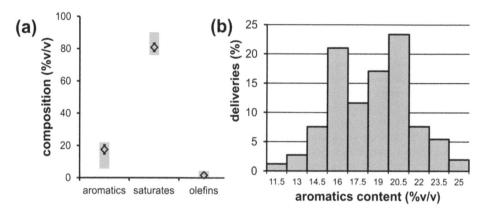

Figure 11.5.   Hydrocarbon content in jet fuels by class from (a) the World Fuel Sampling Program, adapted from Hadaller and Johnson (2006); and (b) all jet fuel deliveries to the US military in 2004, reproduced from Colket *et al.* (2007) with permission from the American Institute of Aeronautics and Astronautics.

content in the final blended product. In most of the synthetic F-T fuels, up to 50% synthetic is blended with conventional jet fuel, with all of the aromatics coming from the petroleum stream (Moses, 2008).

The WFSP analyzed the hydrocarbon content of the sample fuels using a number of ASTM-specified protocols, some of which gave somewhat conflicting results, particularly at low concentrations. Figure 11.5a shows the composition of the jet fuel analyzed *via* ASTM D1319. It is separated into categories of aromatics, saturates (alkanes in the form of *n*-, iso- and cyclo-paraffins) and olefins (alkenes, or unsaturated hydrocarbons with one or more double carbon-to-carbon bonds). Figure 11.5b shows a histogram of the aromatics content in all of the deliveries of Jet-A to the US military in 2004. Note that each of the numbers on the abscissa (*x*-axis), cover a 1.5% range in value, i.e., about 24% of the fuel deliveries had aromatics content in the range of 20.5 ± 0.75%v/v. On average, the aromatics content of fuels was in the range of 20%, although it varied from below 10% to about 24%. All of these fuels met the standards for aviation fuel, which indicates that, from a practical standpoint, the requirements allow substantial flexibility in composition. However, there are will be some differences in performance between the fuels with the highest and lowest aromatics content.

Napthenes (cycloalkanes) are other fuel components of interest. They are also based on carbon rings, like aromatics, except that, rather than some double bonds, they have only single carbon-to-carbon bonds, i.e., they are saturated cyclic hydrocarbons.

Hydrogen content measured in nondimensional units as a percentage of hydrogen by mass to total fuel mass [%m/m]: Fuels with higher hydrogen content burn more cleanly and produce more energy per unit mass. The disadvantage of high hydrogen content fuels is the relatively lower energy content per unit volume. The WFSP found that hydrogen levels in jet fuel were in the range of 14%m/m, and that synthetic Fisher-Tropf fuel had higher hydrogen content (Hadaller and Johnson, 2006).

Lubricity commonly measured by wear scar diameter in units of *L*, millimeters [mm]: Lubricity refers to a fluid's capacity to reduce friction between moving parts. It is characterized by measuring the wear on a fixed steel ball after a specific length of time after contact with a rotating cylindrical ring that is partially immersed in fuel (Chevron, 2006). The depth of the wear scar increases with decreasing lubricity. This property is important for jet fuels, as aircraft components rely on the fuel itself to lubricate moving parts, e.g., in fuel pumps and control units. For a discussion of wear scar testing on a wide range of samples, see Knothe (2008b).

Thermal stability commonly measured by the pressure drop, Δp, across a filter measured in units of *L*, millimeters of mercury [mm Hg]: Specialized test equipment called the Jet Fuel

Thermal Oxidation Tester (JFTOT) exposes the fuel to a heated aluminum alloy tube in a controlled way and passes it through a filter to collect any particulates that have formed. After the test is complete, the pressure drop across the filter and a visual inspection of the aluminum tube for discoloration are used to evaluate thermal stability. It may look unusual to measure a pressure drop in terms of length when it is actually a force per unit area. The height of the fluid column is related to the pressure though its density and gravitational acceleration (g), specifically $\Delta p = \rho g h$. By convention, standard atmospheric pressure at ground level is 760 mm Hg = 101.325 kPa. This property represents the capacity for maintaining fuel properties due to thermal exposure. It is particularly important for aviation fuels because high-performance engines use the fuel as a mechanism for heat exchange, e.g., for cooling certain engine components or hydraulic fluid. High temperatures can accelerate oxidation reactions in the fuel, leading to gum and particulate formation. Antioxidants are used as additives to improve thermal stability.

Storage stability: refers to a fuel's capacity for retaining its essential properties while in storage. This is essential for reserve stocks that may be stored for several years, but a more typical storage time for aviation fuel is six months. Fuel properties can be altered by oxidation processes over time, which are a function of air exposure, elevated temperature, contaminants in the fuel or absorbed from the container walls, and the presence of peroxides and antioxidants (Knothe, 2007).

Electrical conductivity measured in units of $M^{-1}L^{-2}T^3A^2$, picoSiemens per meter [pS/m]: The rate at which electrical energy dissipates is proportional to the fuel's capacity to conduct electricity. A fuel made up of pure hydrocarbons would not be electrically conductive, but trace elements, specifically, the presence of polar molecules, could bestow the capability of holding a charge. During pumping operations, the fuel can be in contact with a variety of materials in hoses, pipes, fittings, and valves. Contact between dissimilar materials can cause the formation of an electrical charge in the fuel. Given time, the static electricity would dissipate. However, if the time constant for the dissipation is long, vaporized fuel could be at risk of burning should it come into contact with a stray ignition source. Naturally derived petroleum products have more contaminants than synthetic fuels and will tend to have a higher conductivity. Jet A-1 and JP-8 require the incorporation of a static dissipater additive, which increases electrical conductivity, thereby decreasing the time required for dissipation and reducing the risk of unplanned combustion.

Vapor pressure measured in units of $ML^{-1}T^{-2}$, kilopascals [kPa]: For a given substance at thermodynamic equilibrium in a closed system, the liquid and gaseous phases (and solid phase, if any is present) have no driving force to change phase, so they will remain in the same proportions unless the system is perturbed. The proportionality is a function of temperature and pressure. At a fixed temperature, liquids will remain stable at all pressures above the vapor pressure, but they will boil when the pressure drops below the vapor pressure. Kerosene-type jet fuel, such as Jet A, has a vapor pressure of about 1 kPa at 38°C. Wide-cut fuel, such as Jet B, is more volatile and it has a higher vapor pressure. The specifications in CGSB-3-22 require that the maximum permissible vapor pressure is 21 kPa at 38°C. Jet A carries no vapor pressure requirement.

Heat capacity, measured in $L^2T^{-2}K^{-1}$, Joules per gram per degree C [J/(g°C)]: Heat capacity is not specified in the requirements, but it is a property that can affect engine efficiency. More efficiency can be squeezed out of an engine if the fuel itself is used to cool engine components, while it is simultaneously heated up en route to the combustor. If temperature is too high (over 480°C or so), thermal/catalytic cracking can occur. In WFSP, the average value of the heat capacity at 0°C was 1.582 J/(g°C) and showed a linear dependence on temperature.

Cetane number, a dimensionless term: The cetane number is related to the ignition delay time between the injection of the fuel and the onset of ignition. It is particularly important for piston-driven engines with periodic combustion spurts, for which precise control over combustion timing is critical. Higher cetane number correlates to shorter ignition delay times, so especially for automotive applications, a higher cetane number will result in cleaner, more complete combustion (Refaat, 2009).

For additional reading on jet fuel standards and testing methods, see Chevron (2006), Knothe (2008b); on jet fuel properties and composition, see Hadaller and Johnson (2006), Moses (2008,

2009), and Moses and Roets (2009); and on thermal, oxidative and storage stability, see Jain and Sharma (2010), and Knothe (2007).

11.2.2   *Alternative jet fuel specifications*

For the near term, alternative fuels must be "drop-in" fuels, which can be used directly in current aircraft turbine engines, usually by blending jet-grade petroleum and bio-jet fuel. Note that there are standards for aviation gasoline (Avgas) for small piston-driven aircraft. Since that represents a small percentage of aviation fuel, we will focus on the heavier aviation fuel designed for use in turbine-powered aircraft. A variety of alternative jet fuels have been thoroughly tested on the ground and have been used in demonstration flights by commercial airlines as well as the US military. See Hendricks *et al.* (2011), and Kinder and Rahmes (2009) for some of this background and Edwards (2007) for a description of the historical evolution of aviation fuel standards. The challenge lies in making these fuels commercially viable, environmentally friendly, and broadly available. The basic fuel types for alternative fuels for aviation are:

- *SPK*: Synthetic Paraffinic Kerosene derived from natural gas, such as Syntroleum's S-8, Shell's GTL, and Sasol's GTL-1 and GTL-2. (GTL stands for Gas To Liquid, a reference to its Fischer-Tropsch (F-T) manufacturing process.) SPK fuels are functionally similar to Jet A, but they have negligible aromatics content.
- *IPK*: Synthetic paraffinic kerosene derived from coal, with properties that are comparable to SPK.
- *FSJF*: Fully Synthetic Jet Fuel, a synthetic paraffinic kerosene developed by Sasol, which consists of coal-derived fuel with synthetic aromatics. This is the only alternative fuel at this time which does not require blending with conventional fuels (Moses and Roets, 2009).
- *HRJ*: Hydrotreated Renewable Jet fuel, which is derived from renewable fuel sources. Vegetable oils are one such source, but they require hydrotreatment to condition the oil to jet-fuel quality. This fuel has also been called *bio-SPK* to underscore its functional similarities to SPK. In 2011, it was re-named *HEFA* for Hydrogenated Esters and Fatty Acids. This fuel was certified by ASTM in July 2011 for use as jet fuel in a 50/50 blend with conventional Jet A fuel under alternative fuel specification D7566.

An analysis by Hilemann *et al.* (2010) concludes that these fuels would reduce the net fuel energy consumed by the aviation industry, improving overall fleet-wide energy efficiency. With the exception of HEFA/HRJ, these fuels are derived from coal and natural gas. They use a variant of the F-T process, which is a liquefaction technique. In 2009, F-T processing was the first manufacturing method to be approved by the ASTM under its alternative jet fuel specification ASTM D 7566. The SPKs were the first approved synthetic blending component under this standard in 2009, and revisions were made in 2011 to also include HEFA.

"Neat" fuels are not mixed or diluted with other fuels. (For pure biofuels, the designation "B100" may also be used, in which the "100" indicates that 100% of the fuel is bio-derived.) One study found that an array of SPK and IPK F-T fuels met all jet fuel specifications except for density (Moses, 2008). To mitigate this property, a 50/50 blend of F-T and conventional fuel was certified for commercial aviation. It has been widely tested on the ground and in civilian and military aircraft.

The composition of the F-T synthetic fuels can vary in hydrocarbon content, but they tend to peak at a slightly lower carbon number than petroleum-based fuels. They also have little aromatics content. The primary advantages and disadvantages of such synthetic fuels relative to conventional fuels are:

- *Advantages:* cleaner burn; reduced carbon monoxide (CO), sulfuric gases (SO_x) and particulate emissions; better thermal stability; potentially carbon-neutral; and
- *Disadvantages:* lower energy density, poor lubricity, higher freeze point, higher viscosity, carbon capture and sequestration is required to be considered sustainable.

Commercial production of F-T fuels is well established in South Africa, Qatar, and Malaysia. By 2012, global production is scheduled to be more than 30,000 barrels per day (Bartis and Van Bibber, 2011). Due to the high temperatures and pressures required for the F-T process, the refining process is energy-intensive and releases higher quantities of carbon dioxide (CO_2) during manufacture than petroleum refining. Consequently, although this fuel is the most commercially viable alternative fuel, it requires capture and sequestration of the carbon emissions in order to be considered a sustainable fuel (Bartis and Van Bibber, 2011; Kinsel, 2010). With sequestration, the environmental impact of F-T fuels is comparable to that of petroleum (Bartis and Van Bibber, 2011). However, the capture/sequestration steps can add substantially to the investment cost for such fuels. Nevertheless, F-T fuels are the most commercially ready solution for current energy needs. For the longer term, the CO_2 emissions of petroleum-based F-T fuels render it less attractive than biofuels, which have the potential to be CO_2-negative. In this case, substituting biofuels for petroleum-based fuel would reduce the level of greenhouse gases in the atmosphere.

While the molecules in petroleum-derived fuel are primarily composed of pure hydrocarbons, vegetable oil and biodiesel are comprised of hydrogen, carbon and oxygen. The long hydrocarbon chains that are typical of paraffins are present in biodiesel as a component of a fatty acid, but they are coupled at one end of the chain to other functional groups. The molecular composition is $R\text{-}(CH_2)_n\text{-}COOH$, where R indicates an alkyl group, which looks like a hydrocarbon chain with a missing hydrogen at one end of the chain. This formula describes an ester, which is obtained through the condensation of an alcohol and a fatty acid. Most often, the alcohol used is methanol, because it is inexpensive and provides good fuel properties. Ethyl esters, based on ethanol, are common in areas like Brazil in which ethanol is abundant. Since the ethanol in Brazil is largely derived from sugar cane, this type of biodiesel is completely based on renewable sources. In contrast, methanol is usually derived from natural gas. Butanol or propanol can also be used to form butyl and propyl esters, respectively, but this is much more expensive and less common. When the ester has only a single fatty acid, the biodiesel blend is comprised of mono-alkyl fatty esters, which are termed Fatty Acid Methyl Esters (FAME) somewhat generically, or more inclusively, Fatty Acid Alkyl Esters (FAAE). At one time, they were under consideration as a jet fuel-blending component, but their poorer cold-weather properties and specific energy are inadequate for jet fuel. The new standards for bio-derived jet are only expected to include HEFA fuel. In this work, we will refer to vegetable oils and greases as "green crude", to the methyl ester blends created from green crude as "FAAE", to the blends of FAAE and conventional diesel which conform to the standards of EN 14214 and ASTM 6751 as "biodiesel", and biofuel consisting of mixtures of pure hydrocarbons derived from renewable sources as "HEFA". Only the latter is under consideration as a blending component for jet fuel.

Some requirements on biodiesel properties in its neat state are identified in Table 11.2. Note that these requirements are not targeted toward HEFA, but they are provided here as a reference.

Although the European EN14214 specifications place a range limit on biodiesel density for FAME and both standards identify a permissible range on viscosity, most of the requirements have to do with fuel composition. The presence of excess mono-, di- and triglycerides, methanol, and glycerol would indicate an incomplete reaction in the refining process or flaws in the separation process or both. Relative to the specifications for jet fuel in Table 11.1, one important difference is that the standards for FAME/FAAE in Table 11.2 specify a range on kinematic viscosity at 40°C, while jet fuel specifications are referenced to a much lower temperature, −20°C. This will be discussed in more detail in the next section. Relative to petroleum fuel, biodiesel derived from FAAE have the following:

- *Advantages*: cleaner burn; high lubricity; reduced carbon monoxide (CO), sulfuric gases (SO_x) and particulate emissions; better thermal stability; carbon-negative (provided the manufacturing process is sustainable); biodegradable, and
- *Disadvantages*: higher freeze point; lower energy density; higher NO_x emissions; poorer oxidative and storage stability; high cost and low availability of feedstock and refining plants.

Table 11.2.   Key fuel specifications for biofuels based on methyl esters for automotive applications.

Property	FAME	FAAE
Density at 15°C [kg/m^3]	860–900	–
Viscosity at 40°C [mm^2/s]	3.5–5.0	1.9–6.0
Methanol [%m/m]	<0.20	<0.20 or flash temperature <130°C
Sulfur [mg/kg]	<10	<15
Total esters [%m/m]	>96.5	–
Monoglyceride [%m/m]	<0.8	–
Diglyceride [%m/m]	<0.2	–
Triglyceride [%m/m]	<0.2	–
Free glycerol [%m/m]	<0.02	<0.02
Total glycerol [%m/m]	0.25	<0.24
Total acidity [mg KOH/g]	<0.5	<0.5
Specification	EN 14214	ASTM D 6751

Typically, green crude has a higher proportion of heavier carbon components than jet fuel, resulting in a fluid that is more viscous than diesel (Fig. 11.4b). Although vegetable oil can be blended with conventional fuel for use in some piston-driven diesel engines without treatment, it can form deposits, plugging up filters and injectors. Plant-derived fuels have very little sulfur or aromatic content, so, in general, its emissions have less environmental impact.

A recent review of material compatibility (Fazal *et al.*, 2011) finds that, relative to petroleum-based diesel, some biodiesel results in greater potential for plugged filters, sticking parts, and corrosion of some metals. The corrosion process can also degrade fuel properties such as density and viscosity. This study was primarily concerned with biodiesel for automotive applications, but its findings could also apply to aviation. In any event, the disadvantages of biodiesel can largely be mitigated by additives and blending with conventional fuels, with the exception of its inherently lower energy density.

Biorefining processes such as transesterification are used to break down the vegetable oils to alkyl esters to form FAAE. Vegetable oils, free fatty acids and FAAE can be used as precursors for HEFA. In order to meet freeze point requirements and boost the specific energy, they must undergo hydroprocessing to remove the oxygen, become saturated with hydrogen, and reduce chain length (section 11.5.3). The composition of HEFA fuel will be essentially the same as that of petroleum-derived fuels, but will have much less cycloparaffin content and only traces of olefins, aromatics and sulfur. As a result, it will have the following:

- *Advantages*: better stability and blending properties due to absence of double bonds, oxygenated molecules, and heteroatoms (here, primarily nitrogen and sulfur); cleaner burn, lower sulfur and particulate emissions relative to petroleum jet; lower freeze point and higher specific energy than FAAE/diesel blends, and
- *Disadvantages*: lower lubricity than biodiesel, so it will likely require additives or blending. Since natural antioxidants may be lost during hydroprocessing, this fuel may also need additives for stability.

For further background on biodiesel, see Blakey *et al.* (2011), Knothe (2010a,b); and on SPK composition, properties, and standards, see Moser (2009).

11.3   FUEL PROPERTIES

11.3.1   *Effect of composition on fuel properties*

The hydrocarbon composition of any particular tank of jet fuel is dependent on the properties of the crude oil, refining operations and the use of additives. Petroleum-based jet-grade fuels are mixtures of various hydrocarbons, including alkanes (paraffins) and cycloalkanes (naphthenes), aromatics, and alkenes (olefins), as well as much smaller amounts of benzene and polycyclic aromatic hydrocarbons, and trace amounts of sulfur and other materials. Note that, for the most part, the standards described in Table 11.1 did not specify hydrocarbon composition. Instead, the fuel properties are primarily specified in terms of performance. As a practical matter, the jet fuel performance specifications (section 11.2.1) limits the carbon content of jet fuels to the carbon numbers from C8 to C16 (Edwards, 2010), but the molecules can appear as straight, branched or linked chains. Most of the fuel components are saturated, meaning that only single carbon-to-carbon bonds appear in the mixture.

When crude oil is refined, it undergoes a distillation process to separate out the hydrocarbon components by weight. As heat is applied, the lightest hydrocarbons will start to vaporize first and will move up a distillation column, followed later by heavier hydrocarbons. Jet B is a wide-cut fuel, which has a more diverse hydrocarbon blend with chain lengths from about C4 to C16. The presence of the light carbon components gives Jet B its cold temperature properties. Jet A-1 is required to have a lower freeze point than Jet A (Fig 11.3c), which makes it desirable for operation in colder climates. However, it can be less expensive to refine oil to Jet-A standards, since a broader temperature range can be used in the distilling process, permitting recovery of a wider range of hydrocarbons.

The thermophysical properties of a pure hydrocarbon fluid are dependent on its carbon number and its structural complexity. The fluid can be saturated (holding the maximum possible number of hydrogen atoms with only single bonds between all atoms), or unsaturated (replacing some of the single carbon-to-carbon bonds with double bonds). The most abundant forms are that of simple, straight-chain, saturated n-paraffins; branched, saturated isoparaffins; cyclic, saturated paraffins (cycloalkanes/naphthenes); unsaturated olefins with at least one double bond; and unsaturated cyclic aromatics (see Appendix A for a description of the underlying organic chemistry). Figure 11.6a shows that density increases with carbon number for saturated compounds. The densities of n-paraffins and isoparaffins at 20°C are indistinguishable, but further structural complexity and double bonds increase the density at a given carbon number. Discussion of density correlations as a function of composition can be found in Alptekin and Canakci (2008), Alptekin and Canakci (2009), Refaat (2009), and Saravanan and Nagarajan (2011).

Kinematic viscosity increases with chain length (Fig. 11.6b), degree of saturation, and branching (Knothe, 2005; Refaat, 2009). The boiling point increases with increasing carbon number (Fig. 11.6c), and is dependent to a lesser extent on structural complexity. The specific energy decreases with increasing carbon number and the presence of naphthenes and aromatics (Fig. 11.6d), while the opposite holds true for the energy density (Fig. 11.6e). Note that the n-paraffins and isoparaffins share nearly the same dependence, while naphthenes are more strongly influenced by carbon number from C8 to C10. Since the energy density increases as the carbon number increases, in a situation with a fixed volume, such as a completely filled fuel tank, more energy can be released by a fuel with higher carbon number. High carbon numbers are advantageous in this regard, although that must be balanced against the deleterious effects of increased viscosity (Fig. 11.6b) and increased freeze temperature (Fig. 11.6f).

Lighter, low-carbon fuels have a lower freeze temperature than high-density high-carbon fuels (Fig. 11.6f). This provides some insight as to the desirability of using a wide-cut blend for better cold-weather performance, since a wide blend tends to have more hydrocarbons with low carbon numbers. Also note that the only graph in Figure 11.6 in which n-paraffins and isoparaffins separate out appreciably is in this property. The freeze temperature decreases with increased structural complexity, such as increased branching and reduced level of saturation. Relative to straight-chain, ladder-like alkanes, such molecules exhibit reduced packing efficiency, so that lower temperatures are needed to reach an entropy level at which crystallization can occur (Refaat, 2009). The freeze temperature of naphthenes from C8 to C10 increases more strongly than the rate of increase for n-paraffins and isoparaffins. Considering the specific energy and freeze temperature, increasing isoparaffin content will produce better cold weather properties, without a substantial hit on the available energy content. If the concentration of isoparaffins can

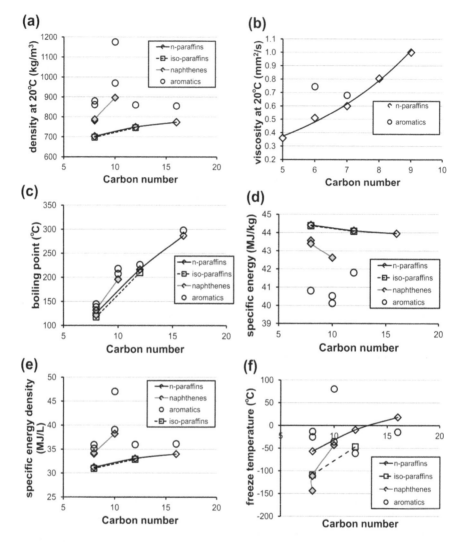

Figure 11.6.   Compositional dependence of thermophysical properties of pure hydrocarbons. (a) Density at 20°C; (b) kinematic viscosity at 20°C; (c) boiling point; (d) specific energy; (e) specific energy density; and (f) freeze temperature. Data derived from ASTM (2011), Touloukian *et al.* (1975) and TRC (2011).

be increased, even at the same carbon number, the fuel will have lower susceptibility to cold temperatures, as well as a lower viscosity (Moses, 2008). The presence of naphthenes may deliver better specific energy density with increasing carbon number from C8 to C10, but it comes at the penalty of reduced tolerance to cold temperatures.

Other properties that are affected by carbon number include vapor pressure and lubricity. For alkanes, vapor pressure increases with decreasing carbon number. Low-density hydrocarbons, such as methane and propane, appear as vapors at standard pressure and temperature at ground level (by convention, 101.325 kPa, 20°C), while higher density hydrocarbons appear as liquids. In general, increasing carbon number correlates to better lubricity, although that may also be linked to the presence of certain polar molecules in the fuel (Refaat, 2009). Fuels with a bias toward lower carbon number will burn more cleanly. Consequently, there is less deposition of unburned hydrocarbons in the fuel system, which can improve engine combustion quality and simplify maintenance.

For precise control of an intermittent combustion process, such as that of a piston-driven engine, it is advantageous to have a narrower distribution of fuel rather than a wider one so that all of the fuel components ignite almost simultaneously. A short, consistent ignition delay time allows for more efficient operation of such engines, as well as cleaner burns with fewer emissions. Even for turbine engines (which is based on a continuous combustion process), if the distribution of components is too narrow, there may be difficulties in high-altitude relight (Blakey *et al.*, 2011).

In Figure 11.7, the composition of many fuel samples are shown as percentages of the fuel composition by carbon number, and, where available, structural class. Despite the obvious differences in the fuel blends, all of these fuels meet jet fuel specifications with one exception. Note that most of the graphs are presented in terms of mass fraction. However, the measurement for Jet A in Figure 11.7a is given as a molar percentage, although the units are still dimensionless. Fuel composition on a molar basis is useful for computational chemistry, because the equations for chemical kinetics are naturally expressed in those units. The mole fraction is related to the mass fraction through the atomic mass, which is expressed in grams per mole [g/mole]. Mass-based carbon composition in a sample of JP-4 (the military equivalent of Jet B) are shown in Figures 11.7b,c. The shale-derived Jet B in Figure 11.7c is much richer in the C7–C9 range, relative to the other petroleum-derived fuels, and is strongly influenced by an abundance of isoparaffins in the fuel mixture. Based on the discussion of hydrocarbon composition effects on fuel characteristics above, one should expect that these fuels would have good cold weather properties, as is required for JP-4.

The Synthetic Paraffinic Kerosene (SPK) fuels in Figures 11.7d–i are manufactured from natural gas (Fig. 11.7d–g) and coal (Fig. 11.7h,i), in a process which compresses the fuel from the gaseous state into liquid. In order to meet the minimum density requirements and aromatics content for jet fuel, they must be blended in a 50/50 mixture with conventional petroleum jet fuel (Moses, 2008). The S-8 fuel (Fig. 11.7d) is a wide cut fuel and was used in a 50/50 blend with JP-8 by the US Air Force. The composition of the GTL-1 (Fig. 11.7e) was tuned for automotive diesel, not jet fuel, and it did not meet the freeze temperature requirements. (Note that this fuel is dominated by *n*-paraffins.) The GTL-1 fuel was further processed to create GTL-2 (Fig. 11.7f), which increased the isoparaffin content, although it also shifted and broadened the hydrocarbon distribution. This manipulation permitted GTL-2 to meet freeze temperature requirements. The Shell GTL fuel (Fig. 11.7g) is a natural gas-derived synthetic fuel that exhibits significantly different composition from the prior fuels but still meets the requirements for jet fuel. The ITK fuel (Fig. 11.7h) is a coal-derived product, which has been used at the OR Tambo International Airport in Johannesburg since 1999. An example of the first fully synthetic F-T fuel is shown in Figure 11.7i. It is manufactured from coal, and enhanced with synthetic aromatics in order to meet jet fuel standards. This is the only synthetic fuel that does not need to be blended with conventional fuels at this time.

One example of a Hydrogenated Esters and Fatty Acids (HEFA, also called bio-SPK or HRJ) is shown in Figure 11.7j, derived from oil extracted from jatropha and algae by Boeing. This fuel has a relatively high percentage of isoparaffins, which would be expected to improve its cold-weather properties while retaining the advantageous energy density of higher carbon content. Boeing has also characterized other bio-derived HEFA, for which the green crude was comprised of oils from camelina, jatropha and algae in varying proportions. All of these blends yielded similar hydrocarbon composition in terms of carbon number distribution and the ratio of *n*-paraffins to isoparaffins (Kinder and Rahmes, 2009).

While developing new aircraft designs and vetting new alternative fuels, we would like to improve the combustion efficiency through smart design of combustors, inlets and nozzles and to minimize pollutant output. Computational fluid dynamics has been used for decades to model

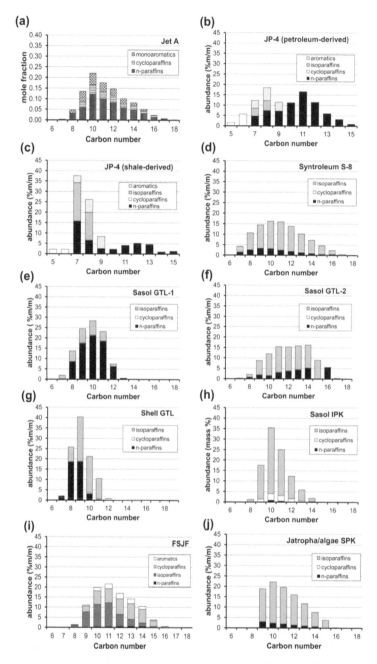

Figure 11.7.   Petroleum and synthetic fuel composition by carbon number for (a) Jet-A fuel (note that this figure is in terms of mole fraction, unlike the other histograms); (b) and (c) from Jet-A derived from traditional petroleum reservoirs and from shale-derived sources; (d)–(g) are fuels derived from compressed natural gas; (h) is derived from coal; (i) is a fully synthetic coal-derived fuel that incorporates synthetic aromatics; and (j) is an example of Hydrogenated Esters and Fatty Acids (HEFA) jet fuel made from a blend of bio-SPK and petroleum jet fuel. Adapted from (a) LeClercq and Aigner (2009); (b)–(c) ATDSR (1995); (d)–(h) Moses (2008); (i) van der Westhuizen *et al.* (2011); and (j) Kinder and Rahmes (2009).

both fluid flow and the chemical kinetics that govern fuel combustion in aircraft components. To properly model the chemical kinetics requires an understanding of the composition of the blended fuel components, as well as appropriate chemical kinetic models that describe the cascade of chemical reactions governing the combustion process. Once the constituents of the fuel are determined, blending rules are applied to calculate the thermophysical properties of the hydrocarbon mixture, such as density, viscosity, specific energy, and others as needed. During a simulation, these properties can evolve over time as a result of chemical reaction, which changes the composition of the fuel, or through changes to the local temperature and pressure. Considering the variability shown in Figures 11.3, 11.5 and 11.7, it would be a disheartening task to attempt to comprehensively model the entire range of possible fuel mixtures. This leads to the desire for identifying representative fuel blends, i.e., jet fuel surrogates (Anand *et al.*, 2011; Colket *et al.*, 2007; Dooley *et al.*, 2010; Herbinet *et al.*, 2010; Huber *et al.*, 2010; LeClercq and Aigner, 2009; Mensch *et al.*, 2010; Pitz and Mueller, 2011; Singh *et al.*, 2011; Westbrook *et al.*, 2009). Surrogate fuels provide a standardized reference composition for comparison of different designs and operating conditions. Both experimental and numerical experiments have employed surrogate fuels to simplify the chemistry while maintaining the key elements of the combustion process to keep all of the essential physics in place. C16 (*n*-hexadecane) is a primary reference fuel for experimental work on diesel engines. The Lawrence Livermore National Laboratory (LLNL) developed a suite of chemical kinetic models for alkanes from *n*-octane to *n*-hexadecane for low and high temperatures (Westbrook *et al.*, 2009). Their work showed that all *n*-paraffins in the range of C8–C16 exhibit nearly the same ignition behavior. If the right behavior can be appropriately captured in a numerical study by substituting a lower carbon number fuel for *n*-hexadecane, this vastly reduces the size of chemical kinetic model. As a reference point, LLNL's model for the combustion of *n*-hexadecane has 8000 reactions and 2100 species, in contrast to *n*-decane (C10), which has 3900 reactions and 950 species (Westbrook *et al.*, 2009).

The identification of the right blend of fuel components for surrogate fuels is currently an active area of research. Databases for petroleum fuel components have been laboriously built up over decades. In contrast, the characterization of the more complex chemistry of biofuels is relatively recent (Kohse-Hoinghaus *et al.*, 2010). Methyl decoanate is sometimes used as a surrogate for biodiesel fuels. It includes reduced chemistry for the oxidation of *n*-heptane that is accurate for both low and high temperatures (Herbinet *et al.*, 2008) More recent work has incorporated two large, unsaturated esters to show the influence of the double bond (Herbinet *et al.*, 2010). Another study used two of the five major components of biodiesel to examine toxic emissions (Kohse-Hoinghaus *et al.*, 2010). They studied methyl stearate and methyl esters using LLNL's pre-existing classes and rules for the chemical reactions, but added additional mechanisms (3500 chemical species and >17,000 reactions) to represent some of the unique characteristics of methyl esters. For example, the monounsaturated compound, methyl oleate (also called oleic acid methyl ester, with a lipid number of C18:1 and a chemical formula of $C_{19}H_{34}O_2$) is slightly less reactive than methyl stearate (stearic acid, C18:0, $C_{19}H_{36}O_2$). The double bond in methyl oleate inhibits some of the reaction pathways that produce chain branching at low temperatures (Naik *et al.*, 2011).

It is also useful to develop a framework to compute fuel properties as a function of composition in order to find the most efficient geometries and operating conditions. Reliable models for properties such as density and viscosity will permit computational analysis of other aircraft systems, such as calculating the power required to pump a given volume of fuel from its storage reservoirs to the combustor. Figure 11.8 shows the effect of composition and temperature on properties for certain fatty acid esters that are typical of biodiesel. At this point, we move into lipid notation for the fuel's fatty acid subunit, C*x*:*y*, in which *x* represents the carbon number and *y* represents the number of unsaturated carbon-to-carbon bonds. Figure 11.8(a) presents experimental data representing the temperature dependence of density for saturated methyl esters (C*x*:0) in the range of C16 to C22, as well as unsaturated methyl esters with one or three double bonds (C*x*:1, C*x*:3) (Pratas *et al.*, 2011a). As with the pure hydrocarbons, density decreases linearly as the temperature increases. The density also increases with increasing chain length and increasing levels of unsaturation (Refaat, 2009). The data in Figure 11.8(b) show the temperature

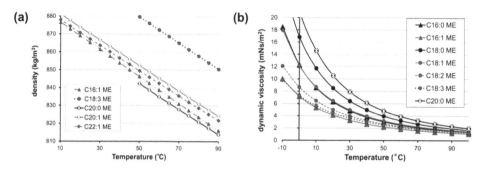

Figure 11.8.   Temperature dependence of saturated and unsaturated compounds for (a) density and (b) viscosity. Data from Freitas *et al.* (2011) and Pratas *et al.* (2011a).

dependence of viscosity, which were computed from exponential functions based on the revised Yuan model developed by Freitas *et al.* (2011). As the carbon number for the methyl esters increases, the viscosity also increases. As the level of saturation increases, however, the viscosity decreases for the methyl esters shown here. The curve for monounsaturated methyl oleate (C18:1 ME) is nearly indistinguishable from that of saturated methyl palmitate (C16:0 ME). This behavior can also be identified in the analysis of methyl ester blends (Saravanan and Nagarajan, 2011).

However, the situation is a little more complicated than this. The methyl esters shown in Figure 11.8 have *cis* double bonds, meaning that the physical structure is such that the same functional groups are located on the same side of the double carbon-to-carbon bond. Some of these compounds have isomers with *trans* double bonds in which like functional groups are located on opposite sides of the double bond. For the unsaturated compounds in Figure 11.5b, each carbon molecule that is joined in a double bond to another carbon is also linked *via* single bonds to one hydrogen and a third carbon atom. It is the placement of these latter C and H atoms relative to the double bond that provides the distinction between *cis* and *trans*. Although *trans*-isomers are much less common in biodiesel, they can be introduced during the refining process through catalytic partial hydrogenation (Moser, 2009). Once the hydrocarbon chain length increases beyond a handful, there are a number of possible structural configurations for unsaturated compounds based on where the double carbon-to-carbon bond(s) are located and their orientation. These variations can introduce changes to the ester's properties. Furthermore, the double bonds add rigidity to the molecule that can introduce kinks into the molecular structure, which can affect intermolecular interactions. For clarity, various forms of identification for the essential methyl esters characteristic of biofuels are shown in Table 11.3. The common name for the FAAE is listed in the first column, followed by the lipid number associated with the fatty acid subunit. The chemical formula shows that the number of H atoms drops by 2 for each C-to-C double bond. All of these esters have 2 O atoms except for methyl ricinoleate with 3 O atoms. This compound is the primary fatty acid present in an important vegetable crude source, the castor bean, but it is rarely found in other vegetable oils.

Even when including the molecular weight, the first three identifiers in Table 11.3 are not specific enough to point to a single molecular structure due to the wide variety of possible structural configurations. The CAS number denotes a system of identification that can avoid ambiguity when referring to these compounds. There are a number of other naming systems and conventions that have been developed for these molecules in different disciplines, and two of the more common alternate designations are found in the last column. The CAS identifier is associated with these and other synonyms for these esters.

Biodiesel from canola (rapeseed) and soy demonstrated differences in combustion properties that were related to the relative amounts of five methyl ester components (Westbrook *et al.*, 2011). Specifically, the properties of the long-chain alkyl group dictated the ignition delay time, and

Table 11.3.   Fatty acid methyl esters that are important in vegetable oil-derived biofuels (ME denotes methyl ester).

FAAE	Lipid number	Chemical formula	MW (g/ mol)	cis/trans configuration	CAS number	Alternate names
methyl laurate	C12:0	$C_{13}H_{26}O_2$	214.34		111-82-0	Lauric acid ME, methyl dodecanoate
methyl myristate	C14:0	$C_{15}H_{28}O_2$	242.4		124-10-7	myristic acid ME, methyl tetradecanoate
methyl palmitate	C16:0	$C_{17}H_{34}O_2$	270.45		112-39-0	palmitic acid ME, methyl hexadecanoate
methyl palmitoleate	C16:1	$C_{17}H_{32}O_2$	268.43	cis	1120-25-8	palmitoleic acid ME, methyl cis-9-hexadecenoate
methyl stearate	C18:0	$C_{19}H_{38}O_2$	298.5		112-61-8	stearic acid ME, methyl octadecanoate
methyl oleate	C18:1	$C_{19}H_{36}O_2$	296.49	cis	112-62-9	oleic acid ME, methyl cis-9-octadecenoate
methyl elaidate	C18:1	$C_{19}H_{36}O_2$	296.49	trans	1937-62-8	elaidaic acid ME, methyl trans-9-octadecenoate
methyl ricinoleate	C18:1	$C_{19}H_{36}O_3$	312.49	cis	141-24-2	ricinoleic acid ME, methyl 12-hydroxy-9-octadecenoate
methyl linoleate	C18:2	$C_{19}H_{34}O_2$	294.47	cis	112-63-0	linoleic acid ME, methyl cis-9-cis-12-octadeca-dienoate
methyl linoelaidate	C18:2	$C_{19}H_{34}O_2$	294.47	trans	2566-97-4	linolelaidic acid ME, methyl 9-trans-12-trans-octadecadie-noate
methyl linolenate	C18:3	$C_{19}H_{32}O_2$	292.46	cis	301-00-8	linolenic acid ME, methyl all-cis-9,12,15-octadecatrienoate
methyl arachidate	C20:0	$C_{21}H_{42}O_2$	326.56		1120-28-1	eicosanoic acid ME, methyl eicosanoate
methyl gadoleate	C20:1	$C_{21}H_{40}O_2$	324.54	cis	2390-09-2	cis-11-eicosanoic acid ME, methyl cis-11-eicosanoate
methyl behenate	C22:0	$C_{23}H_{46}O_2$	354.61	cis	929-77-1	behenic acid ME, methyl docosanoate
methyl erucate	C22:1	$C_{23}H_{44}O_2$	352.09	cis	1120-34-9	eruric acid ME, methyl cis-13-docosenoate

not the carbon number. An examination of thermophysical properties as a function of carbon number and level of saturation (Fig. 11.9) can illuminate a number of other trends. In Figure 11.9a, the kinematic viscosity at 40°C clearly shows that viscosity increases with carbon number. It also decreases with increasing unsaturation; monounsaturated methyl palmitoleate (C16:1) is less viscous than the saturated methyl palmitate (C16:0). For the *cis*-type unsaturated compounds at C18, the viscosity reduces as saturation decreases. On the other hand, the viscosity of methyl elaidate (C18:1 *trans*) and methyl stearate (C18:0) are indistinguishable on this figure just below 6 mm^2/s. The viscosity of methyl linoelaidate (C18:2 *trans*, open circle at 5.33 mm^2/s) is substantially higher than methyl linoleate (C18:2 *cis*, 3.65 mm^2/s). This provides evidence that the beneficial reduction of viscosity offered by *cis*-type unsaturated compounds may not exist for *trans*-type isomers.

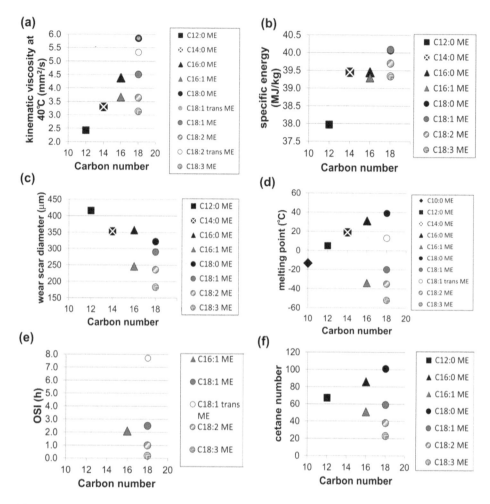

Figure 11.9.　Dependence of the thermophysical properties of methyl esters (MEs) on carbon number and saturation for (a) kinematic viscosity; (b) specific energy; (c) wear scar diameter, which is a measure of lubricity; (d) melting point; (e) Oxidative Stability Index (OSI); and (f) cetane number. Melting point for C10:0 ME from Knothe (2008a); remaining data from Moser (2009).

The trends are different for other properties. For the specific energy (Fig. 11.9b), the value for methyl oleate (C18:1) falls cleanly on top of methyl stearate (C18:0) at about 40 MJ/kg, and the value for methyl palmitoleate (C16:1) is nearly the same as that of methyl palmitate (C16:0).

Another feature in favor of methyl esters over ethyl esters for biofuel is that they exhibit greater lubricity (Holser and Harry-O'Kuru, 2006). For lubricity (Fig. 11.9f), the wear scar diameter decreases with increasing chain length (which indicates increasing lubricity). The lubricity also increases with increasing saturation (i.e., wear scar diameter decreases for the C18:y MEs as y increases). At the molecular scale, oxygen-containing compounds like methyl esters can adsorb or react on surfaces that rub together, reducing the friction between asperities (microscale or smaller hills and valleys on a surface). This can reduce wear and the tendency toward seizure (Fazal et al., 2011). This is, however, highly dependent on maintaining a water-free environment due to corrosive potential.

A review of material compatibility of diesel with respect to biodiesel (Fazal *et al.*, 2011) has identified some common themes among biodiesels: although they have better inherent lubricity than diesel (Holser and Harry-O'Kuru, 2006; Knothe and Steidley, 2005), they are hygroscopic and tend to absorb any moisture that may be present in their surroundings. This can promote the growth of microbes that produce corrosion-producing metabolic products, as well as render electrochemical corrosion more likely upon exposure to certain metals. In field-testing for automotive applications, biodiesel generally produces similar or less wear than conventional diesel. However, elemental testing suggests that certain metals should not be paired with other particular metals. Biodiesel is, in general, particularly corrosive to copper compounds as compared to ferrous compounds, but corrosion is dependent on both the parent feedstock and the type of metal (Fazal *et al.*, 2011).

Figure 11.9d shows that melting point increases with chain length and with increasing saturation for *cis*-type compounds. However, *trans*-type compounds exhibit reduced melting point relative to saturated compounds, but the melting point is substantially higher than their sister *cis*-type esters. Note that, for most pure substances, the melting point is nearly equivalent to the freeze point, i.e., the temperature at which solids and liquids co-exist. Some substances can be supercooled below this value when there are no nucleation sites, so the use of "melting point" may be considered more precise by some than "freeze point". Moreover, some substances like agar do have different solid-to-liquid and liquid-to-solid transition temperatures. For the methyl esters discussed here, the freeze point and the melting point are nearly equivalent.

In any event, the high melting/freezing points of methyl esters become problematic for creating an adequate jet fuel blend, since the jet fuel specifications call for a freeze point of −40°C or below. We have seen that the freeze point of a fuel is a function of its composition. Unlike pure hydrocarbons, some FAAEs may not exhibit a simple, linear dependence on blend ratio, making it more difficult to predict the freeze point of the final blend.

Examination of the melting point data points out another issue with respect to jet fuel suitability. Jet fuel specifications use a reference temperature of −20°C for measurement of kinematic viscosity. At this temperature, Figure 11.9d shows that the saturated methyl esters from C10:0 and up are solids. Most vegetable oils are heavily weighted toward compounds in the range of C16–C18, and consist of both saturated and unsaturated components. The *cis*-type unsaturated compounds at C16–C18 may exist naturally as liquids at −20°C in their pure state, but the blends that are typical of vegetable oil will include solid components at this temperature. Even at −12.5°C, solids were were identified in three commercial biodiesels due to freezing methyl palmitate (C16:0) and methyl stearate (C20:0) (Coutinho *et al.*, 2010). Unsurprisingly, most measurements of viscosity as a function of temperature for FAAE-based biodiesel stop short of –20°C, but they tend to exceed the 8 mm^2/s upper bound for jet fuel well above this temperature (see e.g., Freitas *et al.*, 2011).

The Oxidative Stability Index (OSI), measured in hours, is used to quantify susceptibility to fuel degradation through exposure to oxygen. Increased OSI indicates increased oxidative stability (Fig. 11.6e). The OSI increases with chain length and decreases with the presence of double bonds. Finally, the cetane number (Fig. 11.6f), which increases as ignition delay time decreases, exhibits an increase with carbon number and a decrease with level of saturation. See Moser (2009) for further discussion.

The monounsaturated compounds (Cx:1) thus present a marked advantage in terms of reduced viscosity and melting point with little or no penalty in terms of energy content. However, the polyunsaturated compounds (Cx:2 and Cx:3), while they may be beneficial in terms of reduced viscosity, may increase the density to undesirable levels, as shown in Figures 11.9a and 11.8a for methyl linolenate (C18:3). Polyunsaturated esters are also less stable with a reduced storage stability (Ramos *et al.*, 2009).

For further reading on the effects of structural composition on biofuel properties, see Knothe (2005, 2008a), Moser (2009), and Yuan *et al.* (2003). For a good description of the analytical methods used for evaluating biodiesel composition, see Chapter 5 of Knothe *et al.* (2005), also van der Westhuizen *et al.* (2011).

11.3.2   Emissions

The emissions from an airplane are a function of fuel properties, the amount of fuel used, operating conditions and the combustion efficiency. To evaluate environmental impact, the following emissions are important for their effects on climate change and public health:

- *Carbon dioxide (CO_2)* is a greenhouse gas that contributes to the warming of the planet. Airlines account for ~2.5–3% of global CO_2 emissions. Under conditions of complete combustion, the CO_2 emissions are ~3160 ± 60 g per kilogram of fuel [g/kg] (Lee *et al.*, 2010). Carbon monoxide (CO) is also a regulated gaseous emission. Other greenhouse gases are also of concern with respect to climate change.
- *Water vapor (H_2O)* is emitted at a rate of 1230 ± 20 g/kg when fuel is completely burnt (Lee *et al.*, 2010). For supersonic aircraft, water vapor is the primary concern for gaseous emissions. At high altitudes, it can form contrails and promote formation of cirrus clouds. Contrails are formed when the hot, moist exhaust of the aircraft mix with sufficiently cold ambient air. Although the detailed mechanisms are not fully established at this time, contrail formation may have an effect on climate change. Other hydrogen-containing compounds of concern are hydroxyl radicals (OH) and hydrogen peroxide (H_2O_2). The former is particularly important in chemical processes that produce sulfuric acid (H_2SO_4), NO_x, and ozone (O_3).
- *Sulfur oxide (SO_x)* generation is directly proportional to sulfur content in the fuel. The subscript x is meant to indicate that this is a class of compounds that contain sulfur and oxygen. Sulfur dioxide (SO_2) is the most abundant sulfur-containing species. It is directly proportional to the sulfur content in the jet fuel and is mostly produced under the high-temperature conditions of the combustor (Lee *et al.*, 2010). These emissions may play a role in the development of acid rain. The specifications permit an upper limit on sulfur content to 3000 ppm, although jet fuels are typically in the range of 500–1000 ppm currently (Chevron, 2006). Such compounds may also play a role through the formation of contrails and affect particulate emissions.
- *Nitrous oxide (NO_x)* is formed through reaction with atmospheric nitrogen under the high-temperature conditions in the combustor. Trace quantities of nitrogen bound to the fuel will also form NO_x. At ground level, NO_x is of concern because it is toxic and is a precursor to chemical smog. It is linked to the creation of ozone in the troposphere (ground level to approximately 12,000 m). When emitted above the troposphere, NO_x is implicated in the destruction of the stratospheric ozone layer.
- *Particulate matter (PM) and unburned hydrocarbons (HCs)* are the result of incomplete combustion. Emissions of this pollutant strongly depend on engine design and operating conditions, but fuel characteristics are also important. At altitude, PM can act as nucleation sites for contrail formation. Since the aromatic content in alternative fuels is nearly negligible, the generation of PM is also significantly reduced in terms of particle size, number density, and total PM mass (Timko *et al.*, 2011).

The effects of these pollutants on the atmosphere are compared through their impact on radiative forcing (RF) of the climate. The mean global surface temperature is directly related to RF. A positive RF corresponds to a warming effect, which is characteristic of CO_2 and soot emissions. A negative RF contributes to a cooling effect on the atmosphere, which is typical of sulfate particle emissions. NO_x emissions result in a positive RF due to formation of tropospheric ozone (O_3), while it can induce a negative RF contribution from the destruction of ambient methane (CH_4). This framework provides a way to link the pollutants, so that the net effect on the atmosphere can be computed as a summation of all of the contributions.

Near ground level, air quality is an issue. At the low power conditions that are typical of airplane takeoff and landing, fuel is less likely to undergo complete combustion. This leads to higher emissions of CO, NO_x, HC and PM, which may induce smog formation and haze. PMs are also linked to respiratory distress and increased mutagenic potential (Krahl *et al.*, 2005) that could be a factor in the development of lung cancer.

In testing for automotive emissions, the presence of FAAE reduces all types of emissions, except for NO_x. However, one study found that the release of this pollutant can be controlled by modification of injection timing (Krahl *et al.*, 2005). For aviation purposes, the presence of FAAE or F-T fuel in a jet fuel blend reduced all categories of the emissions discussed here (Timko *et al.*, 2011).

For a thorough description of the complex interplay by which gaseous and particulate emissions influence the climate, see the excellent review by Lee *et al.* (2010). For a discussion of particulate matter, see Kumar *et al.* (2010), and Timko *et al.* (2010); biofuel combustion chemistry and the generation of toxic emissions Kohse-Hoinghaus *et al.* (2010), and Krahl *et al.* (2009); systems analysis of emissions reduction strategies for Europe, see Dray *et al.* (2010).

11.4   BIOFUEL FEEDSTOCKS FOR AVIATION FUELS

There are three commonly used categories of biofuels that indicate their readiness for commercialization. Although there is no universal agreement on the specifics, we will categorize them by the sophistication of the conversion technology as follows:

- *First-generation biofuels*: Easiest to bring to market using current technology. This category uses fermentation processes to produce bioethanol. Commonly used feedstocks include corn in the US and sugar cane in Brazil. The corn-based bioethanol in particular is widely criticized for producing a larger environmental cost than petroleum fuels. The low energy density of bioethanol is inadequate for aviation fuel (Hileman *et al.*, 2010).
- *Second-generation biofuels*: Requires process improvements in refining technology to commercialize. This category uses biomass or coal as feedstocks in gasification/liquefaction processes, such as Fischer-Tropsch, to produce biofuel. Biomass sources include switchgrass, agricultural waste, wood chips and other forest residue.
- *Third-generation biofuels*: Requires cost-effective, sustainable means of producing the feedstock, as well as efficient harvesting, oil extraction, and conversion. This category includes fuels from oils derived from vegetables and microbes, as well as greases from animal fat.

Second-generation biofuel processing is discussed elsewhere in this book and in other sources. See, e.g., for general background Kinsel (2010), Kreutz *et al.* (2008), Sims *et al.* (2010), Sivakumar *et al.* (2010); for cellulosic genomics, see Rubin (2008); and for lignocellulosic co-products, see Mtui (2009). In this section, we will focus on feedstock production for first- and third-generation biofuels. The latter is an important area, since the high cost of feedstock has the biggest impact on the fuel price. For sustainability, the feedstock choice may change from one geographic region to another, depending on factors such as climate, land use, water availability, and the location of the nearest processing facilities.

11.4.1   *Crop production for oil from seeds*

Many plants have oil-rich seeds that can be converted to fuel, such as soybeans, camelina, canola (rapeseed), and sunflower. Other feedstocks with good oil content are cotton seed, babassu, palm, and coconut. Much of the research to date on biodiesel has been performed on these crops, particularly soybeans (see, e.g., Akbar *et al.*, 2009; Freitas *et al.*, 2011; Yuan *et al.*, 2005, 2009). The lipid composition of soy methyl esters is quite consistent across different studies, with a preference for C16–C18 range, and a high proportion of monounsaturated and polyunsaturated C18 compounds, as shown in Figure 11.10a. (Yuan *et al.*, 2005) presented data that showed genetically modified (GM) soy, which had a more desirable fatty acid profile with a higher concentration of monounsaturated methyl oleate (C18:1) (Fig. 11.10b). In Europe, canola has dominated the scene as a biofuel feedstock. It has a high lipid content of up to 50% and a composition that is mostly in unsaturated methyl esters at C18 and saturated ME at C16 (Fig. 11.10c). Many vegetable oils exhibit a preference for the C16–C18 range, such as cotton (Fig. 11.10d), jatropha (Fig. 11.10e), sunflower (Fig. 11.10f), palm (Fig. 11.10g), milkweed, and camelina. There are also some vegetable oils that

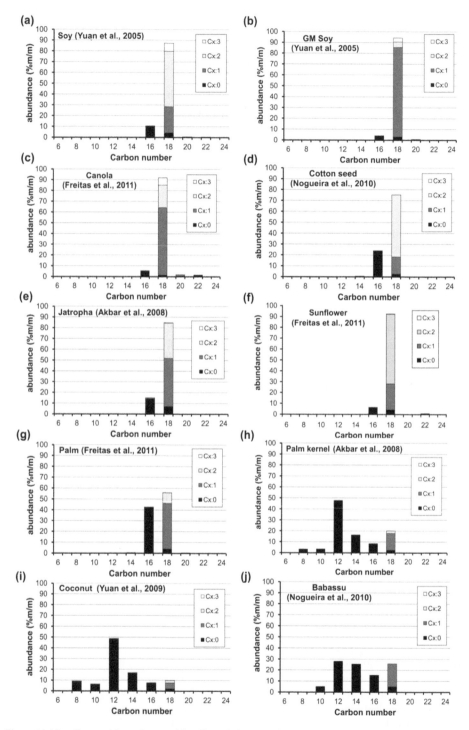

Figure 11.10.   Composition of vegetable oil methyl esters for (a) soy; (b) genetically modified soy; (c) canola; (d) cotton seed; (e) jatropha; (f) sunflower; (g) palm; (h) palm kernel; (i) coconut; and (j) babassu. Data were obtained from Akbar *et al.* (2009), Freitas *et al.* (2011), Nogueira *et al.* (2010) and Yuan *et al.* (2005, 2009).

present a broader range of carbon numbers, such as the palm kernel (Fig. 11.10h), which is the seed of the palm fruit, coconut (Fig. 11.10i), and babassu palm (Fig. 11.10j). For hydrotreatment of vegetable oil into renewable jet fuel, it is advantageous to have a carbon distribution that is heavily weighted in the C10–C14 range, which is closer to conventional jet fuel. In this case, the biorefining process might be able to avoid the energy-intensive hydrocracking stage to break down long chain-length molecules, as is needed for C16–C18. However, most of the feedstocks shown in Figure 11.10 are not desirable feedstocks because they may be used for food and/or have limited availability. In addition, cultivation techniques, such as that used to create palm plantations, have been linked to deforestation.

For sustainability, we would like to avoid competition with food crops or nutritional supplements, so much of the emphasis in feedstock development has shifted to other oil-producing crops that can be grown on marginal land which is unsuited for farming. The state of Montana in the US has instituted a program to grow camelina in intercropping, i.e., in crop rotations when the land would otherwise have been left fallow. Camelina is related to canola, has 37–45% lipid content (~40% monounsaturated, ~50% polyunsaturated) and is a good source of ω3 fatty acid and vitamin E. It is one of the feedstocks that have been successfully used to create HEFA, and it has been tested extensively in civilian and military aircraft. However, it has already been approved as an animal feed supplement in the US, and it may be approved for human consumption, which will limit its desirability as a biofuel feedstock. Two inedible beans that have generated much interest are *Jatropha curcus* (Fig. 11.7f) and castor beans.

A decade ago, *Jatropha curcas* appeared to be the perfect answer to sustainable biofuel production. Jatropha is a small tree (5–7 m tall) that produces fruit after its first year and matures at 3–5 years, with a lifetime of about 50 years. It is native to Mexico, Central America and parts of South America. Jatropha seeds are oil-rich but inedible. It was believed that the trees were pest-, disease- and drought-resistant and could thrive without irrigation on marginal land. Its large central tap root and shallow lateral roots are still widely believed to protect against wind and water erosion. Aggressive campaigns to increase land devoted to jatropha production were instituted in India, Africa and elsewhere.

Admittedly, jatropha was known to have some undesirable properties that needed to be addressed. Some jatropha varieties are quite toxic, and their seeds contain toxins, such as phorbol esters, curcin, trypsin inhibitors, lectins and phylates. To mitigate this drawback, some have tested out the concept that the leftover meal after oil extraction could be detoxified with heat treatment and used for animal feed (Xiao *et al.*, 2011) or the seed/kernel cake could be used as a fertilizer or as biomass in an anaerobic digester. Except for its role as biomass, the other uses have not yet been proven in a commercial setting.

The original optimism has been tarnished somewhat. Since it has not yet been domesticated, crop production is subject to wide variability. In a compilation of data from 1–9 year-old plants in South America, India, and Africa, the annual yields ranged widely from 313 to 12,000 kg oil/ha (Achten *et al.*, 2008). In general, trees produce seeds at a rate of 0.2–2 kg per tree (Achten *et al.*, 2008; Yang *et al.*, 2010), although there are reports of higher values. The oil content of the seeds is in the range of 27–44% oil by mass (Achten *et al.*, 2007). Under controlled conditions for 2 year-old trees developed from wild varieties throughout southern China, the maximum oil yield per tree per year was 15 times higher than that of the minimum oil yield (Yang *et al.*, 2010). (Note that it is not appropriate to extrapolate from biomass yield per tree to oil yield per hectare; the biomass yield depends on plant spacing, canopy management and other production parameters, and the conversion between biomass and oil yield can vary substantially.)

Yields have not been impressive on marginal lands, and are highly dependent on rainfall, soil type, soil fertility, genetics, plant age, spacing and management methods. While jatropha can grow on a wide range of soils, for best biomass production, it requires an infusion of nitrogen and phosphorus as fertilizer (Foidl *et al.*, 1996) and water (Achten *et al.*, 2008; Yang *et al.*, 2010). Seeds do not mature all at once, which makes harvesting a labor-intensive process. Damage due to pests or disease have been noted in continuous monocultures in India (Achten *et al.*, 2008). In non-native localities, jatropha may be an invasive species, and it has already escaped into the wild in Florida (Gordon *et al.*, 2011). In spite of these criticisms, there may be a role for jatropha as a biofuel feedstock in some geographical locations, in part because plantations have already been established.

Castor beans are related to jatropha, but are less toxic, and still have a lipid content of 40–60%. Castor-derived biodiesel blends may provide better lubricity than that of other vegetable oils, even at very low concentrations of less than 1% (Goodrum and Geller, 2004). This is likely due to the fact that castor oil contains high quantities of ricinoleic acid ($C_{18}H_{34}O_3$), an unsaturated ω9 fatty acid, and trace amounts of dihydroxystearic acid ($C_{18}H_{36}O_4$), which have more hydroxyl groups (1 and 2, respectively) than most vegetable oils (Refaat, 2009). Evogene, an Israeli company, is systematically breeding castor plants for good oil production potential. Biodiesel created from Evogene's castor has been produced by UOP, with testing by the US Air Force and NASA. Preliminary characterization of castor-based biodiesel indicates that the hydrocarbon composition has substantial content in the C9–C11 range (Bruno and Baibourine, 2011) and looks promising for meeting the standards required for aviation fuel.

For further reading on other types of vegetable oils under consideration as biofuel feedstocks, see Balat (2011b), Holser and Harry-O'Kuru (2006), Kumar *et al.* (2010), Razon (2009), and Singh and Singh (2010).

11.4.2   *Crop production for oil from algae*

Third-generation feedstocks may be obtained from algae, cyanobacteria and halophytes. Algae are photosynthetic organisms that span length scales from just a few microns (unicellular micro-algae) up to 50 m (multicellular macroalgae, such as kelp). Cyanobacteria are also photosynthetic organisms, but, unlike algae, these microbes lack a membrane-bound nucleus. Halophytes are salt-tolerant plants, such as salt marsh grass, that can thrive in saltwater. There is a huge environmental benefit to growing saltwater-tolerant halophytes and microbes (Yang *et al.*, 2011), because they can be nourished from seawater or brackish water rather than freshwater. Algae are also an attractive crop because they can sequester carbon by using the flue gas from power plants as a nutrient source and remediate wastewater. In this section, we will limit the discussion to algae. For reading on cyanobacteria, see Bouriazos *et al.* (2010), Quintana *et al.* (2011), Tan *et al.* (2011); and for halophytes, see Hendricks (2008), Hendricks *et al.* (2011), and McDowell Bomani *et al.* (2009).

Oleaginous (oil-producing) algae have been hailed as the most efficient producers of green crude over all feedstock types – potentially. The cost-effective growth of algae for conversion to biofuel in a production-scale setting has not yet been achieved, but there are many companies that are currently building such facilities. For a listing of commercial facilities for algae growth, see Singh and Gu (2010) as a starting point, but note that their data, based on a 2009 study, is already obsolete (!), and many more are in process.

From 1978 to 1996, the US Department of Energy funded the Aquatic Species Program (ASP) to quantitatively explore the concept of producing biodiesel from algae. The program analyzed over 3000 strains of microalgae and diatoms (algae with a cell wall of silica), which were narrowed down to the 300 most promising microbes. The intent was not only to understand which species were the best at oil production, but also their hardiness with respect to seasonal temperature variation, pH, and salinity, and the ability to outgrow wild competitors, all of which affect the stability of the culture. Algal growth in industrial-scale open ponds with 1000 m^2 surface area was examined for feasibility of mass production in California, Hawaii and New Mexico (Sheehan *et al.*, 1998). As is typical of open-pond aquaculture, the depth of the ponds was shallow to aid in light penetration, here 10–20 cm. The ASP determined that microalgae use far less water and land than oil-producing seed crops, estimating that 200,000 hectares could produce significantly more energy than seed crops, about one quadrillion BTUs of energy ($\sim 1 \times 10^{18}$ Joules, or roughly 1% of global energy consumption). Nevertheless, the ASP concluded that biofuel from algae would not be cost-competitive with petroleum fuel. In their 1995 evaluation, they projected the cost of algae-based biofuel to be 59–186 US$/bbl *versus* petroleum at $20/bbl. Since then, the gap has likely narrowed due to adjustments on both types of fuel.

Biological productivity has been identified as the single largest influence on fuel cost (DOE, 2010; Yang *et al.*, 2011). This includes such parameters as growth rate, metabolite production, tolerance to environmental variables, nutrient requirements, resistance to predators, and the culturing system. As with jatropha, the reported yields for algal biomass have been wildly disparate because the technology for mass production is not mature.

The two major categories of growth facilities for photosynthetic production are: (i) open ponds, typically in a racetrack configuration with flow driven by a paddlewheel; and (ii) closed, transparent photobioreactor systems that consist of an array of cylindrical tubes or an enclosure formed by two flat plates. Open ponds are the simplest design, and they are by far the least expensive to set up, operate, and clean after a growth cycle (about 2 weeks). However, they must contend with environmental conditions, such as large temperature swings, evaporative losses, and the threat of contamination by opportunistic wild species and predators. The algae selected for this type of operation must be hardy and fast-growing to outperform any competitors that are present in the environment. There are many examples of commercial success with this type of aquaculture for higher-value products, such as β-carotene and ω3 fatty acids. Photobioreactors are, in comparison, expensive to build, run, and maintain. On the other hand, they can operate at higher algal concentrations, so that the water that must be extracted at the end of the process is reduced. Depending on the design, the algae may have better access to light. They are also better equipped to handle less robust organisms, such as algae that have been genetically modified for improved oil production. Scale-up remains an issue; there is a limit as to the length of the cylindrical tubes if it is desirable to maintain a uniform temperature, pH, and dissolved gas content in the growth medium. Other design challenges are that heat and oxygen must be removed from the reactor, and a carbon source such as carbon dioxide must be replenished. The design must also have a protocol for preventing or removing biofilm buildup on the transparent surfaces. Both types of systems have strong advocates. For further discussion, see Brennan and Owende (2010), Carvalho *et al.* (2006), Cheng and Ogden (2011), and Jorquera *et al.* (2010). Within a few years, production facilities that are under construction or in development should provide more quantitative data, and there may well be a complementary role for open and closed systems.

In open-pond raceways that are supplied with nutrients from the flue gas of power plants, many strains of algae in an autotrophic growth regime can consistently increase its biomass on average by 20 grams (dry weight) per square meter of pond surface area per day ($g/m^2/d$), (Ben-Amotz, A., *personal commun.*, 2009). Overall biomass production from the Aquatic Species Program averaged $10\,g/m^2/d$, but at times achieved up to a maximum of $50\,g/m^2/d$ (Sheehan *et al.*, 1998). Theoretically, the average maximum yield could be substantially higher than that, but its attainment is elusive in a commercial-scale setting.

For algae that generate enough oil to be considered for biodiesel production, they are usually about 15–45% lipids when grown in the lab, depending on the strain and growth conditions, although some studies report lipid contents of up to 70%. Weyer and co-workers (Weyer *et al.*, 2010) examined the theoretical maxima for oil production in many geographical locations with an optimistic algal lipid content of 50% and with efficiencies for photon transmission, photon conversion, and biomass conversion of 95, 50 and 50%, respectively. With those assumptions, the annual oil yield ranged from the worst case in Kuala Lumpur of 40,700 L/ha/y (4350 gal/acre/y) to the best case in Phoenix, AZ of 53,200 L/ha/y (5700 gal/acre/y)

This work is consistent with another study that examined theoretical maxima. Cooney and co-workers (Cooney *et al.*, 2011) assumed that the maximum solar irradiance normal to the earth's surface at ground level is 1000 watts per square meter [W/m^2], multiplied by correction factors that correspond to a sunny location with occasional clouds and account for the changing angle of the sun during the day. Since algae can only use a specific spectrum of the incoming sunlight (about 45% of it), that number is also multiplied by a correction factor of 0.45. Then, they used the highest photosynthetic conversion efficiencies to represent the maximum biomass that the algae could produce from that incoming energy. By relating the specific energy of the biomass constituents (protein and carbohydrates at 16.7 MJ/kg, lipids at 37.4 MJ/kg, and ash with zero energy content), they could derive an expression for biomass production as a function of lipid

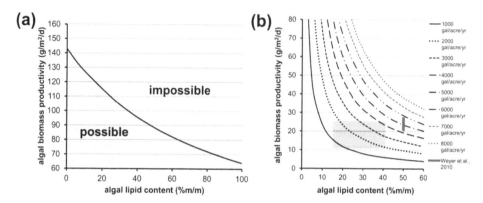

Figure 11.11.   Maximum theoretical algal production. (a) Average daily biomass productivity as a function of lipid content; (b) Curves of constant annual oil yield as a function of lipid content and average biomass productivity, where 1000 gal/acre/y = 1531.9 L/ha/y. Vertical line in (b) represents data derived from (Weyer *et al.*, 2010). The gray box represents a rough envelope of current production capabilities. The remainder of the data is adapted from Cooney *et al.* (2011), reprinted with permission from Elsevier, Inc.

and ash content. Assuming that the facility would grow algae 365 days a year (another optimistic assumption!), Figure 11.11a depicts the average maximum biomass yield per day, which ranges from 143 g/m²/d (lipids = 0%, ash = 100%) to 63 g/m²/d (lipids = 100%, ash = 0%). In these extreme cases, there are no proteins or carbohydrates produced, so the limits are not physically realistic, but in between those limits, all components (lipids, proteins carbohydrates, and ash) are represented. Interestingly enough, note that there is a tradeoff between lipid production and biomass yield. We will have more to say about this later. In an economic sense, whether or not a given number for lipid or biomass production is "good" depends on the value of the lipids and other co-products from the biomass, as well as the cost of oil extraction and refinement, and any other processing of co-products that is needed.

   Assuming that 100% of the oil can be recovered from the biomass, the oil yield is the product of the biomass yield and the lipid fraction (i.e., the lipid content on a percentage basis divided by 100). Figure 11.11b shows a family of curves representing annual oil yield that relates lipid content to biomass productivity. Weyer *et al.*'s (2010) annual oil yield of 4350–5700 gal/acre/y at a fixed 50% lipid content corresponds to an average biomass yield of about 21–27 g/m²/d in Cooney *et al.*'s (2011) figure, shown as a vertical gray line. The gray box in Figure 11.11b represents an envelope that encompasses a generous estimate of current capabilities on a commercial scale. Some reviews are able to critically evaluate the overly optimistic assumptions that are used in projections of algal oil output (Li *et al.*, 2008), based on understanding of the growth process. Others simply report the bloated figures without much thought. These studies on the absolute upper limit that is physically possible can help assess claims that seem too good to be true.

   Most algae can grow autotrophically (using photosynthesis and organic compounds to create food, also called phototrophically), heterotrophically (using complex organic compounds such as sugar for food without the need for sunlight), or mixotrophically (a combination of growth regimes). Contamination is a concern for heterotrophic growth since opportunistic bacteria also grow well on organic carbon. The type of growth regime can significantly affect the lipid (oil) composition. Under heterotrophic growth in the dark, the lipid content of *Tetraselmis* shifted dramatically toward lower carbon number and increased saturation. The abundance of palmitic acid (C16:0) increased from ~20% in mixotrophic/phototrophic growth to 79% for heterotrophic growth. The growth rate was somewhat slower, with a doubling time of 22 h for phototrophic and 25 h for heterotrophic growth (Day and Tsavalos, 1996). Figure 11.12(a) shows the effect of phototrophic *versus* heterotrophic growth regime for *Dunaliella tertiolecta* (Tang *et al.*, 2011). Less

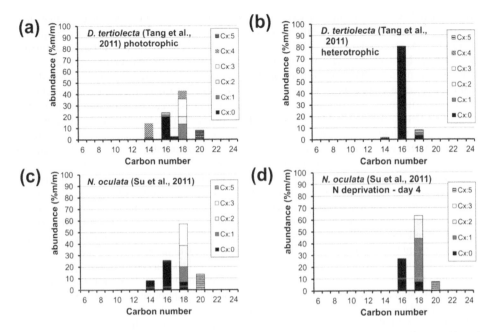

Figure 11.12.   Effect of growth regime on lipid content. *Dunaliella tertiolecta* under (a) phototrophic and (b) heterotrophic growth. Data from Tang *et al.* (2011). *Nannochloropsis oculata* under conditions of (c) nitrogen sufficiency and (d) nitrogen deficiency. Data from Su *et al.* (2011).

dramatic, but similar compositional trends were reported for *Chlorella prototfecoides* (Santos *et al.*, 2011). The ideal oil would be in the C10–C14 range, which is closer to conventional jet, and would require less energy for refining. With this in mind, the shift toward lower carbon number is beneficial for the manufacture of jet-grade fuel, although the hydrocarbon chains are still longer than desirable. Since the C16 components are fully saturated, they will be more difficult to iso-merize (section 11.5.3) to bring down the freeze temperature to acceptable levels. Consequently, this oil may still require energy-intensive hydrocracking to bring it to jet-range quality.

When under stress such as conditions of nutrient deficiency, algae do not waste energy on cell division. Cells become larger, and lipid production increases, resulting in a net accumulation of lipids. The Aquatic Species Program concluded that growing algae in a nutrient-deficient mode was counterproductive since the gain in lipid production was more than offset by the decrease in growth rate (Sheehan *et al.*, 1998). On the other hand, it could reduce production cost directly since nutrients do not have to be supplied during the deprivation phase, and indirectly by reducing the amount of biomass that must be processed. For example, one study found that after four days of nitrogen deprivation in the lab, *Dunaliella tertiolecta* cells accumulated five times more lipids than the control cells (Chen *et al.*, 2011). Similarly, *Nannochloropsis oculata* increased its lipid content from $35.0 \pm 1.2\%$ to $44.5 \pm 2.2\%$ after four days of nitrogen deprivation (Su *et al.*, 2011). The lipid composition also exhibited a shift toward saturated compounds at C18, as shown in Figures 11.11(c) and (d). Many other studies have examined the utility of mixotrophic growth, initially setting growth conditions for high biomass production, followed by several days of nutrient deficiency to increase the lipid content; see, e.g., Ben-Amotz (1995), Levine *et al.* (2010), and Xiong *et al.* (2010).

Another interesting concept is the growth of multi-species algal communities in open ponds. Recent work studied the impact of species diversity using cultures of one to four algal strains with six different species compositions that were drawn from a collection of 22 standard algal strains. The algae represented all major algal classes, including chlorophytes, cyanobacteria, cryptomon-ades, crysophytes and diatoms. In addition, eight samples of naturally occurring phytoplankton with a shared evolutionary history were collected from the wild. This work suggests that such communities may store significantly more solar energy as lipids than monocultures in photo-bioreactors (Stockenreiter *et al.*, 2011). It would also be useful to know if naturally occurring communities have other beneficial properties, such as better stability than monocultures.

Currently, open-pond design is based on deep familiarity and experience with such systems, making it in some sense a black art. Is there a way to move toward the theoretical maximum in a more scientific, cost-effective way? We know that in high-density cell cultures, the cells nearest to the surface absorb most or all of the available light (Chisti, 2007) due to self-shading. An engineering solution is to improve the vertical mixing in the system, so that more cells have access to light. One design choice in pond construction is whether to use a rectangular cross-section, or an angled cross-section with a wider top than bottom. Recent work at NASA Glenn Research Center showed that the addition of passive mixing devices is more effective in providing access to light than the cross-sectional geometry in dense suspensions. This knowledge permits the choice of channel geometry to be based on other factors, such as ease of construction or maintenance. Other, more sophisticated approaches are to examine in greater depth the stochastic effects of hydrodynamic mixing on cell growth (Chait *et al.*, 2012). This requires better understanding of the time scales that are relevant to industrial production. There is a great deal of research on the short time scales associated with photosynthesis, such as the amount of time it takes a cell to absorb a photon, convert it into food, and be ready to absorb another one, which is on the order of milliseconds. There are similarly small hydrodynamic time scales associated with turbulence, as well as larger ones linked to pond traversal. In current designs, the primary locations for vertical mixing occur at the paddlewheel and at the circular bends at either end of the pond, so that time scales associated with the unmixed state are on the order of tens of minutes. Algae that are near the surface will be continuously in the light during that stage, while the those near the bottom may be in the dark. There are also photoadaptation time scales that appear to be on the order of hours. Exposure to too much light can lead to photoinhibition. There are growth time scales with doubling time on the order of days. The complex, species-specific interplay among all of these factors with the biokinetics is an area that is ripe for exploration and may provide a scientific basis for effective pond design.

See DOE Algal Biotechnology Program Roadmap for a good description of all aspects of algae growth for biofuel (DOE, 2010); for background on basic photosynthetic processes (Nelson and Yocum, 2006); for studies on bioprospecting/screening, see Araujo *et al.* (2011), Doan *et al.* (2011), Gouveia and Oliveira (2009), Griffiths and Harrison(2009), Lee *et al.* (2011), Rodolfi *et al.* (2009); for screening for biodiesel and sewage remediation see Sydney *et al.* (2011); for algae and CO_2 remediation see Ho *et al.* (2011).

11.5   MANUFACTURING STAGES

Extraction processes from vegetable oils, algae and greases release oil that is high in triglycerides and fatty acids with hydrocarbon subunits which are primarily in the range of C16–C18. The goal of biorefining is to convert this green crude (also called "green diesel") into a less viscous fuel with better cold-temperature properties.

Biodiesel is defined as a fuel that is made up of mono-alkyl esters of long-chain fatty acids, which is derived from vegetable oils or animal fats. In petroleum fuel, straight-chain, saturated hydrocarbons are called *n*-paraffins. For biodiesel, paraffinic subunits contained in fatty acids are combined with alcohols to form esters that can be blended with conventional diesel and used in vehicles (section 11.5.2). This form of biofuel is referred to as Fatty Acid Methyl Esters (FAME), or equivalently Fatty Acid Alkyl Esters (FAAE). Transesterification reduces the viscosity of the vegetable oil feedstock to a range that is acceptable for biodiesel, as shown in Figure 11.13.

For renewable jet fuel, the freeze point and cold-temperature properties must be improved further. To manufacture Hydrotreated Esters and Fatty Acids (HEFA) fuel, the fatty acids and triglycerides in the crude oils undergo a different set of reactions to yield CO_2, H_2O and long-chained *n*-paraffins. This is followed by a second reaction (hydrocracking) that breaks

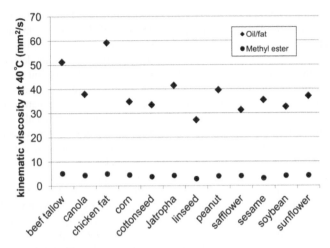

Figure 11.13.   Transesterification reduces viscosity of vegetable oils and greases by conversion to methyl esters. Data obtained from compilations in Jingura *et al.* (2010), Knothe (2008a), Lang *et al.* (1992) and Refaat (2009).

apart the dense hydrocarbon chains to form smaller, highly branched hydrocarbons, shorter-chain *n*-paraffins, and a small amount of cycloparaffins (section 11.5.3). As a result, the molecular makeup of HEFA is very similar to conventional jet fuel with a blend of pure hydrocarbons in the range of C9–C15 (see Fig. 11.7i).

11.5.1   *Dewatering, crude oil extraction and pre-processing*

To prepare for refining, oil-rich beans or seeds are dried, cleaned, cracked and compressed into flakes. The oil is extracted chemically by exposing it to a solvent or, less efficiently, mechanically through pressing. A combination of pre-heating and mechanical extractions can increase oil yield (Mahmoud, Arlabosse and Fernandez, 2011). For algae, the choices become more varied. The algal biomass can be dried and oil extracted by chemical, biochemical and thermochemical approaches, often with some mechanical assistance.

Dewatering can be an energy-intensive stage in oil production from algae. Even at the end of a growth cycle, the algal broth is about 5–17% biomass in industrial-scale open ponds. Although the biomass concentration could be much higher when grown in photobioreactors, there is still significant water content. The biomass can be removed by a number of methods, such as sedimentation, flotation, filtration, or centrifugation. The size and density of the algal cells are a primary consideration for choosing a separation technique. Most algae cells are heavier than water, so sedimentation is one possible, inexpensive strategy, but it is slow and is better for large, dense cells, roughly those larger than 50–100 µm. Many of the most promising algae for biodiesel production are about 1–30 µm in size. Flocculants reduce or neutralize the negative charge of microalgae so that they can form clumps, which effectively increase the particle size for better sedimentation. Flocculants, sometimes coupled with ultrasonic forcing, may be used as a pre-treatment step to encourage clumping. In flotation, air is bubbled up through the culturing system. As algae (or algae clumps) attach to the bubbles, they are carried up to the liquid surface, where they can be removed by skimming. Membrane microfiltration is highly efficient at separating water from biomass in small batches (Zhang *et al.*, 2010), but it is not a good fit for large-scale production due to high energy requirements for pumping and filters that require lots of attention and maintenance. Centrifugation is also very efficient in the lab, but it is prohibitive in an industrial setting due to the high energy costs of running this equipment. If the biomass needs to be dried for oil extraction, techniques range from the low-tech, slow process of drying in the sun to fast, expensive processes such as bed drying, freeze drying, drum drying, and spray drying.

The most common method for lipid extraction from within the algal cell occurs by disrupting the cell walls, releasing algal contents into solution from which they can be separated. This can be accomplished chemically by solvent extraction. The most common solvent in the lab is *n*-hexane because it results in high oil yield, but it is also a slow process, which is not desirable for a large-scale manufacturing facility. Also, the separation and recovery of the solvent and lipids through, e.g., distillation, adds an additional step that requires significant energy input. Finally, there are fire and safety hazards that come with this method. Another option is enzymatic extraction, which uses water as the solvent, with the enzymes acting to break down the cell walls, such as alkaline protease for Jatropha (Achten *et al.*, 2008). Enzymatic extraction is a less effective technique, however, and pretreatment may be helpful, such as ultrasonication. Assistance may also be given by other thermal or mechanical means, such as autoclaving, bead-beating, or microwaves, which are all viable in lab-scale processing. At large scale, mechanical pressing has been suggested as a cost-effective means of lipid extraction, perhaps in combination with solvent extraction (Gong and Jiang, 2011). Thermally assisted mechanical dewatering is discussed by Mahmoud *et al.* (2011).

Some algae, such as *Dunaliella*, have a high extraction efficiency because they do not have a thick cell wall, but other species of interest, such as *Nannochloropsis* or *Chlorella* are more challenging because of their hard cell walls. Other cells, such as diatoms, share the property of hard cell walls. Other extraction methods may be more attractive for these microbes, such as chemically extracting lipids through the cell wall, or manipulating the hydrophobicity to encourage algae to secrete the lipids in a process known as "milking". In the latter case, the algae remain viable after the extraction and can be returned to production as an active cell culture. An algae growth facility in Hawaii is currently being built by Phycal that relies on this technique.

After the extraction step is complete, the oil must be filtered and cleaned before going on to the next manufacturing stage.

For further reading on dewatering of algae, see Uduman *et al.* (2010); for extraction processes, see Grima *et al.* (2003), and Mercer and Armenta (2011).

11.5.2   *Transesterification*

Vegetable oils, or green crude, are highly viscous compounds based on saturated and unsaturated fatty acids, often concentrated in the range of C16–C18. As such, their viscosity is high as is their freeze point, which make them inadequate except for use in some piston-driven diesel engines. The goal of biorefining is to reduce viscosity and improve cold flow properties. The best biodiesel blends will have high levels of monounsaturated and saturated fatty acid esters. It will also be low in polyunsaturated fatty acid esters due to their poor oxidative stability and high density (see Figs. 11.9e and 11.8a). These processes are effective strategies for production of biodiesel regardless of the feedstock.

Most of the fatty acids in vegetable oils are bound up in triglyceride molecules (or triacylglyc-erols or TAGs). These compounds are comprised of a glycerol backbone that is connected to three fatty acids. For some algae, triglycerides are the main carbon storage molecule, but algal oil may also include free fatty acids.

The fatty acids are stripped off the glycerol sequentially, until the result is three separate molecules of fatty acid esters and a glycerol molecule. In the finished product, the presence of monoglycerides or diglycerides (glycerol bound up with one or two fatty acids, respectively) is a sign of incomplete transesterification. In order for this reaction to occur, the triglyceride must be exposed to an alcohol, usually methanol, in the presence of a catalyst. In practice, the reaction is usually carried out with an excess of methanol. The liberated fatty acids react with the alcohol to form esters. This is depicted in Figure 11.14 in which *R1*, *R2* and *R3* represent three alkyl groups, which look like alkanes (paraffins) with one missing hydrogen at one end of the carbon chain. Esters are a molecular alliance of an acid with an alcohol, in which at least one of the acid's hydroxyl (OH) functional groups is replaced with an oxygen atom bonded to a straight-chain saturated hydrocarbon (an alkyl group). For mono-alkyl esters, there is exactly one replacement within the acid by an O-alkyl group, as in Figure 11.14.

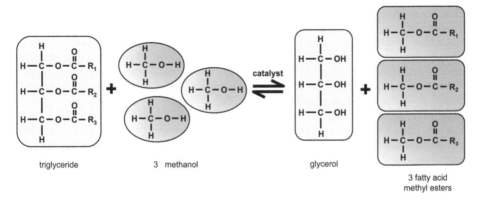

Figure 11.14.   Transesterification is the reaction of a triglyceride and alcohol to form glycerol and alkyl esters.

The process is reversible, in theory, but in actual practice, it is unlikely since the glycerol is not miscible (i.e., it does not mix) with the product, although excess methanol can slow the separation. The result is a two-phase system, with glycerol sedimenting to the bottom of the reaction vessel.

For biodesel, methanol is usually the alcohol of choice due to its low cost and ready availability, although ethanol is used in places such as Brazil in which that fuel is abundant. Some studies have examined the use of genetic modification of *E. coli* to produce higher chain alcohols, such as isobutanol (Atsumi *et al.*, 2008), but this process is not yet ready for prime time.

A catalyst is a substance that speeds up, or otherwise assists, the speed or likelihood of a chemical reaction, without itself being altered. The catalyst can be alkaline (such as sodium hydroxide and sodium methylate), acidic (such as sulfuric acid), or enzymatic. Strong acids donate a proton to the carbonyl group, whereas bases remove a proton from the alcohol (Cordeiro *et al.*, 2011). Acidic catalysts usually provide the most complete reaction, although they are excessively slower than alkaline catalysts, and they require higher temperatures. If free fatty acids are present in the reactant (roughly >0.5–1%), reaction with alkaline catalysts will lead to a saponification reaction, which turns the triglycerides into soaps rather than alkyl esters. In this case, a two-step procedure is needed to use the quicker base catalysts. This method starts with a pretreatment step with an acid catalyst which converts both free fatty acids and triglycerides to alkyl esters. After this reaction is neutralized, it is followed by another conversion step with a base catalyst for the remaining triglycerides. The additional energy and time requirements of acidic catalysis may be more economical than the extra steps that are needed in a two-stage procedure for conversion and catalyst separation. Lipase enzymatic catalysts are more expensive and usually require longer reaction time but are more environmentally friendly. Alkaline protease was used in aqueous enzymatic oil extraction for best results from jatropha (Achten *et al.*, 2008). The conversion of palm oil from lipase enzymes produced by three bacterial strains that support transesterification was tested, but it exhibited low conversion efficiency (~20% *versus* commercial lipases at 90+%) (Meng and Salihon, 2011).

After the reaction is complete, the FAAEs must be separated from the catalyst, excess alcohol, water, free fatty acids, and the glycerol. The glycerol is relatively simple to divide from the products, as it sediments to the bottom of the tank and does not mix with the product. For the others, additional steps, such as distillation, would be needed.

Notice that, in Figure 11.14, the alkyl groups *R*1, *R*2 and *R*3 maintain their identities as functional groups in the reactant and the product. Consequently, the fatty acid profile of biodiesel corresponds in large measure to that of its feedstock. As seen in section 11.4.1, vegetable oils are concentrated at higher carbon number (mostly at C16–C18), relative to petroleum jet fuel (roughly C8–C16), as shown in section 11.3.1.

For sustainability, the overall process must be streamlined to reduce the number of energy-intensive steps and replace any toxic or hazardous compounds with more environmentally friendly ones. Consequently, the development of effective catalysts that are nontoxic, inflammable and recyclable are important areas of research. A feasibility study of lipid extraction from cyanobacteria was able to extract 97% of the lipids with liquefied dimethyl ether, a nontoxic compound. Furthermore, this operation was performed on wet biomass, eliminating the need for the drying and cell wall disruption steps (Kanda and Li, 2011).

The acidic and basic catalysts require a neutralization step to end the reaction. To avoid this extra step, much work has been devoted to extraction processes using solid catalysts that are more selective, safe and environmentally friendly. Zeolites and mesoporous compounds can meet these requirements, in addition to having a high concentration of active sites, high thermal stability, and better shape selectivity (Carrero *et al.*, 2011; Cordeiro *et al.*, 2011; Perego and Bosetti, 2011). Conversion efficiencies using these unique materials have improved sufficiently to be considered for commercial biodiesel production (Perego and Bosetti, 2011; Verma *et al.*, 2011).

Other techniques that avoid using toxic solvents include supercritical gas extraction (Edwards, 2006; Levine *et al.*, 2010; Li *et al.*, 2010; Soh and Zimmerman, 2011) For further reading on the effects of process variables on FAAE yield, see (Alptekin and Canakci, 2011; Rashid *et al.*, 2009), and on production-scale transesterification, see (Van Gerpen and Knothe, 2005).

11.5.3 *Hydroprocessing*

Hydroprocessing is a technique that uses catalysts in the presence of hydrogen to convert a variety of free fatty acids, triglycerides, alkyl esters and other compounds into paraffinic hydrocarbons by removing oxygen and saturating it with hydrogen. This process can also be used to drive contaminants like sulfur, nitrogen and trace metals from a hydrocarbon. Hydrotreatment occurs at relatively low temperatures and pressures, which provide sufficient driving force to break the molecular bonds with S, N or O and replace it with a hydrogen molecule. The residual S, N, and O atoms can combine with hydrogen to form stable compounds. This process works most efficiently on unsaturated oils.

The deoxygenation reaction is shown in Figure 11.15, which is carried out at low temperature (around 300°C, depending on the specifics of the process) using a di-metallic catalyst, such as nickel-molybdenum (Ni-Mo) or cobalt-molybdenum (Co-Mo). (For better readability, note that the hydrogen atoms have been removed from the display of the triglyceride molecule.) The R_1, R_2 and R_3 still denote alkyl groups (paraffinic subunits) of a fatty acid. At the completion of the reaction, the terminal carbon in the alkyl group is saturated with hydrogen, rather than bonded to oxygen. One of the alkyl groups has gained an additional link in the hydrocarbon chain (R_1 CH_3), so that its carbon number increases by one, but the other alkyl groups have become alkanes (paraffins) that maintain the same carbon number. Another product of the reaction is propane (C_3H_8), formed from the glycerol backbone, which can be recovered by fractional distillation. The remaining carbon and oxygen atoms in the triglyceride have been converted to CO_2 (or CO, depending on the reactant) and H_2O. If any carbon atom possesses double bonds in the reactant, these components will become saturated with hydrogen, so that the product consists of long-chain *n*-paraffins.

The benefit of converting the triglyceride to paraffins rather than methyl esters is that the stability, specific energy, and cold-temperature and blending properties of deoxygenated hydrocarbons are better suited for jet fuel.

Since most of the alkyl groups in the reactant vegetable oil are in the range of C16–C18, there are two strategies for reducing the freeze point: (i) convert the dense, straight-chained paraffins into more highly branched hydrocarbons; or (ii) crack the dense hydrocarbons into shorter-chained molecules (∼C12–C14). Recall that Figure 11.6f showed that the freeze temperature decreases as carbon number decreases. The freeze temperature also decreased for isoparaffins as compared to *n*-paraffins, due to their more complex, branched structure. Higher carbon numbers can be better tolerated if the hydrocarbons are isomerized.

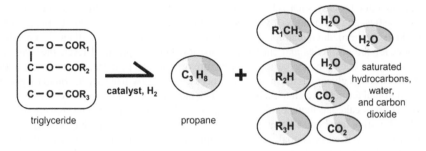

Figure 11.15.   An example of hydrotreatment: deoxygenation of a triglyceride into saturated hydrocarbons, water and carbon dioxide.

There are trade-offs to be made here. The isomerisation process is similar to the deoxygenation reaction described above in that it operates at moderate temperatures (perhaps 250–350°C) and pressures (typically less than 5 MPa). It uses excess H_2 and catalysts as reactants, and it will produce the most jet fuel when the feedstock molecules are in the range of C10–C14. If the feedstock has carbon chains that are either shorter or longer, the resulting hydrocarbon mix can undergo fractional distillation to separate out the heavier and lighter hydrocarbons. The less desirable components can be used for other purposes, such as green diesel or cooking fuel, but it will supply less jet-grade fuel. Isomerisation is also more effective when the feedstock is not fully saturated.

For the case of vegetable oil that is heavily weighted in the C16–C18 range and/or is fully saturated, the more extreme hydrocracking option can be considered. At higher temperatures (~350–420°C), much higher pressures (7–14 MPa), excess H_2, and the right catalyst, the carbon-to-carbon bonds of long-chained *n*-paraffins are ripped apart to form shorter *n*-paraffins and branched isoparaffins. The resulting product will have a higher proportion of hydrocarbons in the range needed for jet fuel, but, due to the large energy input, it will come at a cost, both in terms of fuel cost and in environmental impact.

For a good reference on hydroprocessing, see Robinson and Dolbear (2006); on hydrogenization of unsaturated methyl esters, see Bouriazos *et al.* (2010). Refining vegetable oils removes most of the natural antioxidants such as tocopherols (Holser and Harry-O'Kuru, 2006). For a discussion of additives and blending to improve cold-flow properties see Chastek (2011), Coutinho *et al.* (2010), Joshi *et al.* (2011), Kerschbaum and Rinke (2004), Kerschbaum *et al.* (2008), Moser (2009), Wang *et al.* (2011) and emissions see Moser (2009).

11.5.4   *Other strategies*

Another option for creating jet fuel is to start with relatively low-weight alcohols, such as butanol, and perform an oligomerization step. In this reaction, the short-chained hydrocarbons undergo a reaction to extend the length of the hydrocarbons, thus building up from C3 towards jet fuel range hydrocarbons.

Some studies have examined the use of microbes to produce these compounds directly from sunlight (Atsumi *et al.*, 2008; Tan *et al.*, 2011), fatty acid feedstocks (Dellomonaco *et al.*, 2010) or through fermentation of sugars from lignocellulosic decomposition (Ha *et al.*, 2010). This technology is still in the early stages and is not a near-term solution.

11.5.5   *Co-products*

One of the most critical aspects of sustainable process development and economic viability will be the identification of value-added chemicals, energy, and materials from the remnants of the biofuel production process, such as biogas, animal feed, fertilizers, industrial enzymes and chemicals, bioplastics, and surfactants. It will also require finding creative new applications for these "co-products" as needed. Glycerol, a byproduct of transesterification, is still cited as a valuable co-product for the cosmetics and chemical industries, but at a practical level, the market has become saturated due to the increasing manufacture of biodiesel (Yazdani and Gonzalez, 2007). Since the creation of glycerol is intrinsic to the process, new uses will have to be found for glycerol, such as microbial fermentation into fuels and marketable chemicals (Yazdani and Gonzalez, 2007).

Jatropha press-cake has hemicelluloses, cellulose and lignin, which can be converted through anaerobic digestion into biogas (Demirbas, 2011) or through pyrolysis into bio-oils, gas and char. Its biomass can be used as animal feed (with appropriate detoxification), or as fertilizer. For a detailed description of jatropha fruit, shell, husks and wood that could be used to produce energy, see Jingura *et al.* (2010).

For algae, there are three major components of biomass: lipids, carbohydrates and proteins. Lipids and carbohydrates can be converted into fuel, while the proteins can become co-products, such as animal feed. Another option: anaerobic digestion of algal biomass and cellulose can be used as a means of H_2 production (Carver *et al.*, 2011). For other sorts of co-products from algae, see Cardozo *et al.* (2007). The remediation of wastewater (Aresta *et al.*, 2005; Park and Craggs, 2011; Park, Craggs and Shilton, 2011) or toxic compounds (Petroutsos *et al.*, 2007) or sequestration of CO_2 (Aresta *et al.*, 2005; Sayre, 2010) can be considered co-products for the purposes of calculating environmental impact in life cycle analysis.

11.6   LIFE CYCLE ASSESSMENT

In general terms, sustainability requires that our activities do no harm to the planet or the life that depends upon it. This means that people take priority in the competition for food and water. Balat passionately argues that there is a direct link between using edible oils for biofuel and starvation (Balat, 2011a,b). Thus, the search for energy production should avoid the use of food crops, arable land, or fresh water. The huge preponderance of scientific evidence indicates that we are foolhardy to continue recklessly pumping greenhouse gases (GHG) into the earth's atmosphere. As a result, we seek energy production techniques that are at least carbon-neutral, i.e., the net effect of all parts of the process maintains the same level of GHG as currently exists. However, the smarter strategies will result in a net reduction of GHG in the atmosphere. In considering environmental impact, it is not sufficient to consider only the emissions resulting from fuel burn. The environmental cost of materials, energy and changes in land use that are required to produce the fuel must also be part of the equation. Particularly for the aviation industry, noise and particulate emissions are of concern for public health and safety and could legitimately be considered a criterion for sustainability. To quantify the net impact of any particular process or set of processes, a relatively new field of Life Cycle Assessment has emerged to provide solid guidance for determining which processes are in fact sustainable.

The heart of sustainability evaluations resides in the Life Cycle Assessment, which is a means of quantitatively evaluating a process system from start to finish for the purposes of comparing a number of options. The goal of an LCA may be to evaluate environmental impact, inform market strategies, and/or analyze socioeconomic impact of a particular set of choices. As an example, we will consider the outline of an LCA for the environmental impact of conventional *versus* alternative aviation fuels.

The first task is to decide on the purpose of the LCA and thereby decide on the "functional units" that will be used in the analysis. Conversion factors will be used to relate everything in the analysis to this functional unit, which could be almost anything – but it has to relate to the problem of interest. If we would like to analyze how much you could save over the next five years by changing from incandescent light bulbs to compact fluorescent light bulbs in your living room, your functional unit might be related to the amount of light that is needed to read the paper at night. In this case, we might choose "lumens per square foot per Euro" as a functional unit.

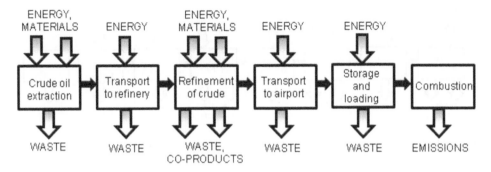

Figure 11.16.   Schematic of life cycle assessment for conventional petroleum fuel.

If we want to compare the change in greenhouse gas emissions due to changing from conventional jet fuel to an alternative fuel while flying a fully loaded Boeing 787 from New York to Tokyo, we might choose grams of CO_2 emissions per kg fuel (g CO_2/kg). If we know how many kilograms of fuel are burned on such a trip, we can calculate the mass of CO_2 emitted during this flight. For this case, everything in the problem must be converted to something representing the mass of CO_2 that is emitted for every kg of fuel that is burned. This becomes a little tricky, because we know that CO_2 is not the only greenhouse gas of interest, and we would like to include the effects of other emissions. Most environmental LCAs incorporate other greenhouse gas emissions by introducing the concept of equivalent CO_2 emissions ($CO_{2,eq}$). The equivalence is developed through conversion factors, such as global warming potential (GWP) or alternatively radiative forcing (RF), which account for the manner in which the pollutant affects the buildup of heat in the atmosphere. When the mass of the greenhouse gas is multiplied by its corresponding conversion factor, the result is in units of g $CO_{2,eq}$, and we can sum the effects of all of the greenhouse gases to find a number that represents the mass of CO_2 equivalent that was pumped into the atmosphere by the flight.

Knowing something about the chemistry of the fuel combustion process, the aircraft performance and operating conditions, it may be relatively straightforward to estimate how much CO_2, CO, NO_x, SO_x, H_2O, and particulate matter is created by burning a kilogram of jet fuel. The really tricky part is that we would also like to include the entire history of CO_2-equivalent emissions that were generated by manufacturing that kilogram of fuel. So now we have to back up and look at the process that brought us this jet fuel.

The life cycle of conventional fossil fuel is depicted in Figure 11.16. The LCA must capture the environmental impact of all of the materials and processes that are required to create aviation fuel, as well as all of the things that are left behind after each step of the process is performed. In the simplest type of analysis, each of these things is expressed in terms of our functional units. It may not be immediately obvious how to do that, but let us begin by identifying all of the stuff along the life cycle path.

The first step is to extract the crude oil, noting all of the energy and materials needed to perform this step, such as chemicals, lubricants and water or other coolants used for drilling. We must also take care in identifying all of the waste that is generated during the extraction. For example, there will be gaseous emissions associated with using fuel to power machinery.

Next, the crude oil is transported to the refinery in a pipeline, in a supertanker, by truck and/or by rail. Each of these modes of transport requires energy (e.g., fuel to run motors or pumps) and generates waste (e.g., gas emissions). At the refinery, the crude oil is processed to yield aviation-grade fuel. Inputs for this step include electricity and/or fuel to heat reaction chambers and pump fuel through distillation columns, as well as the environmental cost of input gases, chemicals, and catalysts. Crude oil is a mixture of organic hydrocarbons, only some of which are suitable for aviation fuel. The remaining (non-aviation) fuel can be sold for other purposes and would be considered a co-product of the refining process. Co-products are materials, energy, or other

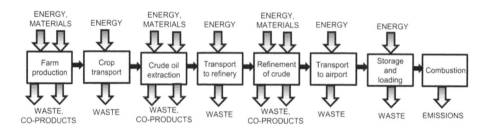

Figure 11.17.   Schematic of the life cycle of a first-generation biofuel.

benefits that are created during the processing step, and they add value in the overall life cycle assessment. Inputs that are consumed and waste that is generated yield a value that represents their environmental cost. Co-products thus offset some of the net environmental cost, so if the inputs and waste are expressed as positive values, then the co-products are given negative values.

After refinement, the aviation fuel is transported to the airport, where it is stored and eventually some proportion of that fuel is loaded onto an aircraft. When the aircraft takes off, the fuel is consumed and the combustion products are exhausted into the atmosphere. Our "system boundaries" in this LCA then extend from the oil well where the crude was extracted to the exhaust gases left behind in the wake of the aircraft, leading to the widely used description of this sort of LCA as "well to wake".

A similar LCA for an alternative fuel, such as castor-derived Hydrotreated Esters and Fatty Acids (HEFA) fuel, might look quite similar to Figure 11.16, but it requires additional steps to describe it. In Figure 11.17, we have added a couple of steps at the front end to represent the production of the castor feedstock. The first box represents the growth of the castor in the fields, harvesting, and the separation of seeds from the rest of the plant material. Inputs might include the environmental costs of seeds, fertilizer, herbicides and pesticides, irrigation, gasoline and electricity for farm machinery, and packaging materials. Output waste might consist of gaseous emissions, fertilizer runoff and plant debris. However, perhaps the plant debris left over on the farm could be transformed into a useful co-product by running it through an anaerobic digester to create methane fuel or turned into compost to improve soil quality. In that case, some of the waste stream might be turned back in toward the crop production process and reduce the amount of input energy or fertilizers needed to produce the crop.

Once the castor seeds are delivered to the processing plant, depicted in the third box of Figure 11.17, they could be crushed to extract vegetable oil. The debris left over from the crushing process could be used for animal feed, so this would be tallied as a co-product. The vegetable oil must be further processed and blended to yield qualified aviation fuel, generally done in a different facility, so it travels on to the crude oil refinery. Beyond this point, the fuel's journey has the same stages as that of petroleum-based jet fuel.

LCAs are meant to systematically evaluate processes, and since processes change over time, the LCAs must also be capable of evolving over time. These changes might include new farming techniques for feedstocks that can increase yield or reduce fertilizer or irrigation requirements, more efficient equipment, development of markets for co-products, changes in water utilization, land use, and waste management. Clearly, there is some degree of subjectivity in defining the components of an LCA, and published studies may not always agree because they include different sets of processes, or different inputs and outputs, or they may not assign consistent values to each component. To address this issue, the inputs and outputs to the LCA can be considered part of a Life Cycle Inventory (LCI), which is a database that provides the numbers that are assigned to each constituent process/material/energy usage in the functional units of interest. Standardized LCIs have become available; some are free (such as the European Reference Life Cycle Database[1])

---

[1] See http://lca.jrc.ec.europa.eu/lcainfohub/datasetArea.vm

while others are proprietary (such as ecoinvent[2]). With such databases, one can drill down to the smallest details and quantify the environmental impact of using material *A versus* material *B* for a pipeline of length *L*.

LCAs are relatively new methodology, so that a literature survey of aviation fuel LCAs can be somewhat confusing due to conflicting results and methodologies. In 2006, ISO 14040 (ISO, 2006a) and 14044 (ISO, 2006b) were established to provide a uniform approach to environmental LCAs. By 2012, the field has matured considerably. GREET (Greenhouse Gases, Regulated Emissions, and Energy Use in Transportation) is a software package developed by Argonne National Laboratories in the US to perform LCAs for vehicular fuels. This code is also commonly used in LCAs for aviation fuels. Another option is the evolving OpenLCA, which is freeware developed for generalized LCAs.

Startup costs, or capital expenditures, are often ignored in economics-focused LCAs, but they can make all the difference in predicting whether a specific venture will be profitable or not, or that a particular fuel will be environmentally friendly or not. If the extraction of the petroleum fuel is at a remote location and requires a road to be built to get there, this can be a significant cost. Not only is there the direct material cost due to asphalt and other building materials, and the energy cost associated with road-building equipment, there may also be indirect costs associated with land use change, such as deforestation if that is needed to build the road. If we slash and burn forest to create farmland on which to grow biofuel feedstocks, the economic cost could be significant, but the environmental impact could be even worse.

In the process of photosynthesis, plants turn CO_2, a greenhouse gas, into O_2. As such, forests are living, breathing systems that store, or sequester, carbon in their limbs and roots. When trees die, fall to the forest floor and decompose, the sequestered carbon in the wood becomes nutritious soil for the next generation of plants. When we clear forestland to create farmland, not only do we lose the benefit of this natural process, but also the carbon that was stored in the plant matter may be released into the atmosphere, typically by burning. The upfront greenhouse gas emissions due to clearing a hectare (ha, with 1 ha = 10,000 square meters) of forest can result in the release of 604–1146 megatons of greenhouse gases (MT) into the atmosphere (Searchinger *et al.*, 2008). Clearing a savannah or grassland can release 74–305 MT/ha into the atmosphere. Since the net environmental benefit of growing corn for ethanol is measured in at best a couple of megatons per hectare per year (MT/ha/y), it can take decades to recover the environmental cost of preparing untouched land for farming (Searchinger *et al.*, 2008). In fact, some have argued that land-use issues end up being *the* dominant issue in sustainability (Melillo *et al.*, 2009; Righelato and Spracklen, 2007).

More subtly, the effect of land use change can be indirect. Suppose that a farmer in Iowa decides to sell the corn harvest to an ethanol refinery rather than a grocery store chain in order to benefit from agricultural subsidies provided in the US. This corn will no longer be available to feed human beings or animals, and the price of corn rises. The globally rising price of corn encourages another farmer in Mexico to clear virgin land to produce corn. The net effect on the environment is to increase the amount of greenhouse gas pumped into the atmosphere due to land use change. Some LCAs do not attempt to quantify indirect land use change, since it is difficult to formulate a meaningful model of the process that adequately describes the socioeconomic factors that are involved. Nevertheless, this is a real phenomenon, and this is an active area of research, particularly for LCAs geared towards analysis of global issues in sustainability.

Other land-use concerns revolve around the use of nutrients to more efficiently grow crops. Nitrogen fertilizers tend to outgas nitrous oxide, which can lead to smog formation when combined with hydrocarbons and sunlight. Nitrogen and phosphorous fertilizer are also responsible for eutrophication as nutrient-rich runoff from farmland escapes into a lake or ocean and create algal blooms. As the algae die off, their decomposition robs the water of dissolved oxygen that fish and other marine life depend on for survival. Not only are we creating unnecessary problems for

---

[2] See http://www.ecoinvent.ch

ourselves by our cavalier use of fertilizers, the global phosphate supply should be considered a precious resource that may not be readily available in a couple of decades (DOE, 2010; Vaccari, 2009).

Many of the early sustainability analyses for corn-based ethanol ignored the issue of land use change altogether, which was a grievous omission. They often did not include the entire system boundaries, nor did they have a complete inventory (von Blottnitz and Curran, 2007). When properly accounting for land use change, system boundaries, and a more thoughtful assessment of the inventory required, corn-based ethanol production shifts from being a good thing for the environment to a strategy that is ultimately harmful to the environment.

Arguably, the most essential feature of any LCA is to thoroughly document the approach taken, including purpose of the LCA, system boundaries, critical assumptions, key parameters, variability in fuel components and uncertainties (Stratton *et al.*, 2011). In many historical LCAs, the values for a given input or output is given as a single fixed number (such as the yield of jatropha oil/ha/y). However, in fact, there are uncertainties in that value associated with real demographic differences. More uncertainties creep into the analysis from other variable quantities, outdated data, or incomplete problem description. To compare different strategies for alternative fuels, feedstocks need to be compared within an equivalent framework. If the approach is fully documented, then future analyses can harmonize the assessment by updating or supplementing the data (e.g., Sun *et al.*, 2011; von Blottnitz and Curran, 2007; Warner *et al.*, 2010).

Defining the appropriate criteria by which to measure energy sustainability is still a work in progress (McBride *et al.*, 2011). Moreover, the hard data on biofuel production that are needed for such analyses are full of uncertainty (Murphy *et al.*, 2011; Slade *et al.*, 2011), because it is not available, still emerging, or preliminary. In order to accurately estimate the amount of land that would be needed globally for producing some specified amount of biofuel, the model must account for local and regional variations for many variables such as climate, water availability, crop production techniques, land suitability, and refining capabilities. Wigmosta and co-workers have developed a comprehensive framework that can account for most of these variables in great detail for the United States down to a resolution of 30 miles (Wigmosta *et al.*, 2011). As such comprehensive tools mature, they will provide even better insight into modeling various scenarios for biofuel production.

For further reading on LCAs for aviation fuels, see Dray *et al.* (2010), Jorquera *et al.* (2010), Kinsel (2010), Kreutz *et al.* (2008), Stratton *et al.* (2010), and Yang *et al.* (2011).

11.7 CONCLUSIONS

Aviation fuel requirements are demanding because of the extreme environment in which aircraft must operate. Without an alternative to the internal combustion engine in the near term, the industry must rely on drop-in fuels that are compatible with existing engines. Although biodiesel based on methyl esters are adequate for automotive applications, their poor cold-weather properties and reduced specific energy make them unsuitable for aviation fuel. Both synthetic paraffinic kerosene and Hydrotreated Esters and Fatty Acids (HEFA) can meet aviation's needs, but it is necessary to find sustainable ways of manufacturing these fuels. Some of the environmental and economic improvements that will bring closure to the problem will include new efficiencies in strain selection, crop management and production, and more creative approaches to finding uses for co-products.

ACKNOWLEDGEMENTS

The author would like to gratefully acknowledge the support of NASA Glenn Research Center and the alternative fuels component of the Subsonic Fixed Wing Program, as well as the wisdom and guidance of Bob Hendricks and Arnon Chait, and the heroic efforts of the NASA Glenn Technical Library, especially Marcia Stegenga.

ABBREVIATIONS

AFRL	US Air Force Research Laboratory
ASP	US DOE's Aquatic Species Program
ASTM	ASTM International, formerly known as the American Society for Testing and Materials
Bio-SPK	bio-synthetic paraffinic kerosene, same as HEFA
CAAFI	Commercial Aviation Fuels Initiative
DOE	US Department of Energy
EU	European Union
F-T	Fischer-Tropf, a processing method to produce synthetic fuel
HEFA	Hydrotreated Esters and Fatty Acids, renewable jet fuel (a/k/a HRJ and bio-SPK)
HRJ	Hydroprocessed Renewable Jet fuel, same as HEFA
ICAO	International Civil Aviation Organization
IATA	International Air Transport Association
IEA	International Energy Agency
LCA	Life Cycle Assessment
LCI	Life Cycle Inventory
ME	methyl esters
NASA	US National Aeronautics and Space Association
RTK	Revenue Tonne Kilometers
SPK	Synthetic Paraffinic Kerosene jet fuel
US	United States
WFSP	World Fuel Sampling Program (Hadaller and Johnson, 2006)

REFERENCES

Achten, W., Mathijs, E., Verchot, L., Singh, V., Aerts, R. & Muys, B.: Jatropha biodiesel fueling sustainability? *Biofuels Bioprod. Biorefin.* 1:4 (2007), pp. 283–291.

Achten, W., Verchot, L., Franken, Y., Mathijs, E., Singh, V., Aerts, R. & Muys, B.: Jatropha biodiesel production and use. *Biomass Bioenergy* 32:12 (2008), pp. 1063–1084.

Akbar, E., Yaakob, Z., Kamarudin, S.K., Ismail, M. & Salimon, J.: Characteristic and composition of *Jatropha curcas* oil seed from Malaysia and its potential as biodiesel feedstock. *European J. Sci. Res.* 29:3 (2009), pp. 396–403.

Alptekin, E. & Canakci, M.: Determination of the density and the viscosities of biodiesel-diesel fuel blends. *Renew. Energy* 33:12 (2008), pp. 2623–2630.

Alptekin, E. & Canakci, M.: Characterization of the key fuel properties of methyl ester-diesel fuel blends. *Fuel* 88:1 (2009), pp. 75–80.

Alptekin, E. & Canakci, M.: Optimization of transesterification for methyl ester production from chicken fat. *Fuel* 90:8 (2011), pp. 2630–2638.

Anand, K., Ra, Y., Reitz, R.D. & Bunting, B.: Surrogate model development for fuels for advanced combustion engines. *Energy Fuels* 25:4 (2011), pp. 1474–1484.

Araujo, G.S., Matos, L., Goncalves, L.R.B., Fernandes, F.A.N. & Farias, W.R.L.: Bioprospecting for oil producing microalgal strains: evaluation of oil and biomass production for ten microalgal strains. *Bioresour. Technol.* 102:8 (2011), pp. 5248–5250.

Aresta, M., Dibenedetto, A. & Barberio, G.: Utilization of macro-algae for enhanced CO_2 fixation and biofuels production: Development of a computing software for an LCA study. *Fuel Process. Technol.* 86:14–15 (2005), pp. 1679–1693.

ASTM: ASTM DS 4B Physical constants of hydrocarbon and non-hydrocarbon compounds, 2011.

ATDSR: Toxicological profile for jet fuels JP-4 and JP-8. Agency for Toxic Substances and Disease Registry, US Dept of Health and Human Services, 1995, http://www.atsdr.cdc.gov/ToxProfiles/TP.asp?id=773&tid=150 (accessed September 2011).

Atsumi, S., Hanai, T. & Liao, J.C.: Non-fermentative pathways for synthesis of branched-chain higher alcohols as biofuels. *Nature* 451:7174 (2008), pp. 86–90.

Balat, M.: Challenges and opportunities for large-scale production of biodiesel. *Energy Educ. Sci. Technol. A Energy Sci. Res.* 27:2 (2011a), pp. 427–434.

Balat, M.: Potential alternatives to edible oils for biodiesel production — A review of current work. *Energy Convers. Manage.* 52:2 (2011b), pp. 1479–1492.

Bartis, J.T. & Van Bibber, L.: Alternative fuels for military applications. Rand National Defense Research Institute, Santa Monica, CA, 2011.

Ben-Amotz, A.: New mode of *Dunaliella* biotechnology: two-phase growth for βcarotene production. *J. Appl. Phycology* 7 (1995), pp. 65–68.

Blakey, S., Rye, L. & Wilson, C.W.: Aviation gas turbine alternative fuels: A review. *Proceedings of the Combustion Institute* 33, 2011, pp. 2863–2885.

Boeing: Long-term market: current market outlook 2011–2030. 2011, http://www.boeing.com/commercial/ cmo/index.html (accessed September 2011).

Bouriazos, A., Sotiriou, S., Vangelis, C. & Papadogianakis, G.: Catalytic conversions in green aqueous media: Part 4. Selective hydrogenation of polyunsaturated methyl esters of vegetable oils for upgrading biodiesel. *J. Organomet. Chem.* 695:3 (2010), pp. 327–337.

Brennan, L. & Owende, P.: Biofuels from microalgae — A review of technologies for production, processing, and extractions of biofuels and co-products. *Renew. Sust. Energy Rev.* 14:2 (2010), pp. 557–577.

Bruno, T.J. & Baibourine, E.: Comparison of biomass-derived turbine fuels with the composition-explicit distillation curve method. *Energy Fuels* 25:4 (2011), pp. 1847–1858.

Bruno, T.J. & Huber, M.L.: Evaluation of the physicochemical authenticity of aviation kerosene surrogate mixtures. Part 2: Analysis and prediction of thermophysical properties. *Energy Fuels* 24 (2010), pp. 4277–4284.

Byron, D.: Air Force officials tackle current, future energy needs. In: *Air Force Print News Today*, Air Force Space Command, Washington, DC, 2011.

Cardozo, K.H.M., Guaratini, T., Barros, M.P., Falcao, V.R., Tonon, A.P., Lopes, N.P., Campos, S., Torres, M.A., Souza, A.O., Colepicolo, P. & Pinto, E.: Metabolites from algae with economical impact. *Com. Biochem. Phys. C Toxicol. Pharmacol.* 146:1–2 (2007), pp. 60–78.

Carrero, A., Vicente, G., Rodriguez, R., Linares, M. & del Peso, G.L.: Hierarchical zeolites as catalysts for biodiesel production from *Nannochloropsis* microalga oil. *Catal. Today* 167:1 (2011), pp. 148–153.

Carvalho, A.P., Meireles, L.A. & Malcata, F.X.: Microalgal reactors: a review of enclosed systems designs and performances. *Biotechnol. Progr.* 22 (2006), pp. 1490–1506.

Carver, S.M., Hulatt, C.J., Thomas, D.N. & Tuovinen, O.H.: Thermophilic, anaerobic co-digestion of microalgal biomass and cellulose for H(2) production. *Biodegradation* 22:4 (2011), pp. 805–814.

Chait, A., Nelson, E.S., Kassemi, M., Pines, V., Park, J., Cohen, M., Etai & Ben-Amotz, A.: Mixing in algae production: how much is not enough, and how much is too much? Manuscript in preparation, 2012.

Chastek, T.Q.: Improving cold flow properties of canola-based biodiesel. *Biomass Bioenergy* 35:1 (2011), pp. 600–607.

Chen, C.-Y., Yeh, K.-L., Aisyah, R., Lee, D.-J. & Chang, J.-S.: Cultivation, photobioreactor design and harvesting of microalgae for biodiesel production: a critical review. *Bioresour. Technol.* 102:1 (2011), pp. 71–81.

Cheng, K.C. & Ogden, K.L.: Algal biofuels: the research. *Chem. Eng. Progr.* 107:3 (2011), pp. 42–47.

Chevron: Aviation fuels technical review. Chevron Corporation, 2006, http://www.cgabusinessdesk.com/ document/aviation_tech_review.pdf (accessed September 2011).

Chèze, B., Gastineau, P. & Chevallier, J.: Forecasting world and regional aviation jet fuel demands to the mid-term (2025). *Energy Policy* 39:9 (2011), pp. 5147–5158.

Chisti, Y.: Biodiesel from microalgae. *Biotechnol. Adv.* 25:3 (2007). pp. 294–306.

Colket, M., Edwards, T., Williams, S., Cernansky, N.P., Miller, D.L., Egolfopoulos, F., Lindstedt, P., Seshadri, K., Dryer, F.L., Law, C.K., Friend, D., Lenhert, D.B., Pitsch, H., Sarofim, A., Smooke, M. & Tsang, W.: Development of an experimental database and kinetic models for surrogate jet fuels. AIAA-2007-0770 American Institute of Aeronautics and Astronautics, *AIAA Aerospace Sciences Meeting and Exhibit*, Reno, NV, 2007

Convery, F.J.: Reflections — The emerging literature on emissions trading in Europe. *Rev. Environ. Econ. Policy* 3:1 (2009), pp. 121–137.

Cooney, M.J., Young, G. & Pate, R.: Bio-oil from photosynthetic microalgae: case study. *Bioresour. Technol.* 102:1 (2011), pp. 166–177.

Cordeiro, C.S., da Silva, F.R., Wypych, F. & Ramos, L.P.: Heterogeneous catalysts for biodiesel production. *Quimica Nova* 34:3 (2011), pp. 477–486.

Corporan, E., Edwards, T., Shafer, L., DeWitt, M.J., Klingshirn, C., Zabarnick, S., West, Z., Striebich, R., Graham, J. & Klein, J.: Chemical, thermal stability, seal swell, and emissions studies of alternative jet fuels. *Energy Fuels* 25:3 (2011), pp. 955–966.

Coutinho, J.A.P., Goncalves, M., Pratas, M.J., Batista, M.L.S., Fernandes, V.F.S., Pauly, J. & Daridon, J.L.: Measurement and modeling of biodiesel cold-flow properties. *Energy Fuels* 24 (2010), pp. 2667–2674.

Day, J. & Tsavalos, A.: An investigation of the heterotrophic culture of the green alga *Tetraselmis. J. Appl. Phycology* 8:1 (1996), pp. 73–77.

Dellomonaco, C., Rivera, C., Campbell, P. & Gonzalez, R.: Engineered respiro-fermentative metabolism for the production of biofuels and biochemicals from fatty acid-rich feedstocks. *Appl. Environ. Microbiol.* 76:15 (2010), pp. 5067–5078.

Demirbas, A.: Competitive liquid biofuels from biomass. *Appl. Energy* 88:1 (2011), pp. 17–28.

Doan, T.T.Y., Sivaloganathan, B. & Obbard, J.P.: Screening of marine microalgae for biodiesel feedstock. *Biomass Bioenergy* 35 (2011), pp. 2534–2544.

DOE: National algal biofuels technology roadmap. DOE/EE-0332. US Department of Energy, Office of Energy Efficiency and Renewable Energy, Biomass Program, 2010, http://www1.eere.energy.gov/biomass/pdfs/algal_biofuels_roadmap.pdf (accessed September 2011).

Dooley, S., Won, S.H., Chaos, M., Heyne, J., Ju, Y.G., Dryer, F.L., Kumar, K., Sung, C.J., Wang, H.W., Oehlschlaeger, M.A., Santoro, R.J. & Litzinger, T.A.: A jet fuel surrogate formulated by real fuel properties. *Combust. Flame* 157:12 (2010), pp. 2333–2339.

Dray, L., Evans, A., Reynolds, T. & Schafer, A.: Mitigation of aviation emissions of carbon dioxide analysis for Europe. *Transport. Res. Rec.* 2177 (2010), pp. 17–26.

Edwards, T.: Cracking and deposition behavior of supercritical hydrocarbon aviation fuels. *Combust. Sci. Technol.* 178:1–3 (2006), pp. 307–334.

Edwards, T.: Advancements in gas turbine fuels from 1943 to 2005. *J. Eng. Gas Turb. Power* 129:1 (2007), pp. 13–20.

Edwards, T.: Jet fuel composition. In: M.L. Witten, E. Zeiger & G.D. Ritchie (eds): *Jet fuel toxicology.* CRC Press, Boca Raton, FL, 2010, pp. 21–26.

EIA: Monthly US refinery yield of kerosene-type jet fuel. US Energy Information Administration, 2011, http://www.eia.gov/dnav/pet/hist/LeafHandler.ashx?n=PET&s=MKJRYUS3&f=M (accessed July 2011).

ExxonMobil: World jet fuel specifications. ExxonMobil Aviation, 2005, http://www.exxonmobil.com/AviationGlobal/Files/WorldJetFuelSpecifications2005.pdf (accessed July 2011).

Fazal, M.A., Haseeb, A.S.M.A. & Masjuki, H.H.: Biodiesel feasibility study: an evaluation of material compatibility; performance; emission and engine durability. *Renew. Sust. Energy Rev.* 15:2 (2010), pp. 1314–1324.

Foidl, N., Foidl, G., Sanchez, M., Mittelbach, M. & Hackel, S.: *Jatropha curcas L.* as a source for the production of biofuel in Nicaragua. *Bioresour. Technol.* 58 (1996), pp. 77–82.

Freitas, S.V.D., Pratas, M.J., Ceriani, R., Lima, A.S. & Coutinho, J.A.P.: Evaluation of predictive models for the viscosity of biodiesel. *Energy Fuels* 25 (2011), pp. 352–358.

Gong, Y.M. & Jiang, M.L.: Biodiesel production with microalgae as feedstock: from strains to biodiesel. *Biotechnol. Lett.* 33:7 (2011), pp. 1269–1284.

Goodrum, J.W. & Geller, D.P.: Influence of fatty acid methyl esters from hydroxylated vegetable oils on diesel fuel lubricity. *Bioresour. Technol.* 96:7 (2004), pp. 851–855.

Gordon, D.R., Tancig, K.J., Onderdonk, D.A. & Gantz, C.A.: Assessing the invasive potential of biofuel species proposed for Florida and the United States using the Australian Weed Risk Assessment. *Biomass Bioenergy* 35:1 (2011), pp. 74–79.

Gourlay, P., Leak, J. & Wright, T.L.: Managing jet fuel and carbon in Europes's new emissions trading system: an OPIS primer. Oil Price Information Service. Gaithersburg, MD, 2011.

Gouveia, L. & Oliveira, A.C.: Microalgae as a raw material for biofuels production. *J. Ind. Microbiol. Biotechnol.* 36:2 (2009), pp. 269–274.

Griffiths, M. & Harrison, S.: Lipid productivity as a key characteristic for choosing algal species for biodiesel production. *J. Appl. Phycology* 21:5 (2009), pp. 493–507.

Grima, E.M., Belarbi, E.H., Fernandez, F.G.A., Medina, A.R. & Chisti, Y.: Recovery of microalgal biomass and metabolites: process options and economics. *Biotechnol. Adv.* 20:7–8 (2003), pp. 491–515.

Ha, S.-J., Galazka, J.M., Kim, S.R., Choi, J.-H., Yang, X., Seo, J.-H., Glass, N.L., Cate, J.H.D. & Jin, Y.-S.: Engineered *Saccharomyces cerevisiae* capable of simultaneous cellobiose and xylose fermentation.

Proceedings of the National Academy of Sciences of the United States of America 108:2 (2010), pp. 504–509.

Hadaller, O.J. & Johnson, J.M.: World fuel sampling program. CRC Report No. 647. Coordinating Research Council, Alpharetta, GA, 2006.

Hendricks, R.C.: Potential carbon negative commercial aviation through land management. ISROMAC12-2008-202421. Presented at the *12th International Symposium on Transport Phenomena and Dynamics of Rotating Machinery*, Honolulu, HI, 2008.

Hendricks, R.C., Bushnell, D.M. & Shouse, D.T.: Aviation fueling: a cleaner, greener approach. *Int. J. Rotat. Machinery* (2011), Article ID 782969.

Herbinet, O., Pitz, W.J. & Westbrook, C.K.: Detailed chemical kinetic oxidation mechanism for a biodiesel surrogate. *Combust. Flame* 154:3 (2008), pp. 507–528.

Herbinet, O., Pitz, W.J. & Westbrook, C.K.: Detailed chemical kinetic mechanism for the oxidation of biodiesel fuels blend surrogate. *Combust. Flame* 157:5 (2010), pp. 893–908.

Hileman, J.I., Stratton, R.W. & Donohoo, P.E.: Energy content and alternative jet fuel viability. *J. Propul. Power* 26:6 (2010), pp. 1184–1195.

Hirsch, R.L.: The inevitable peaking of world oil production. *Bulletin of the Atlantic Council of the United States* 16:3 (2006), pp. 1–9.

Ho, S.H., Chen, C.Y., Lee, D.J. & Chang, J.S.: Perspectives on microalgal CO_2-emission mitigation systems — A review. *Biotechnol. Adv.* 29:2 (2011), pp. 189–198.

Holser, R.A. & Harry-O'Kuru, R.: Transesterified milkweed (*Asclepias*) seed oil as a biodiesel fuel. *Fuel* 85:14–15 (2006), pp. 2106–2110.

Huber, M.L., Lemmon, E.W. & Bruno, T.J.: Surrogate mixture models for the thermophysical properties of aviation fuel Jet-A. *Energy Fuels* 24 (2010), pp. 3565–3571.

IEA: World energy outlook. International Energy Agency, Paris, France, 2010.

ISO: Environmental management — Life cycle assessment — Principles and framework, ISO 14040:2006. International Organisation for Standardisation, Geneva, Switzerland, 2006a.

ISO: Environmental management — Life cycle assessment — Requirements and guidelines, ISO 14044:2006. International Organisation for Standardisation, Geneva, Switzerland, 2006b.

Jain, S. & Sharma, M.P.: Stability of biodiesel and its blends: A review. *Renew. Sust. Energy Rev.* 14:2 (2010), pp. 667–678.

Jingura, R.M., Musademba, D. & Matengaifa, R.: An evaluation of utility of *Jatropha curcas L.* as a source of multiple energy carriers. *Int. J. Eng. Sci. Technol.* 2:7 (2010), pp. 115–122.

Jorquera, O., Kiperstok, A., Sales, E.A., Embirucu, M. & Ghirardi, M.L.: Comparative energy life-cycle analyses of microalgal biomass production in open ponds and photobioreactors. *Bioresour. Technol.* 101:4 (2010), pp. 1406–1413.

Joshi, H., Moser, B.R., Toler, J., Smith, W.F. & Walker, T.: Ethyl levulinate: A potential bio-based diluent for biodiesel which improves cold flow properties. *Biomass Bioenergy* 35:7 (2011), pp. 3262–3266.

Kanda, H. & Li, P.: Simple extraction method of green crude from natural blue-green microalgae by dimethyl ether. *Fuel* 90:3 (2011), pp. 1264–1266.

Karpovitch, E.A.: The green road to energy independence. In: Surface Warfare, Web exclusive, US Navy, 2011, http://surfwarmag.ahf.nmci.navy.mil/green_road.html (accessed September 2011).

Kerschbaum, S. & Rinke, G.: Measurement of the temperature dependent viscosity of biodiesel fuels. *Fuel* 83:3 (2004), pp. 287–291.

Kerschbaum, S., Rinke, G. & Schubert, K.: Winterization of biodiesel by micro process engineering. *Fuel* 87:12 (2008), pp. 2590–2597.

Kinder, J.D. & Rahmes, T.: Evaluation of bio-derived Synthetic Paraffinic Kerosenes (Bio-SPK). White paper. The Boeing Company, Sustainable Biofuels Research & Technology Program, 2009.

Kinsel, W.C.: *Environmental life cycle assessment of coal-biomass to liquid jet fuel compared to petroleum-derived JP-8 jet fuel.* Master of Science in Engineering Management thesis, Air Force Institute of Technology, Wright-Patterson Air Force Base, 2010.

Knothe, G.: Dependence of biodiesel fuel properties on the structure of fatty acid alkyl esters. *Fuel Process. Technol.* 86 (2005), pp. 1059–1070.

Knothe, G.: Some aspects of biodiesel oxidative stability. *Fuel Process. Technol.* 88:7 (2007), pp. 669–677.

Knothe, G.: "Designer" biodiesel: optimizing fatty ester composition to improve fuel properties. *Energy Fuels* 22:2 (2008a), pp. 1358–1364.

Knothe, G.: Evaluation of ball and disc wear scar data in the HFRR lubricity test. *Lubrication Sci.* 20:1 (2008b), pp. 35–45.

Knothe, G.: Biodiesel and renewable diesel: a comparison. *Prog. Energy Combust. Sci.* 36:3 (2010a), pp. 364–373.

Knothe, G.: Biodiesel: current trends and properties. *Top. Catal.* 53:11–12 (2010b), pp. 714–720.

Knothe, G. & Steidley, K.R.: Lubricity of components of biodiesel and petrodiesel: the origin of biodiesel lubricity. *Energy Fuels* 19 (2005), pp. 1192–2000.

Knothe, G., Krahl, J. & Van Gerpen, J.: *The biodiesel handbook*. AOCS Press, Champaign, IL, 2005.

Kohse-Hoinghaus, K., Osswald, P., Cool, T.A., Kasper, T., Hansen, N., Qi, F., Westbrook, C.K. & Westmoreland, P.R.: Biofuel combustion chemistry: From ethanol to biodiesel. *Angew. Chem. – Int. Ed.* 49:21 (2010), pp. 3572–3597.

Krahl, J., Munack, A., Schroder, O., Stein, H. & Bunger, J.: Influence of biodiesel and different petrodiesel fuels on exhaust emissions and health effects. In: G. Knothe, J. Van Gerpen & J. Krahl (eds): *The biodiesel handbook*. AOCS Press, Champaign, IL, 2005, pp. 173–180.

Krahl, J., Knothe, G., Munack, A., Ruschel, Y., Schroder, O., Hallier, E., Westphal, G. & Bunger, J.: Comparison of exhaust emissions and their mutagenicity from the combustion of biodiesel, vegetable oil, gas-to-liquid and petrodiesel fuels. *Fuel* 88:6 (2009), pp. 1064–1069.

Kreutz, T.G., Larson, E., Liu, G.J. & Williams, R.H.: Fischer-Tropf fuels from coal and biomass. *25th Annual International Pittsburgh Coal Conference* Pittsburgh, 7 October 2008 PA USA, revision, 2008, http://web.mit.edu/mitei/docs/reports/kreutz-fischer-tropsch.pdf (accessed August 2011).

Kumar, P., Robins, A., Vardoulakis, S. & Britter, R.: A review of the characteristics of nanoparticles in the urban atmosphere and the prospects for developing regulatory controls. *Atmos Environ.* 44:39 (2010), pp. 5035–5052.

Lang, W., Sokhansanj, S. & Sosulski, F.W.: Modelling the temperature dependence of kinematic viscosity for refined canola oil. *J. Amer. Oil Chem. Soc.* 69:10 (1992), pp. 1054–1055.

LeClercq, P. & Aigner, M.: Impact of alternative fuels physical properties on combustor performance. In: ICLASS 2009, *11th Triennial International Conference on Liquid Atomization and Spray Systems*, Vail, CO, 2009, pp. 1–6.

Lee, D.S., Pitari, G., Grewe, V., Gierens, K., Penner, J.E., Petzold, A., Prather, M.J., Schumann, U., Bais, A., Berntsen, T., Iachetti, D., Lim, L.L. & Sausen, R.: Transport impacts on atmosphere and climate: aviation. *Atmos. Environ.* 44:37 (2010), pp. 4678–4734.

Lee, S.J., Go, S., Jeong, G.T. & Kim, S.K.: Oil production from five marine microalgae for the production of biodiesel. *Biotechnol. Bioprocess Eng.* 16:3 (2011), pp. 561–566.

Levine, R.B., Pinnarat, T. & Savage, P.E.: Biodiesel production from wet algal biomass through *in situ* lipid hydrolysis and supercritical transesterification. *Energy Fuels* 24 (2010), pp. 5235–5243.

Li, L.X., Coppola, E., Rine, J., Miller, J.L. & Walker, D.: Catalytic hydrothermal conversion of triglycerides to non-ester biofuels. *Energy Fuels* 24 (2010), pp. 1305–1315.

Li, Y., Horsman, M., Wu, N., Lan, C.Q. & Dubois-Calero, N.: Biofuels from microalgae. *Biotechnol. Progress* 24:4 (2008), pp. 815–820.

Li, Y.G., Xu, L., Huang, Y.M., Wang, F., Guo, C. & Liu, C.Z.: Microalgal biodiesel in China: opportunities and challenges. *Appl. Energy* 88:10 (2011), pp. 3432–3437.

Mahmoud, A., Arlabosse, P. & Fernandez, A.: Application of a thermally assisted mechanical dewatering process to biomass. *Biomass Bioenergy* 35:1 (2011), pp. 288–297.

McBride, A.C., Dale, V.H., Baskaran, L.M., Downing, M.E., Eaton, L.M., Efroymson, R.A., Garten, C.T., Jr., Kline, K.L., Jager, H.I., Mulholland, P.J., Parish, E.S., Schweizer, P.E. & Storey, J.M.: Indicators to support environmental sustainability of bioenergy systems. *Ecol. Indicators* 11:5 (2011), pp. 1277–1289.

McDowell Bomani, B.M., Centeno-Gomez, D.I. & Hendricks, R.C.: Biofuels as an alternative energy source for aviation — A survey. NASA/TM-2009-215587. NASA Glenn Research Center, Cleveland, OH, 2009.

Melillo, J.M., Reilly, J.M., Kicklighter, D.W., Gurgel, A.C., Cronin, T.W., Paltsev, S., Felzer, B.S., Wang, X., Sokolov, A.P. & Schlosser, C.A.: Indirect emissions from biofuels: How important? *Science* 326:5958 (2009), pp. 1397–1399.

Meng, L. & Salihon, J.: Conversion of palm oil to methyl and ethyl ester using crude enzymes. *J. Biotechnol. Biomaterials* 1:5 (2011), doi:10.4172/2155-952X.1000110.

Mensch, A., Santoro, R.J., Litzinger, T.A. & Lee, S.Y.: Sooting characteristics of surrogates for jet fuels. *Combust. Flame* 157:6 (2010), pp. 1097–1105.

Mercer, P. & Armenta, R.E.: Developments in oil extraction from microalgae. *Eur. J. Lipid Sci. Tech.* 113:5 (2011), pp. 539–547.

Moavenzadeh, J., Torres-Montoya, M. & Gange, T.: Repowering transport. World Economic Forum, Geneva, Switzerland, 2011.

Moser, B.R.: Biodiesel production, properties, and feedstocks. *In Vitro Cell. Dev. Biol. Plant* 45:3 (2009), pp. 229–266.

Moses, C.A.: Comparative evaluation of semi-synthetic jet fuels: Final report. CRC Project No. AV-2-04a, 2008.

Moses, C.A.: Semi-synthetic jet fuels Addendum: Further analysis of hydrocarbons and trace materials to support Dxxxx. CRC Project No. AV-2-04a, 2009.

Moses, C.A. & Roets, P.N.J.: Properties, characteristics, and combustion performance of Sasol Fully Synthetic Jet Fuel. *J. Eng. Gas Turb. Power* 131:4 (2009), Article 041502, doi:10.11.15/1.3028234.

Mtui, G.Y.S.: Recent advances in pretreatment of lignocellulosic waste and production of value added products. *Afr. J. Biotechnol.* 8:8 (2009), pp. 1398–1415.

Murphy, R., Woods, J., Black, M. & McManus, M.: Global developments in the competition for land from biofuels. *Food Policy* 36 (2011), pp. S52–S61.

Naik, C.V., Westbrook, C.K., Herbinet, O., Pitz, W.J. & Mehl, M.: Detailed chemical kinetic reaction mechanism for biodiesel components methyl stearate and methyl oleate. *Proceedings of the Combustion Institute* 33, 2011, pp. 383–389.

Nelson, N. & Yocum, C.F.: Structure and function of photosystems I and II. *Annu. Rev. Plant Biol.* 57 (2006), pp. 521–565.

Nogueira, C.A., Feitosa, F.X., Fernandes, F.A.N., Santiago, R.S. & de Sant'Ana, H.B.: Densities and viscosities of binary mixtures of babassu biodiesel + cotton seed or soybean biodiesel at different temperatures. *J. Chem. Eng. Data* 55:11 (2010), pp. 5305–5310.

Novillo, E., Pardo, M. & Garcia-Luis, A.: Novel approaches for the integration of high temperature PEM fuel cells into aircrafts. *J. Fuel Cell Sci. Technol.* 8:1 (2011), Article 011014, doi:10.115/1.4002400.

Park, J.B.K. & Craggs, R.J.: Algal production in wastewater treatment high rate algal ponds for potential biofuel use. *Water Sci. Technol.* 63:10 (2011), pp. 2403–2410.

Park, J.B.K., Craggs, R.J. & Shilton, A.N.: Wastewater treatment high rate algal ponds for biofuel production. *Bioresour. Technol.* 102:1 (2011), pp. 35–42.

Perego, C. & Bosetti, A.: Biomass to fuels: The role of zeolite and mesoporous materials. *Micropor. Mesopor. Mat.* 144:1–3 (2010), pp. 28–39.

Petroutsos, D., Katapodis, P., Christakopoulos, P. & Kekos, D.: Removal of *p*-chlorophenol by the marine microalga *Tetraselmis marina*. *J. Appl. Phycology* 19:5 (2007), pp. 485–490.

Pitz, W.J. & Mueller, C.J.: Recent progress in the development of diesel surrogate fuels. *Prog. Energy Combust. Sci.* 37:3 (2011), pp. 330–350.

Pratas, M.J., Freitas, S., Oliveira, M.B., Monteiro, S.C., Lima, A.S. & Coutinho, J.A.P.: Densities and viscosities of minority fatty acid methyl and ethyl esters present in biodiesel. *J. Chem. Eng. Data* 56:5 (2011a), pp. 2175–2180.

Pratas, M.J., Freitas, S.V.D., Oliveira, M.B., Monteiro, S.C., Lima, A.S. & Coutinho, J.A.P.: Biodiesel density: experimental measurements and prediction models. *Energy Fuels* 25:5 (2011b), pp. 2333–2340.

Quintana, N., Van der Kooy, F., Van de Rhee, M.D., Voshol, G.P. & Verpoorte, R.: Renewable energy from cyanobacteria: energy production optimization by metabolic pathway engineering. *Appl. Microbiol. Biotechnol.* 91:3 (2011), pp. 471–490.

Raikos, V., Vamvakas, S.S., Kapolos, J., Koliadima, A. & Karaiskakis, G.: Identification and characterization of microbial contaminants isolated from stored aviation fuels by DNA sequencing and restriction fragment length analysis of a PCR-amplified region of the 16S rRNA gene. *Fuel* 90:2 (2011), pp. 695–700.

Ramos, M.J., Fernandez, C.M., Casas, A., Rodriguez, L. & Perez, A.: Influence of fatty acid composition of raw materials on biodiesel properties. *Bioresour. Technol.* 100:1 (2009), pp. 261–268.

Rashid, U., Anwar, F. & Knothe, G.: Evaluation of biodiesel obtained from cottonseed oil. *Fuel Process. Technol.* 90:9 (2009), pp. 1157–1163.

Razon, L.F.: Alternative crops for biodiesel feedstock. *CAB Reviews* 4:045 (2009), pp. 1–15.

Refaat, A.A.: Correlation between the chemical structure of biodiesel and its physical properties. *Int. J. Environ. Sci. Technol.* 6:4 (2009), pp. 677–694.

Renouard-Vallet, G., Saballus, M., Schumann, P., Kallo, J., Friedrich, K.A. & Müller-Steinhagen, H.: Fuel cells for civil aircraft application: on-board production of power, water and inert gas. *Chem. Eng. Res. Des.* 90:1 (2012), pp. 3-10.

Righelato, R. & Spracklen, D.V.: Environment — Carbon mitigation by biofuels or by saving and restoring forests? *Science* 317:5840 (2007), p. 902.

Robinson, P. & Dolbear, G.: Hydrotreating and hydrocracking: fundamentals. In: C. Hsu & P. Robinson (eds): *Practical advances in petroleum engineering.* Springer, New York, 2006, pp 177–218.

Rodolfi, L., Zittelli, G.C., Bassi, N., Padovani, G., Biondi, N., Bonini, G. & Tredici, M.R.: Microalgae for oil: strain selection, induction of lipid synthesis and outdoor mass cultivation in a low-cost photobioreactor. *Biotechnol. Bioeng.* 102:1 (2009), pp. 100–112.

Rubin, E.M.: Genomics of cellulosic biofuels. *Nature* 454:7206 (2008), pp. 841–845.

Santos, C.A., Ferreira, M.E., da Silva, T.L., Gouveia, L., Novais, J.M. & Reis, A.: A symbiotic gas exchange between bioreactors enhances microalgal biomass and lipid productivities: taking advantage of complementary nutritional modes. *J. Ind. Microbiol Biotechnol.* 38:8 (2011), pp. 909–917.

Saravanan, S. & Nagarajan, G.: Effect of single double bond in the fatty acid profile of biodiesel on its properties as a CI engine fuel. *Int. J. Energy Environ.* 2:6 (2011), pp. 1141–1146.

Sayre, R.: Microalgae: the potential for carbon capture. *Bioscience* 60:9 (2010), pp. 722–727.

Searchinger, T., Heimlich, R., Houghton, R.A., Dong, F.X., Elobeid, A., Fabiosa, J., Tokgoz, S., Hayes, D. & Yu, T.H.: Use of US croplands for biofuels increases greenhouse gases through emissions from land-use change. *Science* 319:5867 (2008), pp. 1238–1240.

Sheehan, J., Dunahay, T., Benemann, J. & Roessler, P.: A look back at the US Department of Energy's Aquatic Species Program — Biodiesel from algae. NREL/TP-580-24190. National Renewable Energy Laboratory, Golden, CO, 1998.

Sims, R.E.H., Mabee, W., Saddler, J.N. & Taylor, M.: An overview of second generation biofuel technologies. *Bioresour. Technol.* 101:6 (2010), pp. 1570–1580.

Singh, D., Nishiie, T. & Qiao, L.: Experimental and kinetic modeling study of the combustion of *n*-decane, Jet-A, and S-8 in laminar premixed flames. *Combust. Sci. Technol.* 183:10 (2011), pp. 1002–1026.

Singh, J. & Gu, S.: Commercialization potential of microalgae for biofuels production. *Renew. Sust. Energy Rev.* 14 (2010), pp. 2596–2610.

Singh, S.P. & Singh, D.: Biodiesel production through the use of different sources and characterization of oils and their esters as the substitute of diesel: a review. *Renew. Sust. Energy Rev.* 14:1 (2010), pp. 200–216.

Sivakumar, G., Vail, D.R., Xu, J.F., Burner, D.M., Lay, J.O., Ge, X.M. & Weathers, P.J.: Bioethanol and biodiesel: alternative liquid fuels for future generations. *Eng. Life Sci.* 10:1 (2010), pp. 8–18.

Slade, R., Gross, R. & Bauen, A.: Estimating bio-energy resource potentials to 2050: learning from experience. *Energy Environ. Sci.* 4:8 (2011), pp. 2645–2657.

Soh, L. & Zimmerman, J.: Biodiesel production: the potential of algal lipids extracted with supercritical carbon dioxide. *Green Chemistry* 13:6 (2011), pp. 1422–1429.

Stockenreiter, M., Graber, A.-K., Haupt, F. & Stibor, H.: The effect of species diversity on lipid production by micro-algal communities. *J. Appl. Phycology*: 24:10 2012), pp. 45-54.

Stratton, R.W., Wong, H.M. & Hileman, J.I.: Life cycle greenhouse gas emissions from alternative jet fuels. PARTNER Project 28 report, version 1.2. Cambridge, MA, 2010.

Stratton, R.W., Wong, H.M. & Hileman, J.I.: Quantifying variability in life cycle greenhouse gas inventories of alternative middle distillate transportation fuel. *Environ. Sci. Technol.* 45:10 (2011), pp. 4637–4644.

Su, C.-H., Chien, L.-J., Gomes, J., Lin, Y.-S., Yu, Y.-K., Liou, J.-S. & Syu, R.-J.: Factors affecting lipid accumulation by *Nannochloropsis oculata* in a two-stage cultivation process. *J. Appl. Phycology* 23:5 (2011), pp. 903–908.

Sun, A., Davis, R., Starbuck, M., Ben-Amotz, A., Pate, R. & Pienkos, P.T.: Comparative cost analysis of algal oil production for biofuels. *Energy* 36 (2011), pp. 5169–5179.

Sydney, E.B., da Silva, T.E., Tokarski, A., Novak, A.C., de Carvalho, J.C., Woiciecohwski, A.L., Larroche, C. & Soccol, C.R.: Screening of microalgae with potential for biodiesel production and nutrient removal from treated domestic sewage. *Appl. Energy* 88:10 (2011), pp. 3291–3294.

Tan, X.M., Yao, L., Gao, Q.Q., Wang, W.H., Qi, F.X. & Lu, X.F.: Photosynthesis driven conversion of carbon dioxide to fatty alcohols and hydrocarbons in cyanobacteria. *Metabolic Eng.* 13:2 (2011), pp. 169–176.

Tang, H., Abunasser, N., Garcia, M.E.D., Chen, M., Simon Ng, K.Y. & Salley, S.O.: Potential of microalgae oil from *Dunaliella tertiolecta* as a feedstock for biodiesel. *Appl. Energy* 88:10 (2011), pp. 3324–3330.

Timko, M.T., Onasch, T.B., Northway, M.J., Jayne, J.T., Canagaratna, M.R., Herndon, S.C., Wood, E.C., Miake-Lye, R.C. & Knighton, W.B.: Gas turbine engine emissions-Part II: Chemical properties of particulate matter. *J. Eng. Gas Turb. Power* 132:6 (2010), Article 061505, doi:10.115/1.4000132.

Timko, M.T., Herndon, S.C., Blanco, E.D., Wood, E.C., Yu, Z.H., Miake-Lye, R.C., Knighton, W.B., Shafer, L., DeWitt, M.J. & Corporan, E.: Combustion products of petroleum jet fuel, a Fischer-Tropsch synthetic fuel and a biomass fatty acid methyl ester fuel for a gas turbine engine. *Combust. Sci. Technol.* 183:10 (2011), pp. 1039–1068.

Touloukian, Y.S., Saxena, S.C. & Hestermans, P.: *Viscosity*. IFI/Plenum Data Company, New York, 1975.

TRC: TRC *Thermodynamic tables edited*. Thermodynamic Research Center, Texas A&M University, TX, 2011.

Uduman, N., Qi, Y., Danquah, M.K., Forde, G.M. & Hoadley, A.: Dewatering of microalgal cultures: a major bottleneck to algae-based fuels. *J. Renew. Sustain. Energy* 2:1 (2010), Article 012701, doi:10.163/1.3294480.

Vaccari, D.A.: Phosphorus: a looming crisis. *Sci. Amer.* 300:6 (2009), pp. 54–59.

van der Westhuizen, R., Ajam, M., De Coning, N., Beens, J., de Villiers, A. & Sandra, P.: Comprehensive two-dimensional gas chromatography for the analysis of synthetic and crude-derived jet fuels. *J. Chromatogr. A* 1218:28 (2011), pp. 4478–4486.

Van Gerpen, J. & Knothe, G.: Basics of the transesterification reaction. In: G. Knothe, J. Krahl & J. Van Gerpen (eds): *The biodiesel handbook* edited. AOCS Press, Champaign, IL, 2005.

Vaughan, N.E., Lenton, T.M. & Shepherd, J.G.: Climate change mitigation: trade-offs between delay and strength of action required. *Climatic Change* 96:1–2 (2009), pp. 29–43.

Verma, D., Kumar, R., Rana, B.S. & Sinha, A.K.: Aviation fuel production from lipids by a single-step route using hierarchical mesoporous zeolites. *Energy Environ. Sci.* 4:5 (2011), pp. 1667–1671.

von Blottnitz, H. & Curran, M.A.: A review of assessments conducted on bio-ethanol as a transportation fuel from a net energy, greenhouse gas, and environmental life cycle perspective. *J. Clean. Product.* 15:7 (2007), pp. 607–619.

Wang, Y., Ma, S., Zhao, M.M., Kuang, L.N., Nie, J.Y. & Riley, W.W.: Improving the cold flow properties of biodiesel from waste cooking oil by surfactants and detergent fractionation. *Fuel* 90:3 (2011), pp. 1036–1040.

Warner, E., Heath, G. & O'Donoughue, P.: Harmonization of energy generation Life Cycle Assessments. (LCA) edited., pp. 1–17, National Renewable Energy Laboratory, 2010.

Westbrook, C.K., Pitz, W.J., Curran, H.J., Herbinet, O. & Mehl, M.: Recent advances in detailed chemical kinetic models for large hydrocarbon and biodiesel transportation fuels. *7th Asia-Pacific Conference on Combustion*, National Taiwan University, Taipei, Taiwan, 2009.

Westbrook, C.K., Naik, C.V., Herbinet, O., Pitz, W.J., Mehl, M., Sarathy, S.M. & Curran, H.J.: Detailed chemical kinetic reaction mechanisms for soy and rapeseed biodiesel fuels. *Combust. Flame* 158:4 (2011), pp. 742–755.

Weyer, K.M., Bush, D.R., Darzins, A. & Willson, B.D.: Theoretical maximum algal oil production. *Bioenergy Res.* 3:2 (2010), pp. 204–213.

Wigmosta, M.S., Coleman, A.M., Skaggs, R.J., Huesemann, M.H. & Lane, L.J.: National microalgae biofuel production potential and resource demand. *Water Resour. Res.* 47 (2011).

Xiao, J.H., Zhang, H., Niu, L., Wang, X.G. & Lu, X.: Evaluation of detoxification methods on toxic and antinutritional composition and nutritional quality of proteins in *Jatropha curcas* meal. *J. Agr. Food Chem.* 59:8 (2011), pp. 4040–4044.

Xiong, W., Gao, C.F., Yan, D., Wu, C. & Wu, Q.Y.: Double CO_2 fixation in photosynthesis-fermentation model enhances algal lipid synthesis for biodiesel production. *Bioresour. Technol.* 101:7 (2010), pp. 2287–2293.

Yang, C.-Y., Deng, X., Fang, Z. & Peng, D.-P.: Selection of high-oil-yield seed sources of *Jatropha curcuas* L for biodiesel production. *Biofuels* 1:5 (2010), pp. 705–717.

Yang, J., Xu, M., Zhang, X.Z., Hu, Q.A., Sommerfeld, M. & Chen, Y.S.: Life-cycle analysis on biodiesel production from microalgae: water footprint and nutrients balance. *Bioresour. Technol.* 102:1 (2011), pp. 159–165.

Yazdani, S.S. & Gonzalez, R.: Anaerobic fermentation of glycerol: a path to economic viability for the biofuels industry. *Curr. Opin. Biotechnol.* 18:3 (2007), pp. 213–219.

Yuan, W., Hansen, A.C. & Zhang, H.: Predicting the physical properties of biodiesel for combustion modeling. *Transactions of the American Society of Agricultural Engineers* 46:6 (2003), pp. 1487–1493.

Yuan, W., Hansen, A.C. & Zhang, Q.: Predicting the temperature dependent viscosity of biodiesel fuels. *Fuel* 88:6 (2009), pp. 1120–1126.

Yuan, W.Q., Hansen, A.C., Zhang, Q. & Tan, Z.C.: Temperature-dependent kinematic viscosity of selected biodiesel fuels and blends with diesel fuel. *J. Amer. Oil Chem. Soc.* 82:3 (2005), pp. 195–199.

Zhang, X.Z., Hu, Q., Sommerfeld, M., Puruhito, E. & Chen, Y.S.: Harvesting algal biomass for biofuels using ultrafiltration membranes. *Bioresour. Technol.* 101:14 (2010), pp. 5297–5304.

APPENDIX A. BASIC TERMINOLOGY AND CONCEPTS IN HYDROCARBON CHEMISTRY

For our purposes, we can think of molecular structure and chemical reactions as being all about putting a stable number of electrons in atomic orbitals. Each orbital can accommodate a discrete number of electrons; in successive rings, that number is 2, 6, 10, and 14. Hydrogen has a single proton in its nucleus and one electron in the first orbital, so that hydrogen can share one electron and one vacancy with another atom. Carbon has vacancies for four electrons in its outer shell, so it can share electrons with up to four hydrogen atoms. With a series of single bonds (each set of atoms share only two electrons and two vacancies total), chains of hydrogen and carbon can also be built up to form stable molecules. The chemical formula for these molecules can be written as C_nH_{n+2} where n is the number of carbon atoms present in the molecule. The extra hydrogen atoms are needed to stabilize the carbon atoms at the end of a chain, as shown in Figures A.1a and b. As the number of carbons in the molecule increase, the straight-chain hydrocarbons have higher viscosities, boiling points, and lubricating indices. However, the molecule does not have to follow this straight-chained structure. The molecule in Figure A.1c is an isomer of butane, i.e., it has the same chemical formula, but it has a different configuration, in this case a branched structure. Hexane has a ladder-like structure with 6 carbon and 14 hydrogen atoms. Cyclohexane, shown in Figure A.1d, is an isomer of hexane. For carbon, 6 is the magic number through which the atoms can form a stable ring. All of the molecules from A.1a through A.1d are called saturated hydrocarbons, meaning that each carbon atom still shares a single electron with one neighboring carbon and hydrogen atoms. Such molecules consisting of single bonds are called alkanes (also paraffins, particularly in petroleum engineering). Alkanes with a cyclic structure (Fig. A.1d) are also called cycloalkanes or napthenes.

Alkanes react with oxygen to produce carbon dioxide (CO_2), water (H_2O) and energy in a chemical reaction of the form:

$$C_nH_{2n+2} + (3n + 1)/2O_2 \rightarrow nCO_2 + (n + 1)H_2O + energy$$

Note that this is the net reaction; the actual chemical reaction will almost always take a series of steps in which intermediate species are created and consumed. Hydrocarbons with double or triple covalent bonds between adjacent carbon atoms are termed "unsaturated". An alkene (also olefin) is an unsaturated hydrocarbon with one or more carbon-to-carbon double bonds. The simplest non-cyclic alkenes have only one double bond and can be described by the chemical formula C_nH_{2n}. such as ethylene, shown in Figure A.2a. There are also ring-shaped alkenes, such as benzene (Fig. A.2b). Aromatics are very stable molecules that contain one or more benzene rings; many of them smell good, hence the name. Their properties are sufficiently different so that they are considered as a separate class from alkenes.

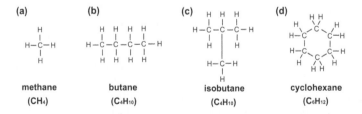

Figure A.1.   Molecular structure of alkanes include straight chain hydrocarbons such as (a) methane; (b) butane; Branched alkanes such as (c) isobutane; and Cyclic hydrocarbons such as (d) cyclohexane.

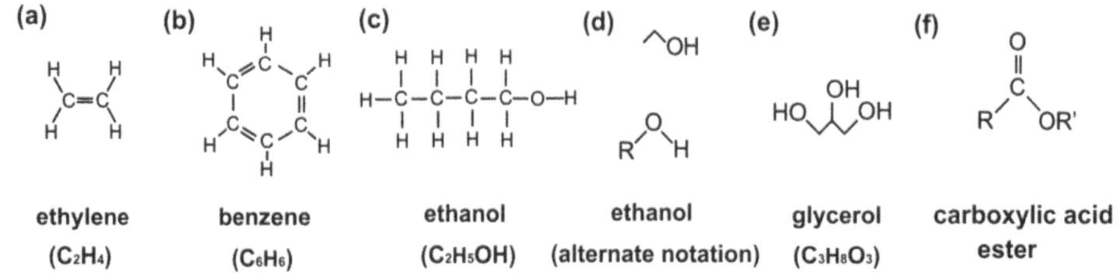

(a)   (b)   (c)   (d)   (e)   (f)

ethylene   benzene   ethanol   ethanol   glycerol   carboxylic acid
(C₂H₄)   (C₆H₆)   (C₂H₅OH)   (alternate notation)   (C₃H₈O₃)   ester

Figure A.2.   Molecular structure of more complex hydrocarbons, including alkenes such as (a) ethylene; and cyclic alkenes such as (b) benzene. Replacing a hydrogen molecule with a hydroxyl group (OH⁻) yields alcohols such as (c) ethanol, which can also be written in alternate notation (d). Other compounds of interest include (e) glycerol and esters, such as (f) a carboxylic acid ester.

Alcohols, such as ethanol in Figure A.2c, have similar structures, but in this case one or more hydrogens are replaced by hydroxyl (OH⁻), which is a functional group that results in a net molecular negative charge. Functional groups are atomic subunits within a molecule that give rise to chemical reactions. Higher carbon alcohols include ethanol, a straight-chain molecule with a terminal hydroxyl functional unit (Figs. A.2c,d). As the complexity of the molecule increases, notation becomes more challenging and a sort of shorthand is used. Some simplifications are shown that represent ethanol in Figure A.2d. Carbon atoms and their supporting hydrogens are represented by kinks and the empty terminus in the version on top. On the bottom, R is used to designate a representative alkyl, in this case the ethyl group C_3H_5, while the other atoms are explicitly shown. Glycerol (Fig. A.2e) is a reaction product from the refining process and mimics half of a sugar molecule. Esters are present in vegetable oils and animal fats, and they may also appear as reaction products. They come from from the reaction of an alcohol with an acid. Carboxylic acid esters (Fig. A.2f) are based on a carbon atom and have the general form $RCOOR'$, where R is an alkyl group such as C_3H_5 and R' is an aryl group. A moiety is a subunit within a molecule that may itself include functional groups within it. There may be several ways of parsing a molecule into its constituent moieties.

Other molecules that use hydrogen, carbon and oxygen atoms but in more complicated configurations include lipids, fatty acids, sugars and carbohydrates (a/k/a saccharides) with formula form $C_m(H_2O)_n$. Smaller carbohydrates (mono- and disaccharides) are sugars. Polysaccharides are polymeric carabohydrates, i.e., they are comprised of simpler units, such as a mono- or disaccharide, in a repeating chain. Fatty acid methyl esters (FAME) are key components of vegetable oils and animal fats. Free fatty acids are carboxylic acids with a long unbranched hydrocarbon chain attached to it. The hydrocarbons may be saturated without any double carbon-to-carbon bonds; monounsaturated with a single double bond; or polyunsaturated with multiple double bonds in the hydrocarbon chain. Glycerol, shown in Figure A.2e, has three ends at which it may couple with fatty acids. When the coupling occurs, the molecules are termed "glycerides". A glycerol with one fatty acid is a monoglyceride; likewise a glycerol with two fatty acids is a diglyceride. A triglyceride (also called triacylglycerol or triacylglyceride or TAG) is comprised of three fatty acids joined to glycerol subunit. TAGs are lipids that some algae use as its main carbon storage molecule. These compounds will be important when we consider biorefining in section 11.5.

More nomenclature: $Cx:y$ is the lipid number, in which x is the number of carbon atoms in the fatty acid and y is the number of double bonds in the fatty acid. $Cx:0$ is a saturated compound. C18:3 denotes an acid with 18 carbon atoms and 3 double bonds, but there are several molecules that fit this description. To be precise, it is appended with notation that indicates where the unsaturated bond is located relative to the carboxylic acid end, $n - x$, $\Delta - x$ or Δ^x (where the xth carbon molecule has a double bond with the $(x + 1)$th carbon molecule), depending on the field of study. Palmitic acid is C16:0; stearitic acid is C18:0; oleic acid is C18:1; linoleic acid (a/k/a ω-6 fatty acid) is C18:2; α-linoleic acid (a/k/a ω-3 fatty acid) is C18:3. Other examples and discussion may be found in section 11.3.1.

CHAPTER 12

Pulp and paper industry – trends for the future

Erik Dahlquist & Jochen Bundschuh

We have seen a strong expansion of pulp and paper production during the last 20 years in especially Asia and South America. Many new pulp mills have been erected in Indonesia and Brazil where relatively cheap and fast growing species like Eucalyptus has become very important. Other species like Acacia and hard wood from natural forests have increased.

At the same time, many old mills are closed in Europe and North America as they get difficulty to compete. The more expensive wood species as well as higher salaries and often old machines are important factors, but the fact that pulp and paper industry have difficulties to compete on capital for investments in e.g. the US, Canada and Europe makes it difficult to enhance the production also in some existing mills. Due to expanding markets in the "new" economies in China, India and Brazil and many other countries, the demand for paper is increasing there. As fiber is a limited resource, recycling of used fibers is becoming even more important.

At the same time, we can see that the use of newsprint paper is decreasing due to competition from electronic media, while packaging material as well as tissue is increasing. In the US newsprint sales have gone down by 48% between 2005 and 2010 (Thorp, 2010). In SPCI (2011), a prediction has been made about when newsprint will be "insignificant". This will be so already by 2017 in the US with the present trend! In UK we will be there in 2019 and in Canada and Norway in 2020. To meet these new demands the large pulp and paper corporations are specializing into more narrow segments of the market. Storaenso for instance is concentrating a lot of their efforts into packaging according to Jouko Karvinen, CEO (Karvinen, 2011). Due to poor economic conditions in North America Storaenso is also leaving this continent at least concerning production units. Instead, they expand in China, Brazil and in the future also in India, where the markets are growing. Borregard has concentrated a lot on dissolving pulp/viscose and has several partners in different countries. One of them is SAPPI who is now also focusing on dissolving pulp, especially in their South African mills and in the US (SAPPI, 2011).

Borregard as well as Domsjö are also producing many different chemicals like lignosulfonate, ethanol and Borregard also Vanilla. As they have sulfite processes this fits very well, and is quite different from many other companies. They are leading the trend towards biorefineries.

SCA is concentrating on fine paper and hygiene products (SCA, 2011) but also wind power production in a large on-land wind power farms in Northern Sweden. Södra, owned by the forest owners in Sweden, is concentrating on long fiber pulp for especially the European market. They have no intention to produce paper, as they primarily want to have someone buying their forests to a good price. Energy is important for them, so thus they cooperate with Statkraft on forest based wind power parks at their forest owners land. Another group, Korsnas, is focusing on liquid board, where Tetrapak is the main customer. Here the strategy is to have a few large plants where the production is highly automated. The owners are actually investing a lot in TV-channels, the METRO journal and telephone companies, as a side activity to the pulp and paper. The company also is selling surplus heat to the local municipality, Gavle. UPM has many mills producing printing paper in a declining European market for this type of products. Thus, mills are closing. For the future integration of biomass use for energy and other applications is of interest (Arve, 2001a,b), and the company owns part of a new nuclear power plant in Finland. Other products are Labels and Plywood that can be formed. The company also is expanding in China, where printing papers is still growing fast. In the future, there will be new products for energy, bio-chemicals,

Figure 12.1.   Ngodwana pulp mill in South Africa, SAPPI.

bio-composites and micro fibrils. UPM also has a goal to reduce own energy use by 30% in primarily the paper mills.

In East Asia APP (Asia Pulp and Paper, part of Sinar Mas group) is building new newsprint machines in China where the market for this type of products is still increasing rapidly (APP, 2011). They also produce packaging and tissue and the focus is not to be narrow and specialized, but more to have very low production cost. They have many environmental organizations taking actions towards their use of rain forests, but claim to have taken strong environmental actions last few years and actually claim they follow EU Ecolabel. The other major South East Asian company based in Singapore is APRIL, which has been primarily focusing on fine paper produced in Indonesia, but coming from dissolving pulp originally. They are today one of the biggest producers of bleached hardwood pulp, with also ECF, Elementary Chlorine Free. They have a broad spectrum of products like liquid packaging, printing and writing paper, tissues, shopping bags, food packaging, magazines and books, which shows that the specialization has not yet been driven that strongly. It is still more important to grow than specializing on a specific segment (APRIL, 2011). Also APRIL is growing in China with a pulp mill starting up 2010 in Rizhao and a fine paper machine in Guangdong and plans for liner board, according to AJ Devanesan, CEO (Asp, 2011).

Visy pulp and paper, with the base in Australia is focusing very much on packages for fruit and heavy products, and is expanding with this globally. From 2006 to 2010, Weyerhaeuser reduced their production by around 50% on structural lumber board and oriented strand board. During the same period, pulp production went down by 20% while paper slightly increased. The major volume was on fluff pulp for tissue (Weyerhaeuser, 2010). A main focus for Weyerhaeuser is to manage the 6 million acres of timberland which may "provide a source of sustainable products, green energy and carbon sequestration as well as a host of other solutions for challenges facing the world" according to the CEO Dan Fulton (Weyerhaeuser, 2011). Real estate business is one business area of major importance, which is integrating the timber forwards in the value chain. International paper (2011), one of the biggest pulp and paper companies globally, is focusing on uncoated papers and industrial and consumer packaging, complemented by xpedx. This was just some examples.

From a biomass perspective it is interesting to note that the major commercial use of biomass was for pulp and paper only some 20 years ago, while today the demand for use of wood for energy use is increasing especially in Northern Europe in countries like Germany, Sweden and Finland and in e.g. Brazil for production of Ethanol but then from sugar cane. Generally, we can foresee that the competition for fibers will increase. This will on the other hand drive the development of new wood and crop species and use of waste for both pulp and paper and many other applications. Biorefineries have already been mentioned in several chapters of this book, but in the future we probably will see many new applications where wood and crops replace oil as the raw material. This will build new conglomerates where pulp and paper will be one product among many different products. Production of chemicals for production of plastics, all kind of chemicals, fuels for vehicles as well as production of electricity and heat for district heating will be complements to the paper.

Christer Segersteen, president for Södra P&P corporation, says that although Sweden is using 105 TWh/y from forests there is a potential to increase the energy outtake with at least another 20 TWh/y by taking out roots, branches and similar. There also must come more energy forests by having hybrid poplar, hybrid aspen, fast growing larch and pine and others. Södra, owned by the forest owners and farmers, see this as a very interesting business for the future. During the last 20 years the share of earnings from energy has increased from a few % to more than 20% for Södra, and the potential to grow further is very high (Krögerström, 2010).

The chemical company Perstorp in Southern Sweden is today replacing almost all fossil raw materials with wood or crop based materials. From this, they produce among others methanol and paints (Perstorp, 2011).

There is also a new trend towards rather smaller paper machines – approximately 6 meters wide – for copy paper and wood free paper. This is happening at the same time as Voith's CEO Hans Peter Sollinger (Voith, 2011) predicts 500 million tonnes of paper to be produced in 2015. He also says that shortage of energy, water and fibers will drive the business. Voiths R&D manager Thomas Wurster believes recycle fiber will be used much more as total energy will be reduced by 40% and electricity by 20%. This will mean some 100 million tonne more recycle fibers produced until 2025. Many paper qualities will have 100% recycle fibers he believes (SPCI, 2010)

In Sweden there is a strong push to save energy and in a program (SPCI, 2009) a few years ago 0.67 TWh$_{el}$/y in a 5-year program (from 30 TWh/y$_{electricity}$) was achieved. It may sound little, and there is a much larger potential with more efficient processes. In China, there was a program in 2008 to reduce the energy utilization by 20% before the end of 2010 for the 1000 highest consuming companies in China. This included several pulp and paper mills.

On the other hand – some companies only focus on high capacity of paper, and don't care much about the quality, while others focus more on high quality products. Today they will mostly not compete directly with each other. Some products like tissue also are not suitable for long transportation as they lose their properties then. Thus, tissue is often produced in smaller machines close to the customers.

In 2009, Ernest and Young presented a report "Global Trends in the Pulp and Paper industry" (Ernst and Young, 2009). In this, we have a historical development of paper and board from 1990 and predictions until 2020. In 1990 the total consumption was approximately 240 million tonnes per year, 2000 325 million tonnes, 2010 430 million tonnes and predicted 2020 550 million tonnes per year. The biggest increase will be in Asia, while it will be stagnant or even slightly reduced in Western Europe and North America. Eucalyptus and akazia grown in South East Asia and South America is used mostly on these new markets in China and India, where fiber is a limiting source. E&Y has made an overview of the situation in the local regions as follows: China – high growth, limited fiber sources, aggressive investments; Western/Nordic Europe – mature market, restructuring underway, technology leadership; North America – mature/declining market, technology leadership; Latin America – fiber sources available, pulp supplies and investments; Russia – huge coniferous fiber sources, sizeable market potential, poor infrastructure, investments needed; India – big potential, limited own capacity, limited fiber sources.

Ingemar Croon (Croon, 2010), a Nestor in the P&P business since many years, says the change in forest industry is stronger now than ever! This is having several aspects:

- Restructuring – closing or buying complete corporations (e.g. MoDo and Weyerhaeuser) or merger between very large global companies (like SCA Forest Ind and Holmen, UPM-Mreal, Aracruz-Votoratim and large Chinese companies).
- New product on new-old machines.
- Biorefineries including energy production, liquor gasification etc.
- Bio combines
- Pellet- bio char- gasification – methanol- biochemical production.
- Ethanol from biomass.

Among others a 600,000 tonne pellet factory has been starting up in Florida 2008 and German RWE has built an 800,000 tonne pellets plant in Georgia, USA.

Large eucalyptus plantations are found in Brazil and Indonesia. The companies doing these are accused of non-sustainable actions. Still, more is planted, and in 2009 450,000 ha eucalyptus was planted only in the sub-state Bahia, with a cycle of 5–7 years compared to 70–100 years in northern forests like in Scandinavia, Russia and Canada. Eucalyptus has deep roots and can be grown also in dry areas, if surviving the first tough time after planting.

Russia has 25% of the global fiber resources (SPCI, 2011). New pulp mills are being erected to increase export of fibers to especially China.

In Thorp (2010) a review of the use of fibers in best way is made. In the paper, the added value from the US$ 150 value of the pulpwood to final products is calculated. The total value multiplier from Pulpwood to pulp is 14–33, to power 3, while from slash to biofuel 8–18 and to power 5. Conversion using different methods is also calculated like hydrolysis, gasification and pyrolysis.

Technology wise several companies like Borregard and Storaenso are starting pilot plants for nanofibers, where Borregard will use in paint first, while Storaenso for stronger packages (Karvinen, 2011).

A trend is also to utilize wood as good as possible. If you use wood for building houses for instance, you both get a renewable building material that binds CO_2 instead of cement that causes release of large amounts of fossil fuels when utilized for building houses. Antti Lagus at VTT (Lagus, 2011) reports how to use logs efficiently in a report 2011. Today typically 50% is made into sawn timber, 10% becomes sawdust and 30% becomes wood chips. This includes 10% as bark. With better technologies, scanning the stems more timber could be achieved, as knots, defects could be detected, and sorting can be made with respect to wood quality.

QUESTIONS FOR DISCUSSIONS

- How do you think the problem with competition for virgin fiber between pulp and paper industry and other usages like as biomass fuel in power plants should be solved?
- Is it the market forces that shall solve the problems and if so, will subsidies for one application be acceptable in relation to other usages?
- How shall we get enough wood for all the demands we have? Should we start to distribute fertilizers in the forests on a large scale?
- What is your opinion about what trees to cultivate? Should we have optimal species in mono-cultures or should we direct forestry towards many different species like in rain forests? Discuss both advantages and disadvantages with both alternatives.

REFERENCES

APP: 2011, http://www.asiapulppaper.com/ (accessed December 2011).

APRIL: 2011, http://www.aprilasia.com/process_products.html (accessed December 2011).

Arve E.: UPM management has found the way towards the future (UPMs företagsledning har hittat vägen framåt). *SPCI Svensk Papperstidning* 8 (2011a), pp. 12–14.

Arve, E.: Interview with Jouko Karvinen, CEO Storaenso. *SPCI Svensk Papperstidning* 1 (2011b), pp. 12–15.

Asp, M.: Interview with AJ Devanesan, CEO APRIL: *SPCI Svensk Papperstidning* 1 (2011), pp. 20–22.

Croon Ingemar: A new global forest industry is developing massively (En ny global skogsindustri växer fram på bred front). *SPCI Svensk Papperstidning* 6 (2010), pp. 16–19.

Ernest and Young: Global trends in the pulp and paper industry. SPCI/Svensk Papperstidning No. 8, 2009, pp. 28–31.

http://www.internationalpaper.com http://phx.corporate-ir.net/phoenix.zhtml?c=73062&p=irol-irhome. 2011.

Karvinen, J.: Packaging and nanofibers. *SPCI Svensk Papperstidning* 7 (2011), p. 8.

Krögerström, L.: Forestry can contribute more (Skogen kan ge stort tillskott). Interview with Christer Segersteen, president for Södra P&P corporation. *Energivärlden* 3 (2010), p. 14.

Lagus, A.: More out of wood. *VTT Impuls* 2 (2011), pp. 42–47.

Perstorp chairman Lennart Holm: Biorefineries. Kemivärlden Biotech med kemisk tidskrift,o 11 (Nov 2011), p. 7.

SAPPI to convert kraft pulp mill. Nordisk papper & massa, 6 (2011), p. 37.

SCA is strengthening to meet an uncertain future (SCA stärks inför osäker framtid). SPCI/Svensk Papperstidning. No. 9, 2011. p. 6.

SÖDRA investments in Russia (Södras satsning I Ryssland): SPCI/Svensk Papperstidning. No. 8, 2001, p. 16–17.

SPCI/Pulp and paper industry reduced energy most efficiently (Massa och pappersindusrin energieffektivis-erade mest). *SPCI Svensk Papperstidning* 10 (2009), p. 6.

SPCI/Svensk Papperstidning. No. 6, 2010, Voith Thomas Wurster R&D manager, recycle fiber will be much more used.

SPCI/Svensk Papperstidning. No. 9, 2011, A prediction has been made about when Newsprint will be "insignificant". See link www.futureexploration.net.

The 2010 Weyerhaeuser fact book. 2011. http://investor.weyerhaeuser.com/phoenix.zhtml?c=92287&p=irol-irhome.

The 2010 Weyerhaeuser annual report and form 10-k. 2011.

Thorp B. A., & Akhtar M.: The best use of wood. TAPPI Paper 360 Jan/Febr 2010, p. 26–29.

Voith's CEO Hans Peter Sollinger predicts 500 million tonne paper to be produced 2015. Svensk Papperstidning, 2011.

CHAPTER 13

Biorefineries using waste – production of energy and chemicals from biomasses by micro-organisms

Elias Hakalehto, Ari Jääskeläinen, Tarmo Humppi & Lauri Heitto

13.1 INTRODUCTION

In nature's ecosystems, energy is flowing and the matter is circulating. Microorganisms are the essential link in maintaining balance in returning the substances into circulation. They are capable of degrading the biomasses. Modern sustainable bio-industries should be based on the biomass and waste utilization by the microbial strains and their enzymes, which carry out the energetically favorable biocatalysis.

The ordinary methods for researching the microorganisms include their propagation in laboratory conditions using:

- solid media,
- liquid media, or
- semisolid media.

The chosen medium has to contain the carbon (C) source, energy source, nitrogen (N) source, other minerals, trace elements and vitamins, according to the requirements of the cultivated organism(s). Any microbe has its optimum and limits regarding such parameters as pH, dissolved oxygen, or production temperature. The optimal conditions may be different for the microbial growth and for the product formation. In this chapter, we concentrate on the production of fuels and bulk chemicals by microbe cultures. In some sense, this also has been the original meaning of the concept "biotechnology". It also provides the humanity vast potential for rearranging the economies to be based on sustainable industrial microbiology. The processes could then be based on the degradation and circulation chains in nature.

In many recent considerations, the word "biorefinery" has been designating technologies, which could use biomass raw materials instead of oil and other fossil deposits for the foundation of the energy maintenance and the chemical industries. Those applications, where the microbes liberate the chemically bonded energies in and from various organic raw materials could give a positive overall energetic balance, for example for waste treatment processes. These natural reactions make possible the flexible use of alternative biomass sources as well as multiple use of the bioreactor plants and installations. We wish to open up a view into these potentials, and challenges, of the old and at the same time extremely modern discipline of industrial biotechnology.

In the research purposes, we have been using a miniaturized cultivator system, PMEU (Portable Microbe Enrichment Unit) for the experimentation. In this device, it is possible to use various gas flows in selected temperatures for the microbial bio-reactions. The resulting activities and changes are measurable optically, or by following up the gas and volatiles emissions from the minireactor system. By using the PMEU, it has been possible to investigate the interactions of several microbes in mixed cultures, as well as optimize the physicochemical conditions for different processes.

In practice, the implementation of microbial bioengineering requires research on biocatalysis. This enzymatic powerhouse of cell systems is offering a basis for industries on a large scale. In order to facilitate the change into more sustainable biomass-based industries we need to improve the efficiencies of the pretreatments, and productivity of the bioprocesses, together with improving the integration of the downstream processing with the production itself. As a result, we learn to exploit nature's own means for circulating matter. The corresponding technologies represent sustainable industries in their purest form. Their usage includes also circulating the waters. In addition, the carbon-containing gases like CO and CO_2 are being effectively collected in the biorefinery solutions. All the waste materials become useful substrates for the future industries.

13.2   SUSTAINABLE PRODUCTION OF FUELS AND CHEMICALS FROM WASTES AND OTHER BIOMASSES

13.2.1   *Circulation of matter and chemical energy in microbiological processes*

In biorefinery industries, wastes are not wastes, but they form another raw material source. Establishing biorefinery "fields" instead of landslides, for example, we are able to fully exploit the circulation of matter described above. During these times of critical oil price fluctuation and risks of inadequate supply, the biomass alternative could offer a replacement and new backbone for our modern societies. This sustainable alternative is based on the enormous potential of microscopic cells and their enzymes in the production of fuels, plastic, textiles, bulk chemicals and other commodities.

The cell theory was established in 1839, and the reproduction of cells by division was affirmed as a basis of biological life in 1858 (Graham, 1982). All metabolic activities of the cells require exchange of substrates with the surroundings. Small bacterial cells have more surface area in proportion to the cell volume than the larger eukaryotic cells. This partially explains the relatively high speed of their metabolic action. The basis of this exchange of substances is the diffusion of gases. In aerobic bioprocesses on optimal substrate availability oxygen is usually forming the rate-limiting step.

The smaller the cells are, the quicker the nutrient molecules are diffused and metabolized inside the cells. The pace of the metabolic events is determining the rate of the production of the biochemical (of course taking the cell's regulatory mechanisms into account). Thus, the overall productivity of the biotechnical reaction is largely depending on the diffusion conditions. The gases are freely diffused into and out of the cells, whereas larger molecules, such as glucose, need facilitated diffusion across the membranes with carrier molecules. Undoubtedly, the concentration gradient is contributing to the rate of transport. The active transport requires energy in the form of the ATP. Naturally, the diffusion moves substances from higher concentrations toward the lower ones. Therefore, exploitation of the diffusion and overcoming it on the other hand play a key role in the cell metabolic efficiency. These principles are offering new grounds for modern biotechnology era.

Waste is a heterogeneous energy source and different types of wastes can be converted into different energy products or bulk chemicals in different conversion processes (Thorin *et al.*, 2011). In all reaction sequences, the convertible source of chemical energy is derived from solar energy bound on the plant material and into the nutritional chains. Microbes carry out the degradation and circulate the matter in nature. This approach has been adopted by the industrial microbiology discipline.

13.3   REPLACING FOSSIL FUELS BY THE BIOMASSES AS RAW MATERIALS

The term "feedstock" refers to raw materials used in biorefinery (Cherubini, 2010). The biomass is synthesized via the photosynthetic process that converts atmospheric carbon dioxide and water into sugars. In a photosynthetic organism, such as an algae or a plant, the cells utilize the sugar molecules to synthesize *via* the biosynthesis the complex materials that are generically named biomass. Then the so-called heterotrophic organisms use the organic biomass as raw materials for their catabolic and anabolic reactions. These further conversions from one entity to another one are establishing also new classes of substrates. As the raw material supply is an important stage in biorefinery system, the provision of a renewable, consistent and regular source of feedstock is a favorable condition. These biomass sources are replacing the fossil deposits.

On the other hand, many biomass substrates are available on seasonal basis after harvesting, for example. Then it comes to the flexibility of the microbial bioprocesses. Since the heterotrophic processing by the microbial cultures and communities is based on the exploitation of the energy contents of the biomass substances, the overall economics of the process is dependent on the energy density of the substrate. If reduced costs of transport, handling or storage are aimed for, the energy densities need to be increased or concentrated. In a properly functioning biorefinery unit, this could be achieved by combining the raw materials, and the above-mentioned unit operations. If this flexibility is associated with the option for multiple uses of the bioreactor facilities, it could be well stated that the establishment of a full-scale biorefinery system is requiring and also bringing along a new kind of industrial logics. On the other hand, this revolution in our thinking is needed for successful management of biological materials and processes, where ordinary skills and knowledge of basic process chemistry often are inadequate. For example, with proper planning we can move from one operational mode to another according to the availability or cost of the raw materials, or based on the demand of the various product substances in the market.

Biomass can be defined as organic matter available on a renewable basis (Lucia *et al.*, 2006). Renewable carbon-based raw materials for biorefinery purposes are provided from four different sectors (Cherubini, 2010):

- agriculture (dedicated crops and residues),
- forestry,
- industries (process residues and leftovers) and households (municipal solid waste and wastewaters),
- aquaculture (algae and seaweeds).

A further distinction can be made between those feedstocks, which come from dedicated crops and residues from agricultural, forestry and industrial activities, which can be available without deprivation from e.g. food production. The main biomass feed stocks can be grouped in three wide categories: carbohydrates and lignin, triglycerides and mixed organic residues. In the plant materials, the lignin is usually occurring together with cellulose and hemicellulose, forming some extra challenges for the process construction (Gressel and Zilberstein, 2007).

The most important energy products, which can be produced in biorefineries, are, according to Cherubini (2010):

- gaseous biofuels (biogas, syngas, hydrogen, biomethane),
- solid biofuels (pellets, lignin, charcoal),
- liquid biofuels for transportation (bioethanol, biodiesel, FT-fuels, bio-oil).

It is important to take into account that in the industrial combustion it is possible to combine the solid fuels and waste materials with bioprocess fluids and suspensions. Ethanol from fermentation is usually used as fuel in transportation. The gas from anaerobic digestion can also be used after upgrading as fuel for transportation. It can also be burned directly in an oven, gas engine or gas turbine for heat and power production (Thorin *et al.*, 2011).

Biobutanol is a promising alternative for vehicle fuels (www.butanol.com). Its feasible industrial production has been so far restricted by the low product yield and productivity in the bioprocesses. The main route for butanol is the acetone-butanol fermentation, which was found about 100 years ago by Chaim Weizman (Jones and Woods, 1986). This reaction is carried out by the anaerobic bacterium *Clostridium acetobutylicum*.

In some process alternatives, *Clostridium beijerinckii* is being explored as promising producer organism for biobutanol production (Kumar and Gayen, 2011). Also genetically modified organisms (GMO) have been suggested and attempted for that purpose. Their usage, however, includes complications, which possibly could be avoided by the sophisticated use of mixed cultures, integrated or consolidated bioprocess (CBP) etc. modern means.

Common substrates in agricultural biogas plants are (Thorin *et al.*, 2011):

- liquid and solid wastes from livestock husbandries (like manure),
- ensiled energy crops,
- harvesting residues,
- residues from waste material market.

The most important chemical and material from the biorefineries products are the following (Cherubini, 2010):

- chemicals (fine chemicals, building blocks, bulk chemicals),
- organic acids (succinic, lactic, itaconic and other sugar derivatives),
- polymers and resins (starch-based plastics, phenol resins, furan resins),
- biomaterials (wood panels, pulp, paper, cellulose),
- food and animal feed,
- fertilizers.

This list is illustrating the fact that bio-industries powered by microbes could be replacing the petrochemical industries gradually.

In anaerobic digestion biogas, mainly consisting of methane, is produced. A large variety of wastes are digested such as sludge from waste water treatment plants, grease trap sludge, manure, bio-waste, municipal solid waste, food wastes, refuse derived fuel and industrial waste water.

Another biological conversion process that can use waste as feedstock is fermentation to ethanol or butanol. Wastes of interest for fermentation are industrial wastewaters and lignocellulosic wastes such as straw and wood wastes. Fermentation of sugary wastes to ethanol is a mature and well-proven technology but fermentation of lignocellulosic waste and fermentation to butanol is still at the research stage (Thorin *et al.*, 2011). One lucrative possibility could be the combination of different wastes, or new materials, as starting point for industrial processes.

The exploitation of fats derived from animal and plant kingdoms is an industrially feasible approach (Metzger and Bornscheuer, 2006). Due to their pure aliphatic structure, the fatty acids can be readily combusted as fuel. They are also potential raw materials for various chemicals (Metzger and Bornscheuer, 2006).

Proteins are abundant in food and are vital to nutrition and biochemical function (Barone and Schmidt, 2006). They are derived from agricultural sources and used in everyday products such as glues and textiles. They are also used as replacements for several petroleum-derived materials. Proteins are the source of polymers for fiber, molded plastics, films, and an array of products. They constitute a sustainable resource and can often be processed in the same way as conventional synthetic polymers.

13.4   MICROBES CARRY OUT THE REACTIONS WITH ENERGETICALLY FEASIBLE BIOCATALYSIS

Traditional biotechnology could be determined by some ideals into technologies where bio-catalysis is applied for the process management. Therefore, such applications as e.g. biodiesel production fall outside this concept. On the other hand, if biocatalysis is involved, it usually is also a sign of energetically feasible manufacturing system.

13.4.1   *Ecological thinking based on understanding microscopic interactions*

Microbial interactions are differing from the ones of zoological ecology, where various individual encounter occasionally. The microbes are continuously living in a "biochemical universe." They are under the influences of the surrounding chemical gradients. If active, they have to deal, besides with the physicochemical circumstances, also with their own metabolic products. In the microscopic world, they are also subjected to the signaling of other cells and populations.

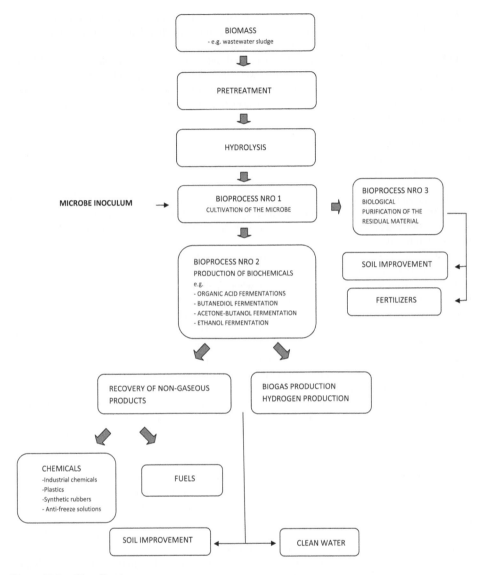

Figure 13.1.   Biorefinery process.

13.4.2   *Air and water pollution diminished by natural processes*

Biorefinery industries need to be established in the proximity of biomass sources. These often are various waste sentiments, sludge, surplus disposals or the overflow or outcome of different treating plants. By handling the organic materials, biorefineries improve the environmental quality as the waste fraction is dealt with according to the schematic flow presented in Figure 13.1.

13.5   TRANSPORT OF FUELS AND CHEMICALS LESS ABUNDANT AND RISKY WHEN LOCAL SOURCES ARE EXPLOITED

Oil is one of the greatest harms from the environment's point of view: it is a non-renewable natural resource (Välimäki, 2005). Oil drilling spoils environment in ecologically sensitive areas.

Oil burning causes a remarkable portion of the world's carbon dioxide emissions. It also causes sulfur and nitrogen emissions, acidifying forests. Gasoline and diesel are the greatest polluters of air in cities. Oil accidents contaminate waters and soil.

Oil spill can cause serious impacts on coastal areas (Gråsten and Kiukas, 2004). Marine organisms may be affected by oil in several ways. For example in Finland, having around 5.4 M inhabitants, roughly 2000 oil accidents take place annually. Majority of these causes only small amounts of oil leaking to the environment. Cases that are more serious amount around 160 per year, and can cause problems:

- as a result of physical contamination
- by toxic effects of chemical components
- by accumulation of substances.

Physical contamination by hydrocarbon components is the main threat to marine organisms and habitats after an oil spill (Hänninen and Rytkönen, 2004). The most endangered organisms are those living near shoreline or sea surface, such as seabirds, marine mammals and fish during their spawning season

Most oil accidents are happening due to various over-fillings, leaks and damaged oil tanks (Veriö, 1990). Over-fillings are usually relating to filling oil tanks in houses or otherwise transferring fuel from a tank to another, such as when fuelling cars. Leaks are caused for example by heat expansion and corrosion of old oil tanks. Oil tank damage happens in traffic accidents. The costs for remediation of the current with oil contaminated land areas during 2005–2025 have been estimated in Finland (Järvinen and Salonen, 2004). Oil contaminated sites, caused by fuel delivery in service stations as well as oil accidents relating to heating of buildings is estimated to cause total costs of 37 M Euro in this time period. In addition, there will be costs from other sites, where oil has been partially causing the contamination.

Options for cleaning up oily disasters may soon be more cutting-edge. New sponges, microbes, and chemicals are in development that could change the ways we respond to oil spills. Some microbial preparations naturally break down petroleum, and several companies are working on oil-eating superbugs, which have been genetically altered to devour a spill more efficiently (Kaufman, 2010). Again, in this context it could be recommendable to choose mixed microbial cultures from e.g. environmental safety reasons.

Oil-eating microbes are selective with respect to their substrates. They go after the most simply shaped oil molecules first, because smoother shapes are easier to nibble on than more complex, jagged oil molecules (Dell'Amore, 2010). Waste oil materials are also potential co-substrates for bio industries.

Many of the bacteria are grown in high-nutrient conditions in the laboratory but when the microbes are added to the salty, lower-nutrient environment of the ocean or coastline, they die immediately. Other research is looking after bacteria that already live in ocean conditions but do not naturally crave crude oil as a substrate. Mixed microbial cultures are able to utilize substances otherwise hardly degradable.

Still, many experts are warning that even when high-tech cleaners become available, oil-spill response will continue to be a messy, hard job. Consequently, spill prevention, not damage control, should always be a top priority (Kaufman, 2010).

13.6   BENEFICIAL IMPACT ON THE SOCIO-ECONOMIC STRUCTURES OF THE NEW, SMALL OR MEDIUM SIZED BIOINDUSTRIES

Research findings reveal a high potential of bio industries for economic development and job growth, especially in the logging sector and in rural regions (Bailey *et al.*, 2011). Bioelectricity production may have relatively large impacts on the energy security in developing countries (Campbell and Block, 2010).

Results show that the use of crop residues in a biorefinery saves GHG emissions and reduces fossil energy demand. For instance, GHG emissions are reduced by about 50% and more than 80% of non-renewable energy is saved (Cherubini and Ulgiati, 2010).

For example, succinic acid co-production can enhance the profit of the overall biorefinery by 60% for a 20 years plant lifetime. Results of Vlysidis *et al.* (2011) indicate the importance of glycerol when it is utilized as a key renewable building block for the production of commodity chemicals. Luo *et al.* (2010) have estimated that the potential worldwide demand of succinic acid and its derivatives can reach 30 million tonnes per year. Thus, succinic acid is a promising high-value product if production cost and market price are substantially lowered. The results of the economic analysis show that a refinery has great potentials compared to the single-output ethanol plant.

13.7   BIOMASS AND RAW MATERIALS

Various potential biomass raw materials are listed in Table 13.1. It is noteworthy that almost all fields of plant or animal materials exploiting industries are producing biomass wastes. Their utilization for bioprocesses should be considered as a self-evident practice. For example, microbial enzymes can be used in tanning and in other processes designed for the treatment of animal hides (Kamini *et al.*, 2010). These enzymes include bacterial and fungal proteinases, specific proteinases like keratinizes, and lipases.

Some conclusions concerning REMOWE project study batch fermentation tests in the Baltic Sea region regarding the municipal wastes (Behrendt *et al.*, 2011):

- "already dumped waste or municipal waste older than six months is not worth the effort to be prepared for anaerobic digestion
- if working with waste water sludge, the high water content has to be considered when thinking about transporting the sludge, or whether an on-site anaerobic fermentation e.g. on the sewage plant site would be more useful
- animal faeces as cow manure can be a suitable co-substrate, but the sole use does not produce very high methane amounts
- biodegradable kitchen and canteen waste has always different compositions especially throughout the seasons and plant design has to be adapted therefore, but methane outcomes deliver promising results and it also does not compete with composting as the digestate from biowaste fermentation can easily be used as fertilizer
- agricultural products or overproduction as hay silage and ley crops silage deliver methane amounts and process behavior as expected, but the methane yields really depend on the age of the material (anaerobic digestion is useful in most of the cases)

Table 13.1.   Potential raw materials for bioconversion to chemicals, solvents and animal feed (Modified from Singh and Mishra, 1995).

Sugar containing	Starch containing	Lignocellulosic	Animal waste
– molasses	– cereal grains	– agricultural residues	– slaughter house wastes
– whey	– corn	– forest residues	– manure
– sweet sorghum	– sorghum	– wood sulfite waste	– leather industry wastes
– sugarbeet	– barley	– other pulp and paper	
– sugarcane	– wheat bran	industry waste	
– fruit wastes	– potato	– waste paper	
– beverage industry	– corn and potato	– municipal solid waste	
wastes	industry wastes	– peat	
		– algae	

- municipal household waste requires quite an effort to be digested as disturbing materials have to be sorted out and sanitation is necessary, but the organic matter left in the household waste makes it worth taking the effort
- industrial organic wastes from alcohol or food industry often contain high energy but most of them also require high attention concerning process stability and they perfectly suit as co-substrates to enhance the methane yield of a mixture.

13.7.1 Enzymatic hydrolysis of macromolecules

One major obstacle in the biocatalysis is formed by the so-called steric hindrances in the molecular structures (Rosenberg et al., 2011). This means the space limitation around the target substrate, and enzyme molecules functional, active sites. These are lowering the reaction speed if not completely preventing the reaction from happening. For example, in the case of plant material, cellulose fibers are not always fully accessible for the hydrolytic enzymes. The hydrolysis of macromolecules is the basis for the effective utilization of organic wastes. Plant materials are often degraded by either amylolytic enzymes in case of starch polymers, or by cellulases and hemicellulases attacking the corresponding cellulose fibers or their hemicellulose bonds. Regarding the slaughterhouse wastes, also such enzymes as proteases and lipases are needed for the pretreatment of the biomasses.

13.7.2 Hemicellulose, cellulose and lignin

There are remarkable physiochemical barriers in the lignocellulosic materials hindering the hydrolysis of cellulose and hemicelluloses to fermentable sugars (Alvira et al., 2010). The pretreatment methods aim at loosening the biomass structures, improving enzyme accessibility and thus making the macromolecular structures more digestible. Such methods are including hot water treatment, steam explosion, ammonia filter explosion (AFEX), wet oxidation, microwave or ultrasound pretreatments and CO_2 explosion. Besides these methods, also traditional acid or alkali treatments, as well as mechanical communition are available. Chemical treatments with ozone, organic solvents and ionic liquids have also been in use for achieving an improved recovery of the chemical energy generating potential of the plant biomass wastes, as well as urban or industrial, paper or board disposals. Increased efficiencies of the pretreatments pave way for the replacement of combustion as the primary waste processing methods. The more is the sustainable microbial biotechnology applied for the refinement of the organic wastes, the more beneficial are the consequences for the air, water and ground. In this fashion, the organic matter is truly circulated based on the natural cycle of carbon and other substances.

Hemicellulases consists of chains of pentose sugars, whose presence is selecting the organisms available for their utilization (Papoutsakis and Meyer, 1985). In the wastewater pools of a paper and pulp mill, it is self-evident to find mixed cultures of hexose and pentose-fermenting bacteria. Among the Nordic wood materials, especially the birch is containing high levels of hemicelluloses (35%) (Alén, 2000). The role of hemicelluloses in wood is to cement the cellulose fibers together. The resulting monosaccharides from the hydrolysis of hemicelluloses include the hexoses glucose, mannose and galactose, and the pentoses xylose and arabinose, and the disaccharide cellobiose (Chandel et al., 2011). Lignin in turn is the most difficult biomaterial in wood for hydrolysis. If it's concentration in wood is rising up it increases the HVVs of the biomass fuels (HHV = higher heating value) (Dermirbas, 2010).

Therefore, the more lignin-containing the material is (and the more difficult is its hydrolysis) the more tempting it is to choose the combustion for the method of choice in treating the particular biowaste. In this context, it is important to keep in mind that the extraction of hemicelluloses and celluloses is increasing the HHVs of the remaining material. Consequently, the microbial bioprocesses could be extremely feasible as pretreatments for the combustible remnants prior to burning. This approach, besides saving the environment, could increase the overall economy of the process. It is clear that as the combustion is currently responsible for about 97%

of the world's bioenergy production, this proportion has to make room for more sophisticated and economically feasible approaches. The microbial biotechnology is one of the most important of these solutions since it is the nature's way and the potential is huge offering practically limitless sources with respect to the biodegradation and circulation of wastes and other biomass sources.

13.7.3   *Starch and other saccharides from food industry by-streams and agriculture*

In food industries, a considerable amount of unused parts of the raw materials of either plant or animal origin, as well as process wastes, cleaning and washing liquids, unsuccessful product rounds, by-products, and several other commodities constitute a fruitful base for biorefinery operations. The substances contain such saccharides as starch, lactose, maltose, and fractions of the sugar industries. These materials provide the bioprocesses an excellent starting point. It is often only a matter of redirecting the focus of industrial production from solely the food products into the direction of balanced multiproduct industries.

Total amount of waste biomasses in the world has been estimated to reach two billion tonnes annually (Gressel and Zilberstein, 2007). The grain crops are required for the food production for the population directly, and as a feed supply for the animal production. It has been a controversial issue whether this biomaterial could be used as feedstock for biotechnological production of fuels and chemicals. In fact, the crops always produce plentiful wastes and by-products, which are an adequate source for bioengineering purposes. In the huge effort to complement the needs for global human nutrition, about a half of the biomasses is wasted in anyway. For example, straw is often being burnt or slowly degraded in the soil in which processes much of the energy value is being lost. As a feed components of the ruminants, the straw cannot be fully exploited because the use by the animal is restricted or limited by a minor component, lignin. Its presence is preventing the complete biodegradation of hemicellulose or cellulose by the rumen bacteria.

Improving the yields of agricultural biotechnologies is highly dependent on the development of hydrolysis processes. Even though the entire process of the grain utilization by the cattle is low in overall efficiency, the liberated chemical energies are enough for the vast production of milk and meat, hides, skulls and bones, and manure during and after the life-span of the animals. We should also not forget the volatiles emitted by the animals resulting from the microbial action in their intestines. This outflow of greenhouse gases is paid a major attention to in New Zealand for example, where new methods are developed for the elimination of this climatological effect. This influence of domestic animals closely equals to the emissions of the industrial production in New Zealand. Likewise, with respect to the water qualities, former Finnish president Tarja Halonen has emphasized the need for effective treatment of agricultural wastewaters caused by the animal production. The PMEU method has been developed for the effective monitoring of the water quality (Hakalehto *et al.*, 2011a).

13.7.4   *Industrial waste biomasses*

Enormous amounts of biomass are processed in the paper and pulp industries. The liquors, process waters, wood wastes such as bark or saw dust, and pooled wastewaters offer a plentiful supply of cellulose, hemicellulose and lignin-containing raw materials. This reservoir is somewhat more challenging as a starting material for bio-industries. For example, waxes and phenolic compounds may restrict or inhibit the required bioactivities. On the other hand, the forest industries possess facilities for transport, storage and pretreatment of wood. Therefore, these industrial plants could be converted from "one product entities" into biorefineries consisting of numerous product lines and processes in the same area. The treatments of the fibrous raw materials into more readily usable sources should be in the scope of their research activities. In this modification of various wood fragments and lignocellulosic wastes microbes offer extended options. Effective hydrolysis is a pre-requisite for a successful process. Considerable amount of research has been directed towards wood-derived materials and their processing for bio-industries since these materials constitute one of the most considerable stores for the solar energy in chemically bonded form. In practice, its exploitation is restricted by the lignin fraction of the biomasses in wood and other plant-derived sources (Gressel and Zilberstein, 2007).

In a novel application, direct conversion of lignin into liquid bio-oil is achieved with a very low oxygen content of the product (Kleinert and Barth, 2008). In this liquefaction process, formic acid serves both as a hydrogen donor and a pyrolysis/solvolysis reaction medium. It has been stated that "Conversion of biomass to chemicals and energy is imperative to sustain our way of life as known to today" (Amidon *et al.*, 2008). A biorefinery plant is "a catch and release" way of using carbon that is beneficial to the environment being also feasible from the point of view of the global economy. In this context, it is important to emphasize that this label of environmentally friendly technologies has to include the recycling of wood-derived products, improved care of the liquid and gaseous disposals, and protection of global forest ecosystems as a diverse reservoir of plants and animals. The multiple uses of forests for recreation, as sources for medicinal plant species and molecules, berries and fruits, mushrooms and other species of relevance for humanity, as well as a valuable "respiratory" organ for the earth, should be taken into account when designing the exploitation of forests or forest ground. It should also be clear that combustion is a wasting process for the wood raw materials. Therefore, improved strategies and methods need to be further developed.

Sugar maple woodchips were subjected to hot water extraction at 160°C for two hours (Amidon *et al.*, 2008). In an extraction pretreatment, the wood material can yield 15% of the mass recovered as wood sugars (2/3) and as acetic acid and other extractives (1/3) (Amidon, 2006). These extraction products have the value being one-half of the value of the wood. Therefore, it could be considered as highly feasible to extract wood or other cellulose-containing biomasses. These revenues could then be further increased by producing valuable chemicals from these materials. The most widely used pulping technology, the Kraft process, is removing about 20% of the wood weight as hemicellulose during cooking. The resulting black liquor has a heating value (13.6 MJ/kg) about 50% that of lignin (25 MJ/kg) (Gullichsen and Fogelblom, 2000). Consequently, it could be more feasible to extract the hemicelluloses prior to pulping. This approach could then lead to such higher value-added products as ethanol, butanol and 2,3-butanediol, or various polymers.

Besides the industrially disposed materials, the municipal waste fractions contain high amounts of carbohydrates, such as paper wastes, mixed wastes, fibrous packing materials, biowaste, and parts of the construction wastes, which could be treated in a biorefinery type of processes (Rättö *et al.*, 2009). Theoretically, the mixed wastes dumped into landslides in Finland could produce annually 260.000 tonnes of ethanol from these materials. For example, in the UK there has been developing a strong interest in producing fuels from municipal wastes instead of combusting them directly (www.engineerlive.com). Critical water (250–300°C under elevated pressure) has been suggested for extractions and separations. When the pressure is then lowered, the water returns to its normal properties. In addition, gas-expanded liquids such as methanol with CO_2 provide a flexible solvent whose properties can be adjusted by changing the pressure. The cellulose polymers can also be decomposed by using supercritical water or methanol extraction methods (Ehara and Saka, 2005; Ishikawa and Saka, 2001). For instance, supercritical methanol treatment at 350°C with 43 MPa for 7 minutes, for instance, converted microcrystalline cellulose (Avicel) to the methanol-soluble form. These types of reactions could potentially be used also as pretreatments for industrial microbiology processes.

In our laboratory in Finland, we have established the use of the PMEU method for the microbial monitoring of the process waters (Mentu *et al.*, 2009). The ecology of the processes and products thereof has a decisive role for the characteristics of the products. Similarly, the bioprocesses originating from wood could be monitored by the PMEU system. This kind of laboratory investigation has revealed a balanced action of common enterobacterial intestinal strains (Hakalehto *et al.*, 2008). These bacteria, which occur also in the wood industry fluids, are forming symbiotic co-cultures. Such interactions are providing huge potential for industrial microbiology processes. For example, we have studied the conversion of cellulose plant wastewaters into bulk chemicals,

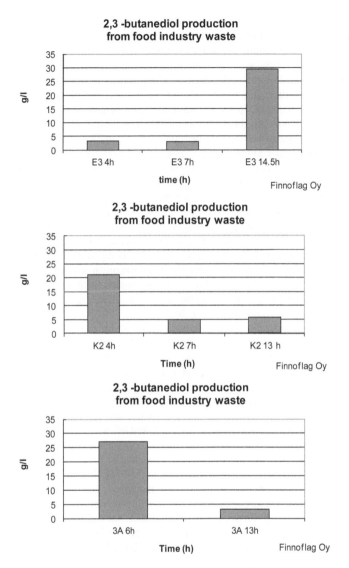

Figure 13.2.   Production of 2,3-butanediol (23BD) isomers from food industry waste material. The determination of 23BD was carried out as described earlier (Hakalehto *et al.*, 2008).

biogas with methane and hydrogen, and soil with substances for plant nutrition. This approach could make all fractions of the wastes useful raw materials as well as promotes the circulation of carbon and organic materials. Approximated productivities for 2,3-butanediol have been estimated from the values presented in Figure 13.2 for food industry wastes.

13.7.5   *Municipal waste and waste water utilization*

The production of fuels or solvents and other chemicals or biogases from biomasses is called *bioconversion*. It is noteworthy that the lignocellulosic wastes, for example, could be treated with microbes prior to combustion, and thus their energy values could be enhanced. If the bioconversion is combined with bioremediation where microbes break down the environmental pollutants into harmless substances, the overall effect of the waste treatment process is achieving a remarkably higher level of sustainability. In order to further improve the ecological (and most often also economical) value of the waste treatment process, it is worth developing actual biotechnological processes based on the understanding of the metabolic activities and diversity of the microorganisms. In practice, this means often increased understanding of the mixed cultures, and mixed substrate utilization. In such conditions, the outcome of the reactions is depending on many parameters. Obviously, the conditions with versatile substances as raw materials in relatively low concentrations of various raw materials favor so-called mixotrophs, organisms with versatile capabilities (Harder and Dijkhuizen, 1982). On the other hand, in the presence of higher concentrations of any particular substrate running the process can lead to the dominance of the microorganisms most furnished for the exploitation of this substrate. In this case, the ecological succession is then determining the selection of other strains further metabolizing the materials. If the waste mixtures contain recalcitrant compounds, which are not easily degraded by any single species, the microbes often develop so called co-metabolism, which extends the capabilities of the mixed cultures with respect to the degradation of several compounds (Dalton and Stirling, 1982). This is a valuable option for the removal of toxic compounds from the wastes. In some environments, it is also possible that the microbes are interacting in a deliberate manner in order to achieve best possible results for their survival as a community. Such interactions, and avoidance of direct competition, could be important in the human digestive tract, in regions where nutrients are quickly absorbed by the host (Hakalehto, 2012).

The integrated waste utilization process could consist of, for example, mushroom production (food protein for human consumption), feeding material (combined solid-state fermentation of agricultural and forest wastes by white rot fungi), biogas and biofertilizers (Chang, 1987). In the case of animal feed, decomposing with fungi increases the rumen digestibility up to 30–60% from only 3% of the undecomposed wood. In case of biogas (methane) production by mixed bacterial cultures, the tendency to produce CH_4 and H_2O is somewhat an alternative to the production of H_2 and CO_2. Besides these processes where the microbial cell mass is either used as single cell protein, and the gases emitted by them are exploited in energy generation, the fermentative cultures produce liquid phase products into the solution or suspension. The usage of this potential has been limited by the view that these anaerobic fermentations are slow reactions in which lack of speed increase the cost of e.g. product recovery. In the PMEU research, we have documented that the capacity of the microbial cells to carry out anaerobic reactions is in fact not slow, but it equals the aerobic reactions (Hakalehto *et al.*, 2007). This is true also with many facultative anaerobic strains. We have applied the principles of the PMEU for the production of 2,3-butanediol, which is valuable compound for the production of synthetic rubber and plastic monomers (after conversion to butadiene). In this experimentation, we achieved the productivity of 8–10 g/L/hour (Fig. 13.2).

Even though the animal manure is containing high nitrogen levels in comparison with the carbon content, it is possible to microbiologically liberate the carbon from the cellulose in the sludge. In the case of piggery sludge, a system with endoglucanase, exoglucanase and cellobiase activities were constructed using five isolated bacterial strains (Ping *et al.*, 2008). The strains were identified as *Pseudomonas citromellolis*, *Sternotrophomorus maltophilia*, *Flavobacterium mizutaii*, and two strains of *Pseudomonas aeruginosa*. In our research experiments in Sotkamo, Finland, with the bioprocess pilot plant (designed by Finnoflag Oy for Biometa Finland Oy), the cow manure mixture (with dry weight of a few percent only) was producing relatively high yields of hydrogen and methane gases, which were resulting from the carbon polymers in the waste material. The hydrogen generation has been shown to increase the energy efficiency of waste treatments. For example, in case of cereal or food processing wastewater, the treatment process was linked with a microbial fuel cell (MFC) (Oh and Logan, 2005). The maximum power density was above 50 mA per a liter of wastewater in this experiment. It is also noteworthy that in an advanced bioprocess system the aliquots are mixed with untreated new wastewater flow in order to maintain adequate substrate levels. It has been also shown that in the PMEU conditions water microbes are recovered fast emitting gases in a short time from appropriate substances (Hakalehto *et al.*, 2011a). This approach of wastewater treatment also links the microbiology with direct electricity generation.

The hydrogen production reactions can be divided into photobiological processes and dark hydrogen fermentation. In the latter ones, many bacteria reduce protons to hydrogen in order to get rid of reducing substances resulting from their primary metabolism. The hydrogen thus replaces the oxygen as an electron acceptor in anaerobiosis. In some reactions, organic substances can act as electron acceptors. Such a case is the conversion of butyrate into butanol. This reaction is also resulting in hydrogen production, and is an important potential route for the treatment of municipal sludges and animal manures, which contain high concentrations of butyrate as a result from activities of the butyric acid bacteria. These strains are important members of the microflora of the cecum and the colon (Hakalehto, 2012). They are also provoked by the CO_2 produced by other bacteria (Hakalehto and Hänninen, 2012). This illustrates the potential power of the mixed cultures in the waste treatment. The members of the *Enterobacteriacae* are not inhibited by high H_2 pressures, which could make them a potential production strain for the energy gas (Tanisho *et al.*, 1987; Kumar and Das, 2000).

In our own experiments in the Finnoflag Oy laboratory we were able to convert the butyric acid produced by some other clostridia strains further into butanol by *Clostridium acetobutylicum* in about four hours in an enhanced mixed fermentation with simultaneous hydrogen production. In fact, in these mixed cultures the sole products are CO_2, H_2, butyric acid and butanol. The typical products for a conventional ABE fermentation result also in acetic, lactic and propionic acids as well as ethanol.

The direct production of electricity from cellulose was demonstrated by a microbial fuel cell (MFC) using a defined binary culture (Ren *et al.*, 2007). In this experiment, the cathode was built from the insoluble substrate cellulose, which was hydrolyzed by a *Clostridium cellulolyticum* strain. This bacterium alone did not produce electricity, but combined with an electrochemically active bacterium, *Geobacter sulfurreducens*, the mixed culture generated remarkable amounts of electricity. Hydrogen, ethanol and acetate were the main residual metabolites of the binary culture, all of which could be further exploited in the energy generation from the cellulose wastes.

Municipal wastewater is a combination of liquid and water-carried solid wastes. Treatment of wastewaters produces massive volumes of excess sludge and other organic material, which require proper treatment. Even one third of operating costs in a wastewater treatment plant may be caused by the treatment of this excess sludge (Laaksonen, 2011). It should be taken into account that this excess sludge is not only waste, but it can also offer many economically feasible possibilities, such as nutrients for agriculture and a source of energy e.g. *via* biogas production or combustion. Moreover, excess sludge contains valuable organic material, such as proteins and enzymes. According to Metcalf and Eddy Inc. (2003), approximately 75% of total suspended solids in wastewater are organic. The organic matter is typically composed of proteins (40–60%), carbohydrates (25–50%) and lipids (8–12%), but the composition may vary very much depending on the wastewater sources. Sludge utilization is of great interest (LeBlanc *et al.*, 2008).

13.7.6   *Removal of harmful substances*

The chemicalized industrial production in many fields, as well as in the agriculture and forestry is creating a chain of unbalanced and distorted microenvironments and biomass build-up. Chemical contaminants then accumulate into the organic matrices and matter, such as waste materials and sludges. During combustion, for example, some of them are concentrated into the ashes, whose deposition can become an environmental problem. Microbial biotechnologies, such as bioleaching, offer solutions for the removal of toxic elements (Fanga *et al.*, 2011).

13.8   FERMENTATION PROCESSES AND BIOREACTOR DESIGN REVOLUTIONIZED

13.8.1   *Increased productivity lowers the cost of bioreactor construction and downstream processes*

Traditionally, industrial microbiology has been aiming for the applications with CSTR's (Continuous Stirred Tank Reactor) (Lakatos *et al.*, 2011). In these systems, the raw material, or substrate, addition is carried out without interruption and the products are removed in an equally continuous liquid suspension flow. In a properly functioning continuous fermentation, chemostat, the specific growth rate of the microbial culture equals the dilution rate. The end-product inhibition is avoided with the removal of product substances. Alternatively, the fermentations could be arranged as batch or fed-batch processes. In the latter ones, the substrate is added in intervals, and this approach has become rather popular in the bioindustries. It is often easier to run a fed-batch system than a bioreactor with continuous flow. Many fermentation products are results from the secondary metabolism, which is switched on after some more easily metabolizable substrates have been exhausted from the batch. This natural division of metabolic functions requires multi-phased arrangement in the fermentation process. In a fed-batch mode, the remaining microbial mass in the bioreactor serves as inoculum, or starter, for the next round of production. Another possibility for arranging the bioprocess initiation could be a feedback loop in the process.

Some drawbacks of the bioreactions accomplished in the stirred tank reactors are related with the shear stress in the liquid (Bhojwani and Razdan, 1996). The stirring equipment is causing material flows and pressures on the biocatalyst cells or enzymes. This physical stress is lowering the production by simply breaking up the cells or molecular structures. These undesired effects are often avoided by immobilizing the biocatalysts (Tischer and Wedekind, 1999). This approach is also eliminating at least partially such phenomena as the end-product inhibition. In addition, the downstream processing is simplified if the biocatalysts are closed within their own compartments.

Regardless of the processing difficulties caused by the shear forces in the liquids, it is of utmost importance for the outcome of the fermentation to have some kinds of techniques for overcoming the so-called diffusion limitation, which is lowering the reaction speed (Zyskin *et al.*, 2007).

Alternative bioreactor modes instead of the tank reactors are e.g. airlift and hollow-fiber fermenters (Shkilnyy *et al.*, 2012). In these facilities, the biocatalysts are often immobilized within some matrices. In order to speed up the diffusion, moderate pressures could be attempted. Thus, the accessibility of the substrate is increased, and the product removal eased up by more rapid flow of the processed fluids or materials. The more concentrated the product is, the more effective (and less costly) is its collection and purification,

13.8.2   *PMEU (Portable Microbe Enrichment Unit) used for process simulation*

The PMEU system was originally developed for the rapid detection of microbial strains in hospitals, industries, and in the environment (Hakalehto, 2009, 2010). It has been used for the detection of salmonellas (Hakalehto *et al.*, 2007), Campylobacteria (Pitkänen *et al.*, 2009), intestinal enterobacterial strains (Pesola *et al.*, 2009; Heitto *et al.*, 2009) and forest industry bacilli (Mentu *et al.*, 2009). The basic idea behind the use of the PMEU equipment is called "enhanced enrichment." In different PMEU versions, the verification of bacterial growth is based on the optical sensing in the PMEU Spectrion® (Fig. 13.3) and on the detection of volatile organic compounds (VOC) in the PMEU Scentrion®. The rapid growth of *Escherichia coli* with various gas flow parameters is demonstrated in the Figure 13.4. In the monitoring of the hospital pathogens with the PMEU Scentrion®, very low bacterial concentrations were detected in blood samples of neonatal septic patients (Hakalehto *et al.*, 2009). In bioreactor applications the PMEU versions could be used for simulating the microbial growth, interactions and product formation.

The effect of different gas flow conditions on otherwise identical cultures in the selective medium has been studied in the PMEU. The cultivation took place in specific PMEU cultivation syringes. Nitrogen-gas was led to syringe 1, and the outcome gas from syringe 1 was then led to

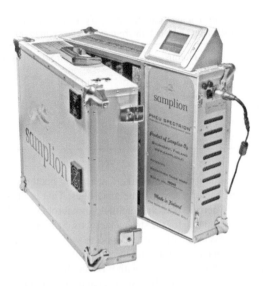

Figure 13.3.   PMEU Spectrion®. All PMEU models are innovated in Finnoflag Oy. Some versions have been manufactured by Samplion Oy. Both companies are from Kuopio and Siilinjärvi, Finland.

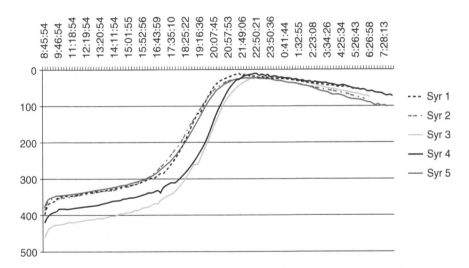

Figure 13.4.   Growth of *E. coli* in PMEU Spectrion® (Samplion Oy and Finnoflag Oy, Kuopio, Finland) using Colilert™ media (IDEXX Laboratories Inc., Maine, USA).

syringe 2, from syringe 2 to syringe 3 and so on. The gas flows from a syringe to the next one as well as the final cell concentrations are presented in Table 13.2. Temperature was set at 37°C. *E. coli* amount was about 2×10^4 cfu/mL in the beginning of the experiment in every syringe.

13.8.3   *Anaerobiosis made efficient*

Pasteur already proposed that the aerobic bacteria were consuming free molecular oxygen (O_2) in mixed cultures, which also gave the opportunity for the anaerobic bacteria to survive beneath the aerobic flora. When *E. coli* was used for the purpose of oxygen removal, an E_h value (oxidation-reduction potential) of $-440\,mV$ was achieved (Porter, 1948). Much higher values up to $+50\,mV$

Table 13.2. *E. coli* amount (cfu/mL) in every syringe at the end of the
experiment and gas flow from syringe to syringe (mL/min).

	cfu/mL	Gas flow from syringe to syringe
Syringe 1	2×10^9	72 mL/min
Syringe 2	4×10^8	60 mL/min
Syringe 3	2×10^9	45 mL/min
Syringe 4	5×10^9	36 mL/min
Syringe 5	3×10^9	

are sufficient for promoting the growth of obligate anaerobic bacteria (Fung, 1988). Actually, in human skin a majority of bacteria are obligate anaerobes. This is illustrating the situation in natural microenvironments, where aerobes and facultative anaerobes rapidly consume the oxygen thus preparing the conditions suitable for the strict anaerobes. Similarly, it could be well supposed that in natural conditions the exchange of substances occur within short distances by various sects of microorganisms. These interactions could in future offer interesting possibilities for exploiting the microbial world for biotechnical purposes with mixed cultures.

Many bioprocess productions require the use of anaerobic microbes (Fung, 1988). In biochemical literature, it is widely emphasized the rapidity of the aerobic reactions, such as glucose utilization, in comparison with the anaerobic fermentation. However, when the diffusion limitation has been overcome, for example, in the PMEU both the aerobic and anaerobic metabolism has been operated at equal speed (Hakalehto et al., 2007). This corresponds with the observation that during the wintertime in Finland many landslides, not only the composts, develop heat, which prevents them from freezing. Therefore, a novel thinking in the bioreactor design and in the exploitation of the anaerobic organisms could provide fruitful improvements in the productivity of the biochemical processes. In our laboratory, we have successfully accomplished e.g. acetone-butanol fermentation in a few hours of time with novel bioreactor arrangements (results not shown here).

13.8.4   *Some exploitable biochemical pathways of bacteria and other microbes*

The fermentative reactions are carried out usually in anaerobic conditions, in which the ultimate electron acceptor is usually CO_2. Simultaneously, the cell metabolism is accumulating some molecular species as end products. In fact, the CO_2 is also initiating the bacterial growth (cell division) by some unknown mechanism. With suboptimal CO_2 concentration, the growth rate of *E. coli* is controlled by the CO_2 concentration in otherwise adequate growth medium (Repuske and Clayton, 1968). It has also been documented that the CO_2 is a necessity for bacterial growth: "The removal of carbon dioxide from an environment, which is otherwise favorable, results in complete cessation of bacterial growth". This illustrates the sensibility of microbial cultures (and biological systems in general) to the environmental conditions in the bioreactor. In practice, this means mainly the circumstances surrounding the cells.

Many bacterial fermentations lead to the accumulation of organic acids. These products can be purified and exploited as such or they can further processed by other microbes, or by chemical reactions. For example, the succinic acid has been suggested as a platform chemical for biorefinery applications (Kamm and Kamm, 2004).

Some other fermentation products, such as ethanol, acetone, butanol or 2,3-butanediol have a wide variety of potential applications in bioindustries. Besides the uses as fuel components, they are typical bulk chemicals for the industries. For example, the 2,3-butanediol is easily converted into butadiene, which serves as new material for synthetic rubber and plastic monomers, as well as anti-icing substances (American Chemistry Council, 2011).

Besides the direct production of bulk chemicals microorganisms produce many hydrolytic enzymes, which are exploited in the biomass pretreatments. Often it has to be decided which type

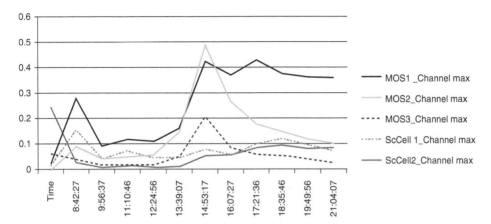

Figure 13.5a.   Experiment carried out with circulated car washing liquid/water using microaerophilic gas flow in PMEU Scentrion®. Syringe 1: sludge with TYG(1:10), gas mixture 2/3 N_2 and 1/3 air. The curves indicate the emissions as detected with different metal oxide sensors (MOS) and Semi-conductor Cells (ScCell). Major microbial breakdown activities were provoked for five hours in this experiment.

of biocatalyst is the most useful for any particular application. Bacterial or mold strains could be used for the production of hydrolytic enzymes in the process fluid or in a pretreatment unit, or alternatively they could be replaced with their enzymes. Nowadays, promising results have been obtained with the so-called consolidated bioprocesses (CBP).

13.8.5   *Mixed cultures in bioengineering*

Microbiological research is largely based on isolated pure cultures. However, in many cases the microbes form integrated communities in which some species act in balanced, or even in a symbiotic fashion. For example, the various types of the small intestinal bacteria, the mixed acid fermenting strains such as *E. coli* and the more neutral substances producing *Klebsiella/Enterobacter* group are maintaining the duodenal pH at 6 (Hakalehto *et al.*, 2008). This dualistic balance was established in simulation with the PMEU, where the acids produced by *E. coli* were neutralized with the 2,3-butanediol fermentation of *Klebsiella mobilis*. The two types of the facultatives were shown to produce equal growth yields in mixed culture and in two separate pure cultures. The strains were shown to establish the balance in almost all cases, and this balance was not shaken by other intruding species (Hakalehto *et al.*, 2010). The dualistic balance between *E. coli* and klebsiellas is one example of the powerful means offered by the mixed cultures in industrial biotechnology.

13.8.6   *Novel principles for the planning of unit operations for bulk production*

In the case of almost any organic biomass, the material itself is containing mixed cultures of various microorganisms. We were studying by the PMEU cultivation method for the circulated car washing water-based solution containing organic substances. When the sterile air flow (aerobic condition) was used for cultivation experiments in the PMEU Scentrion® device, the natural microbial population generated a continuous gas flow (Fig.13.5). The detected emission could partially derive from the washing liquids and the dirt components dissolved or suspended into the fluids. An addition of 1:10 or 1:100 of TYG medium (Tryptone, Yeast Extract, Glucose) was added to boost the microbial growth. However, if laboratory strain of *Klebsiella mobilis* was inoculated into either of the supplemented washing liquids, the emissions of the volatiles peaked quicker than with the natural microbial population. The effect was more readily evoked with sterile air than with the microaerobic gas flow. The results indicate that the naturally developed microbial

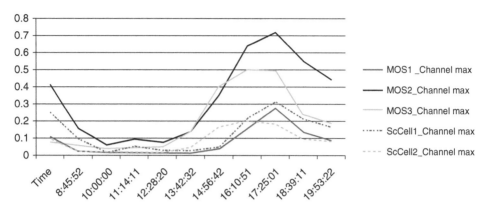

Figure 13.5b.   Syringe 2: sludge with TYG(1:100), gas mixture 2/3 N$_2$ and 1/3 air. See Fig. 13.5a for explanations.

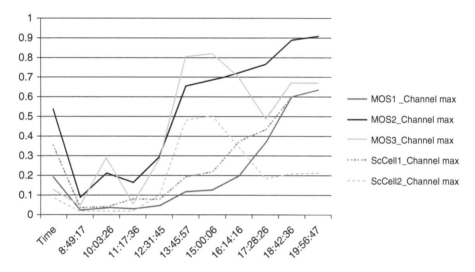

Figure 13.5c.   Syringe 3: sludge with TYG (1:10), *Klebsiella mobilis* 1:1000, gas mixture 2/3 N$_2$ and 1/3 air. See Fig. 13.5a for explanations.

communities were able to exploit the available organic compounds within a longer time range, whereas a rapidly growing and metabolized production strain was more effectively depriving some essential components from the medium.

Interestingly, the natural microbial culture was shown to contain remarkable amounts of klebsiellas as determined by the Chromagar™ medium after 9.5 hours of PMEU cultivation. *Klebsiella mobilis* strain (butanediol fermenting bacterium) was added as pure inoculant culture to the circulated washing liquid (Fig. 13.5a–d). Both in case of naturally "fermented" and in inoculated fluid the proposition of the added TYG medium clearly influenced the outcome as measured by the volatile compounds. In the determination from the raw washing liquid, no klebsiellas could be verified, and the number of *E. coli* cells was around 1000/mL. Moreover, during these 9.5 hours the non-inoculated substrate had been producing 134 µg/mL of 2,3-butanediol, whereas the sample from the *K. mobilis* laboratory strain containing culture at the same time point resulted in 61 µg/mL of the butanediol. This preliminary experimentation shows that such substrate as car washing liquid could form a solution for diluting organic wastes, and it also contains a variety of microorganisms capable of starting the production of biochemicals. It is noteworthy that this

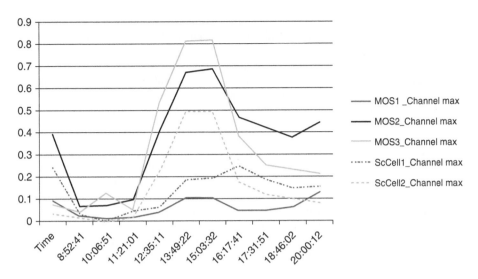

Figure 13.5d.   Syringe 4: sludge with TYG (1:100), *Klebsiella mobilis* 1/1000, gas mixture 2/3 N_2 and 1/3 air. See Fig. 13.5a for explanations.

Table 13.3.   Bacterial densities after 9.5 hours of cultivation in the PMEU Scentrion®.

	Klebsiella sp.		*Escherichia coli*	
	Aerobic	Microaerobic	Aerobic	Microaerobic
a	1×10^8	5×10^7	1×10^8	1×10^8
b	2×10^8	$<1 \times 10^7$	1×10^8	1×10^7
c	1.5×10^8	1×10^8	$<1 \times 10^7$	$<1 \times 10^7$
d	1.5×10^9	4×10^8	$<1 \times 10^7$	$<1 \times 10^7$

used washing water was derived from a circulated system after the biological purification, and still contained organic materials. This kind of application suggests that various wastewaters could be combined with biomass raw materials in order to achieve correct water densities as well as suitable carbon and energy sources for the exploited microflora. The growth of microbes was not inhibited by this liquid matrix, as indicated by Table 13.3 summing up the levels of *Klebsiella* sp. and *Escherichia coli* amongst variable mixed microflora. The a–d refer to cell densities corresponding to Figure 13.5 as measured by colony counts on CromAgar™ medium.

13.9   THERMOPHILIC PROCESSES

As generally understood on the basis of thermodynamics, any chemical reaction is speeded up as the temperature is rising up. In biotechnology, this requires often the selection of new production organisms. However, because of the nature of the biological systems it is not always that simple to achieve the improvements. The entire process has to be considered in the planning of the manufacturing of the fuels or bulk chemicals. This includes the resistance to heat or the product by the facilities, as well as the ability of the organism to exploit the particular raw materials available. In all cases the overall biotechnological principle of three fundamentals has to be taken into account, and none of them should be developed independently, namely (i) the choice of microorganism, (ii) understanding its physiology, and (iii) the reactor design including the downstream processing. In a comparative study by the Finnish Environment Institute regarding the methane production with mesophilic and thermophilic processes, the product yield was 1.3 times higher in the latter case (Kangas *et al.*, 2011). However, the quality of thermophilic process reject water was improved from the sewage treatment plant's point of view. Its COD value was 3.4 times the one deriving from the mesophilic process, and ammonium nitrogen level 1.3 times the one of the mesophilic origin. Thus, the method of choice has to be selected after careful overall comparison.

In addition to mesophilic bacteria, thermotolerant bacterial species such as *Clostridium thermocellum*, *Clostridium thermohydrosulfuricum*, *Clostridium thermosaccharolyticum*, *Clostridium thermosulfurogenes* and *Thermoanaerobacter ethanolicus* can also produce ethanol (Chandel *et al.*, 2011).

For example, coupling the thermophilic organisms with the production of ethanol with low heat of vaporization is resulting in a process where end-product inhibition caused by the product is avoided by its continuous evaporation from the process fluid. In these temperatures, it is also possible to run the process without contaminating microbes occurring. Such case has been demonstrated with *Thermoanaerobacter* BG1L1 bacterium in the production of fuel ethanol from dilute sulfuric acid pretreated corn stover (PCS) (Georgieva and Ahring, 2007). This production organism was able to exploit the xylose effectively from the substrate yielding ethanol by rate of 0.39–0.42 g/g-sugars consumed. This thermophilic anaerobic bacterium was not inhibited by the toxicity of the corn stover hydrolysate in this process.

Another example of ethanol production with thermophilic organisms is the use of anaerobic clostridia. The fermentations of several saccharides derived from lignocellulosics were investigated with a co-culture consisting of *Clostridium thermocellum* and *Clostridium thermolacticum* (Xu and Tschirner, 2011). In this case, mixed cultures of two bacterial species provided a solution for the exploitation of lignocellulose-based raw materials. The co-culture was able to actively ferment glucose, xylose, cellulose and micro-crystallized cellulose (MCC). The ethanol yield observed in the co-culture was up to two times higher than in monocultures. The highest ethanol yield (as a percentage of the theoretical maximum) observed was 75% (w/w) for MCC and 90% (w/w) for xylose. In this case low levels of initial ethanol addition resulted in an unexpected stimulatory impact on the final ethanol productions. Examples of this kind underline the importance of extensive experimentation in the production scale, on the basis of understanding and also challenging the microbial physiology. Especially, using the controlled mixed cultures can be extremely rewarding regardless of their complexity.

In an experiment where the compost of Napiergrass and sheep dung were used as a substrate for selecting effective mixed cultures for degrading lignocellulosic wastes in thermophilic (60°C) conditions, the basic microflora components after stabilization were five main organisms (*Clostridium* strain TCW1, *Bacillus* sp. THLA0409, *Klebsiella pneumoniae* THLB0409, *Klebsiella oxytoca* THLC0409, and *Brevibacillus* strain AHPC8120) (Lin *et al.*, 2011). Acetic acid, ethanol, and butanol were the main biochemical products produced by biological fermentation. Under optimal conditions, ethanol yields from Avicel and Napiergrass reached maxima of 0.108 and 0.040 g/g, representing ethanol productivities of 0.00055 and 0.00028 g/g/h, respectively. This example is again an indication of the increasing development in the field of mixed cultures, which reflect the nature's own methods in generating new substances *via* degradation. In another study, co-cultures of cellulolytic *Clostridium thermocellum* with non-cellulolytic *Thermoanaerobacter* strains (X514 and 39E) significantly improved ethanol production by 194–440% (He *et al.*, 2011). In this reaction, the importance of vitamin B12 for the ethanol production was also documented.

13.10   VOLATILE PRODUCTS

In the rumen, glucose undergoes a bacterial fermentation with the production of volatile fatty acids (VFA): acetic, propionic and butyric, and the gases carbon dioxide and methane (Madigan *et al.*, 2003). In addition, hydrogen results from several fermentations. The same substances are potential products (or raw materials) for biochemical engineering processes. It is possible, as indicated by the wide experimentation with the PMEU units, to develop further applications on the basis of the volatiles. In the refinement of biogas, for example, the gaseous compounds may also cause problems, since upgrading the biogas includes removing carbon dioxide to increase the content of methane and to remove hydrogen sulfide and water, which can cause damage in the gas utilization system (Weiland, 2010). Other possible impurities are nitrogen, oxygen, ammonia, siloxanes, and particles (Petersson and Wellinger, 2009).

Many alternative ways in designing the reactors could be accomplished. One experimental semi-industrial plant has been built in the northern Finland (planned and designed by Finnoflag Oy for Biometa Finland Oy, built by subcontractors YTT Oy and Kaitek Oy). This plant has been constructed in such a way that various sequential runs could be performed simultaneously.

13.11   DIFFERENCES BETWEEN CHEMICAL TECHNOLOGIES AND BIOTECHNICAL PROCESS SOLUTIONS

Several companies are exploiting bioprocesses in their chemical and materials production. Many biotechnological processes are cheaper than traditional chemistry, have higher yields or produce a cleaner product (Venter, 2004). Amino-acid supplements, vitamin supplements, antibiotics, anti-influenza drugs, substances of creams for cosmetics and even the solid rocket-fuel that is used in air-to-air missiles. Several polymeric product applications have been established. Metabolic Explorer Inc., for example, turns glucose into acrylate, a feedstock for the plastics industry. Cargill Inc., a large agricultural company, produces a glucose-derived substance called 3-hydroxypropionic acid. This can be manufactured with modified microbes. Further strains can then produce a dozen chemicals that are precursors for plastics. John Frost of Michigan State University (and the inventor of biotech rocket-fuel) has even worked out how to use bacterial enzymes to make a form of nylon. Cargill-Dow Inc., a joint-venture between Cargill Inc. and Dow Chemicals Inc., produces a cost-effective polymer made from lactic acid that has been produced from maize-derived glucose. DuPont Inc. has Sorona®, a plastic that is half biotech and half-traditional. These are only a few of numerous applications.

The main source of industrial glucose in North America is maize starch, which is relatively costly. However, most of the dry weight of a plant is composed of cellulose. If cellulose residues were used to make glucose, much agricultural waste, such as straw and the leftovers from maize farming, could be turned to account. It can be broken down biologically, and the enzymes to do such a job are found in many bacteria and fungi. The search is now on for the best enzymes and ways of upgrading these into industrial products. Novozyme Inc.'s target chemical is ethanol, which is fermented from glucose. The firm has improved the process of hydrolyzing cellulose into glucose, and then fermenting glucose into ethanol, in ways that have reduced the cost ten-fold. In our own process development, we have found out that such savings are achievable by new planning principles of the biotechnological production. This still means that the enzyme processing for an American gallon of cellulose-derived ethanol costs 50 cents. However, Novozyme hopes to bring that cost down another ten-fold over the next few years, to a point where ethanol derived from cellulose might be cost-competitive with petroleum-derived products. In addition, if ethanol can be made cheaply from cellulose, then this can be achieved with many other bulk chemicals.

In principle, biotechnical industries must often rely on the use hydrolytic enzymes for the pretreatments, which make an additional step into the process when compared with the chemical industries. In the actual process, the biological substances act in a less predictable manner due to the vast amount of parameters controlling their function.

13.12   BIOREFINERY CONCEPT EVALUATION

13.12.1   *New ideas on materials: all process wastes serve as raw materials in nature*

In nature, all organic matter circulates. Besides this network, or multidirectional flow of substances, we have learnt to recognize the importance of water cycle, and carbon, nitrogen, sulfur and phosphorus cycles. All these reaction sequences require interactions between organic and inorganic compartments. One example of the global scale of the microbial life are the gigantic giruses, or giruses, of the photosynthetic ocean algae. The giruses as a group were found only recently due to the fact that they were filtered out of the girus suspensions by the standard 0.2 μm filter (Raoult and Forterre, 2008). That is the reason why these creatures were not detected by the virologists. Nevertheless, in the Pacific Ocean they have shown to participate in the circulation of matter by an interesting success of events (Van Etten, 2011). The giruses dissociate huge blooming algal carpets thus liberating lots of organic biomass for the degradation by other microorganisms. Their action, in turn, causes emissions of an organic sulfur compound DMS (dimethylsulfoxide), which causes huge rains. This example shows the affectivity of the natural circulation of matter in a marine ecosystem.

13.12.2   *Multiple uses of the production equipment*

In fermentation industries, it is a common principle that such basic solutions as the CSTR (Continuous Stirred Tank Reactor) is functional in a wide variety of processes. Variable raw materials can be used in the production of multiple products. The production and the product recovery can be adjusted according to the desired process. Therefore, seasonal variations and raw material costs, as well as market demands can be taken into account within a framework plan of the unit.

13.12.3   *Plant nutrition and agriculture connected with bioindustries*

As already discussed, the biorefinery could be an industrial field where process wastes are combined to new raw materials. This requires efficient substrate flow control and management. By exploiting various energy generating means from or within the biomass, the power requirements of the plant itself can be met and the entire biorefinery becomes an energy field. This goal is achievable by taking into use more effective multistrain production reactions combined with simultaneous downstream processing, implementation of fuel cell technologies, creative exploitments of the photosynthesis as well as material flow management and logistics. Reactor design is serving these goals offering facilities for variable process solutions. After the production of fuels and chemicals, a residual biomass will remain in the plant. One alternative is to refine for use in plant nutrition use after bioremediation, the removal of harmful chemicals on their transformation into harmless form.

The wastes and side products of the food industries can readily be exploited in the biorefineries (EuropaBio and ESAB, 2006). This process design generates positive environmental impacts, such as lowered water pollution. The materials left over from bioprocesses can be directed into the fields after biorefinery processes in the form of plant nutrients.

In Eastern Finland, Biokymppi Oy's business concept is material efficiency *via* versatile waste reception possibilities to be able to broadly serve region's organic waste handling needs. Electricity production is only one branch in Biokymppi Oy's business, others being handling of organic waste streams and recycling of nutrients. Feed materials are hence refined into safe, organic fertilizers for farmers nearby. The plant has two process lines. One line is for raw materials, which are accepted for organic fertilizers such as manure and separately collected biowaste and the other line is for municipal sewage treatment sludge and other materials, which are not accepted for organic fertilizers. The digestion residue from the latter can be utilized, after hygienization, both as solid and liquid recycled fertilizer, also for garden and seedling production. For this product, Biokymppi Oy has received organic fertilizer status, which is the first case among biogas plants in Finland. Overall, with the estimated production rates fertilizers will be produced for 1000–1500 hectares. As its future visions Biokymppi Oy has established the productions of concentrated fertilizer (liquid) and granulated fertilizer (solid) as well as biogas for vehicle use (Juvonen, 2010).

13.12.4 *Local products of microbial metabolism with global impacts*

As a consequence of local energy production, local small-scale production of goods is encouraged. In agriculture, this means safer food production with less transportation. Instead of huge industrial plants destroying the balance in their environment, the local microbial biotechnology facilitates energy and food production with small industries for local communities. Such pattern has turned out successful in Israel, where agricultural kibbutzim have provided grounds for high-tech enterprises.

A bioprocess plant resembles the digestive tract of humans or animals. Ruminants, for example are converting the cellulose-containing plant materials into animal biomasses with high speed. This process is effectively made possible by the microbes and their enzymes. The ruminants have specific microflora in their digestive tract (Madigan *et al.*, 2003). In case of human alimentary tract with a very distinctive flora, being also dependent on the health status of the individual, the reactions resemble the processes carried out in the biorefineries (Hakalehto, 2012).

In the acidic gastric fluids and on the epithelial surfaces the microbes need specific survival strategies as like the *Helicobacter pylori* bacterium, which is protecting its surface colonies on the epithelia by ureolytic enzymes, which balance the pH (Cullen *et al.*, 2011). Even the lactic acid bacteria could survive in the stomach of 13 healthy individuals as indicated by the endoscopic samples enriched in the PMEU (Hakalehto *et al.*, 2011b). After the low pH conditions of the stomach, the chime is pumped in pulses into duodenum and rapidly neutralized (Hakalehto, 2012). There the facultative bacteria constitute a balanced milieu for maintaining their own communities in relatively stable pH and osmotic conditions (Hakalehto *et al.*, 2008, 2013). This is the basis for enzymatic hydrolysis of the food macromolecules and later on the starting point of the column of microorganisms along the intestinal tract. This ecosystem is producing the most of its biomass in the colon, where most of the nutrients required for human consumption have been uptaken earlier (Hakalehto, 2012). The reaction sequences inside our digestive tract could give an inspiration for designing biotechnical applications.

The various ways for different microorganisms to produce their energy are:

- respiration
- fermentative pathways
- anaerobic respiration
- photosynthesis.

Besides the fulfilment of their energy needs, the microbes need also carbon source, minerals, trace elements, growth factors, vitamins etc. Those strains, which get along without organic raw materials are called autotrophs, whereas the strains utilizing organic matter, or biomasses, are heterotrophs. The photosynthetic algae or bacteria, for example, belong to the former group. The heterotrophs, in turn, circulate the biomass substances, acquiring their energy supplies by liberating the energies from the chemical bonds in the organic matter. This is their catabolism. As byproducts, different microbes also produce and emit lots of various substances, which are exploitable in the biotechnological industries. In the waste treatment, these metabolic capabilities of the microorganisms can be exploited.

The vast capacity of the microbial metabolism is multiplied with the use of the mixed cultures. In fact, the term "co-metabolism" means the joint action and outcome of the different microbial species in a co-culture on a particular substance, or raw material, for degrading and utilizing substances otherwise difficult for the processes. The co-metabolism is resulting in enhanced and extended biochemical products. For example, some recalcitrant are most often not degraded by any single microbes, but with a selected mixed cultures (Quan *et al.*, 2004).

The current improvements in the biorefinery technologies are recently reviewed (El-Hawegi and Stuart, 2012). Integrated processes offer more economic feasibility and lower the environmental consequences. As it has been a well-known fact that biotech product facilities could be modified or converted suitable for variable process needs with relatively low cost, this increases flexibility of this industrial discipline. The biorefineries could be seen as fields of different biomass and byproduct and waste piles constituting a high number of raw material mixtures and process alternatives. These various processes and their products could be then manufactured in a flexible manner according to the market demand, for example. Such a "field idea" of biorefinery construction was presented and later put into practice successfully within the framework of BECON 2004 program in Iowa, USA. There the agricultural wastes were and have been in the scope of industrial revolution (Iowa Energy Center, 2013).

13.13   CONCLUSIONS

Modern applications of biotechnological industries are in many countries being directed towards gradual complementation or replacement of the petrochemical industries. This is a consequence of diminishing resources worldwide and potentially limited access to the sources. On the other hand, biomass surplus and wastes are available widely. The process should be developed further into the direction of increased sustainability, productivity and integration. The flexibility of the microbial processes makes them increasingly attractive. Biocatalysis with microbes and their enzymes is also an energetically feasible way to build up production of fuels and chemicals for the future needs. Microbial metabolism offer huge potential for development in this respect. When the capabilities of several microbes are combined in the case of mixed cultures, increasingly better results are to be achieved. In order to have these results actualized as industrial applications, also the reactor design and downstream processing have to develop hand in hand with the process microbiology.

QUESTIONS FOR DISCUSSION

- What is the important aspect of biocatalysis in the biotechnical processes from the energetic point of view?
- Why could the mixed microbial cultures often replace genetic engineering in the biorefinery process design?
- Mention the benefits of using thermophilic organisms in the production of e.g. ethanol.
- Which one of the following claims is untrue? The flexibility of the biotechnological approach in comparison with chemical engineering is based on:
  o producing fewer products from mixed substrates,
  o the possibility of using alternative raw materials in the same production unit,
  o switching the bioreactor into variable processes if needed, or
  o small-scale solutions with local raw materials.
- Efforts to improve the bioprocess can be directed towards selecting the production strain, studying its physiology, bioreactor design, and integrated downstream processing. Describe these alternatives in the commercialization of biobutanol production in an anaerobic sludge reactor.
- Give examples of lignocellulosic biomasses and introduce the problems in their effective utilization.
- Why, when and how should combustion be replaced by bioprocess applications?
- Could the biorefinery field replace the oil refineries, and eventually or step-by-step the petrochemical industries?
- Why could bioprocesses often be considered as an ecologically sustainable solution?

REFERENCES

Alén, R.: Structure and chemical composition of wood. In: P. Stenius (ed.): *Forest products chemistry*. Fapet Oy, Helsinki, Finland, 2000, pp. 11–57.

Alvira, P., Tomés-Peló, E. Ballesteros, M. & Negro, M.J.: Pretreatment technologies for an efficient bio-ethanol production process based on enzymatic hydrolysis: a review. *Bioresour. Technol.* 101 (2010), pp. 4851–4861.

American Chemistry Council: Butadiene product summary. 2011, http://www.americanchemistry.com/butadiene (accessed July 2012).

Amidon, T.E.: The biorefinery in New York: woody biomass into commercial ethanol. *Pulp Pap. Canada* 107 (2006), pp. 47–50.

Amidon, T.E, Wood, C.D., Shupe, A.M., Wang, Y., Graves, M. & Lin, S.: Biorefinery: conversion of woody biomass to chemicals, energy and materials. *JBMB* 2 (2008), pp. 100–120.

Bailey, C., Dyer, J.F. & Teeter, L.: Assessing the rural development potential of lignocellulosic biofuels in Alabama. *Biomass Bioenergy* 35 (2011), pp. 1408–1417.

Barone, J. & Schmidt, F.: Nonfood applications of proteinaceous renewable materials. *J. Chem. Educ.* 83 (2006), pp. 1003–1009.

Behrendt, A., Vasilic, D. & Ahrens, T.: Report on substrate pretreatment, quality and biogas potential of different waste substrates and suitable substrate mixtures for each individual region. REMOWE Report No. O 3.2.3.1, 2011, http://www.remowe.eu (accessed July 2012).

Bhojwani, S.S. & Razdan, M.K.: In: Plant tissue culture: theory and practice. Elsevier. *Studies in Plant Sciences*, 5, 1996.

Campbell, J.E. & Block, E.: Land-use and alternative bioenergy pathways for waste biomass. *Environ. Sci. Technol.* 44 (2010), pp. 8665–8669.

Chandel, A.K., Chandrasekhar, G., Radhika, K., Ravinder, R. & Ravindra, P.: Bioconversion of pentose sugars into ethanol: a review and future directions. *Biotechnol. Molecul. Biol. Rev.* 6 (2011), pp. 8–20.

Chang, S.T.: Microbial biotechnology. Integrated studies on utilization of solid organic wastes. *Resour. Conserv.* 13 (1987), pp. 75–82.

Cherubini, F.: The biorefinery concept: Using biomass instead of oil for producing energy and chemicals. *Energy Convers. Manage.* 51 (2010), pp. 1412–1421.

Cherubini, F. & Ulgiati, S.: Crop residues as raw materials for biorefinery systems – A LCA case study. *Appl. Energy* 87 (2010), pp. 47–57.

Cullen, T.W., Giles, D.K., Wolf, L.N., Ecobichon, C., Boneca, I.G., Trent, M.S.: *Helicobacter pylori versus* the host: remodeling of the bacterial outer membrane is required for survival in the gastric mucosa. *PLoS Pathog.* 7 (2011), e1002454.

Dalton, H. & Stirling, D.L.: Co-metabolism. In: J.R. Quangle & Bull (eds): New dimensions in microbiology: mixed substrates, mixed cultures and microbial communities. *Phil. Trans. R. Soc. Lond.* B 297 (1982), pp. 481–495.

Dell'Amore, C.: Nature fighting back against gulf oil spill. 2010, http://news.nationalgeographic.com/news/2010/05/100507-science-environment-gulf-mexico-oil-spill-cleanup-bacteria/ (accessed July 2012).

Dermirbas, A.: *Biorefineries for biomass upgrading facilities.* Springer-Verlag. London, UK, 2010.

Ehara, K. & Saka, S.: Decomposition behavior of cellulose in supercritical water, subcritical water and their combined treatments. *J. Wood Sci.* 51 (2005), pp. 148–153.

El-hawagi, M. & Stuart, P.R. (eds): *Integrated biorefineries: design, analysis and optimatization.* CRC Press, Boca Raton, FL, 2012.

EuropaBio & ESAB: Industrial or white biotechnology — A policy agenda for Europe. 2006, http://www.bio-economy.net/reports/files/policy_agenda.pdf (accessed July 2012).

Fanga, D., Zhanga, R., Zhouc, L. & Lia, J.: A combination of bioleaching and bioprecipitation for deep removal of contaminating metals from dredged sediment. *J. Hazard. Mat.* 192 (2011), pp. 226–233.

Fung, D.Y.-C.: Methodology of anaerobic cultivation. In: L.E. Erickson & D.Y-C. Fung (eds): *Handbook on anaerobic fermentations.* Marcel Dekker Inc. New York and Basel, 1988.

Georgieva, T.I. & Ahring, B.K.: Evaluation of continuous ethanol fermentation of dilute-acid corn stover hydrolysate using thermophilic anaerobic bacterium Thermoanaerobacter BG1L1. *Appl. Microbiol. Biotechnol.* 77 (2007), pp. 61–68.

Graham, T.M.: *Biology. The essential principles.* CBS College Publishing, Philadelphia, PA, 1982.

Gråsten, J. & Kiukas, I.: Öljyvahingot Etelä-Savon, Kaakkois-Suomen ja Keski-Suomen alueilla. Rekisteri, tutkimussuunnitelma ja toimintamalli. *Etelä-Savon ympäristökeskuksen moniste* 59 (2004). Mikkeli, Finland.

Gressel, J. & Zilberstein, A.: The forgotten waste tonnes for fuel or feed. In: Industrial biotechnology and biomass utilization, prospects and challenges for the developing world. Stockholm Environment Institute and United Nations Development Organization, Vienna, 2007.

Gullichsen, J. & Fogelblom, C.-J.: *Chemical pulping.* Book 6B, Papermaking Science and Technology, Fapet Oy, Helsinki, Finland, 2000.

Hakalehto, E.: Method and apparatus for concentrating and searching of microbiological specimens. US Patent No. 7,517,665, 2009.

Hakalehto, E.: Hygiene monitoring with the Portable Microbe Enrichment Unit (PMEU). *41st R3-Nordic Symposium. Cleanroom Technology, Contamination Control and Cleaning.* VTT Publications 266. Espoo, Finland: VTT (State Research Centre of Finland), 2010.

Hakalehto, E. (ed.): *Alimentary microbiome – a PMEU approach.* Nova Science Publishers, New York, 2012.

Hakalehto, E. & Hänninen, O.: Gaseous CO_2 signal initiates growth of butyric-acid-producing *Clostridium butyricum* in both pure culture and mixed cultures with *Lactobacillus brevis. Can. J. Microbiol.* 58:7 (2012), pp. 928–931.

Hakalehto, E., Pesola, J., Heitto, L., Närvänen, A. & Heitto, A.: Aerobic and anaerobic growth modes and expression of type 1 fimbriae in Salmonella. *Pathophysiology* 14 (2007), pp. 61–69.

Hakalehto, E., Humppi, T. & Paakkanen, H.: Dualistic acidic and neutral glucose fermentation balance in small intestine: Simulation *in vitro. Pathophysiology* 15 (2008), pp. 211–220.

Hakalehto, E., Pesola, J., Heitto, A., Deo, B.B., Rissanen, K., Sankilampi, U., Humppi, T. & Paakkanen, H.: Fast detection of bacterial growth by using Portable Microbe Enrichment Unit (PMEU) and ChemPro100i((R)) gas sensor. *Pathophysiology* 16 (2009), pp. 57–62.

Hakalehto, E., Hell, M., Bernhofer, C., Heitto, A., Pesola, J., Humppi, T. & Paakkanen, H.: Growth and gaseous emissions of pure and mixed small intestinal bacterial cultures: effects of bile and vancomycin. *Pathophysiology* 17 (2010), pp. 45–53.

Hakalehto, E., Heitto, L., Heitto, A., Humppi, T., Rissanen, K., Jääskeläinen, A., Paakkanen, H. & Hänninen, O.: Fast monitoring of water distribution system with portable enrichment unit – Measurement of volatile compounds of *coliforms* and *Salmonella* sp. in tap water. *J. Toxicol. Environ. Health Studies* 3 (2011a), pp. 223–233.

Hakalehto, E., Vilpponen-Salmela, T., Kinnunen, K. & von Wright, A.: Lactic acid bacteria enriched from human gastric biopsies. ISRN Gastroenterol. (2011b), pp. 109–183.

Hakalehto, E., Tiainen, M., Laatikainen, R., Paakkanen, H., Humppi, T. & Hänninen, O.: Rapid bacterial metabolic activity by the production of ethanol and 2,3 -butanediol without population growth protects intestinal flora against osmotic stress. Manuscript in preparation, 2013.

Hänninen, S. & Rytkönen, J.: *Oil transportation and terminal development in the Gulf of Finland.* VTT Publications 547, Espoo, Finland, 2004.

Harder, W. & Dijkhuizen, L.: Strategies of mixed substrate utilization in microorganisms. In: J.R. Quangle & A.T. Bull (eds): New dimensions in microbiology: mixed substrates, mixed cultures and microbial communities. *Phil. Trans. R. Soc. Lond.* B 297 (1982), pp. 459–480.

He, Q., Hemme, C.L., Jiang, H., He, Z. & Zhou, J.: Mechanisms of enhanced cellulosic bioethanol fermentation by co-cultivation of *Clostridium* and *Thermoanaerobacter* spp. *Bioresour. Technol.* 102 (2011), pp. 9586–9592.

Heitto, L., Heitto, & Hakalehto, E.: Tracing wastewaters with faecal enterococci. *Second European Large Lakes Symposium* (Poster), Norrtälje, Sweden, 2009.

Iowa Energy Center: Biomass Energy Conversion (BECON) Facility. http://www.iowaenergycenter.org/biomass-energy-conversion-becon-facility/ (accessed January 2013).

Ishikawa, Y & Saka, S.: Chemical conversions of various celluloses as treated in supercritical methanol. *5th International Biomass Conference of the Americas*, 17–21 December 2001, Orlando, FL, 2001.

Jones, D.T. & Woods, D.R.: Acetone-butanol fermentation revisited. *Microbiol. Rev.* 50 (1986), pp. 484–524.

Juvonen, M.: Presentation of Managing Director Mika Juvonen, Biokymppi Oy, Kitee, Finland. In: *The Second REMOWE Open seminar in Kuopio*, Finland 29.9.2010, 2010.

Järvinen, K. & Salonen, S.: Pilaantuneiden maiden kunnostuskustannukset Suomessa. Memo, Ramboll Oy and Finnish Environment Institute, 2004.

Kamini, N.R., Hemachander, C., Geraldine S.M.J. & Puvanakrishnan, R.: Microbial enzyme technology as an alternative to conventional chemicals in leather industry. 2010, http://www.iisc.ernet.in/currsci/jul10/articles16.htm (accessed July 2012).

Kamm, B. & Kamm, M.: Principles of biorefineries. Mini Review. *Appl. Microbiol. Biotechnol.* 64 (2004), pp. 137–145.

Kangas, A., Lund, C., Liuksia, S., Arnold, M., Merta, E., Kajolinna, T., Carpén, L., Koskinen, P. & Ryhänen, T.: Energiatehokas lietteenkäsittely (Energy efficient management of wastewater sludge). *The Finnish Environment* 17 (2011). Finnish Environment Institute. Helsinki, Finland.

Kaufman, R.: 3 Future oil-spill fighters: sponges, superbugs, and herders. 2010, http://news.national geographic.com/news/2010/05/100511-science-environment-gulf-oil-spill-cleanup-future/   (accessed July 2012).

Kleinert, M. & Barth, T.: Towards a lignocellulasia biorefinery: direct one-step conversion of lignin to hydrogen-enriched bio-fuel. *Energy Fuels* 22 (2008), pp. 1371–1373.

Kumar, M. & Gayen, K.: Developments in biobutanol production: new insights. *Appl. Energy* 88 (2011), pp. 1999–2012.

Kumar, N. & Das, D.: Enhancement of hydrogen production by *Enterobacter cloaceae* HT-BT 08. *Process Biochem.* 35 (2000), pp. 589–593.

Laaksonen, P.: *Production and utilization of municipal wastewater sludges*. MSc Thesis, Tampere University of Technology, Finland, 2011.

Lakatos, B.G., Bárkányi, Á. & Németh, S.: Continuous stirred tank coalescence/redispersion reactor: a simulation study. *Chem. Eng. J.* 169 (2011), pp. 247–257.

LeBlanc, R., Matthews, P. & Richard, R.: Global atlas of excreta, wastewater sludge, and biosolids management: moving forward the sustainable and welcome uses of a global resource. Kenya, United Nations Human Settlements Programme (UN-HABITAT), 2008.

Lin, C.-W. , Wub, C.-H., Tranc, D.-T, Shihd, M.-C., Lie, W.-H. & Wuf, C.-F.: Mixed culture fermentation from lignocellulosic materials using thermophilic lignocellulose-degrading anaerobes. *Process Biochem.* 46 (2011), pp. 489–493.

Lucia, L.A., Argyropoulos, D.S., Adamopoulos, L. & Gaspar, A.R.: Chemicals and energy from biomass. *Can. J. Chem.* 84 (2006), pp. 960–970.

Luo, L., van der Voet, E. & Huppes, G.: Biorefining of lignocellulosic feedstock – Technical, economic and environmental considerations. *Bioresour. Technol.* 101 (2010), pp. 5023–5032.

Madigan, M., Martinko, J. & Parker, J.: *Brock biology of microorganisms*. 10th edn, Pearson Education, Inc./Prentice Hall, New Jersey, 2003.

Mentu, J.V., Heitto, L., Keitel, H.V. & Hakalehto, E.: Rapid microbiological control of paper machines with PMEU method. *Paperi ja Puu / Paper and Timber* 91 (2009), pp. 7–8.

Metcalf & Eddy Inc.: *Wastewater engineering: treatment and reuse*. 4th edn, McGraw-Hill. Boston, MA, 2003.

Metzger, J.O. & Bornscheuer, U.: Lipids as renewable resources: current state of chemical and biotechnological conversion and diversification. *Appl. Microbiol. Biotechnol.* 71 (2006), pp. 13–22.

Oh, S.E., & Logan, B.E.: Hydrogen and electricity production from a food processing wastewater using fermentation and microbial fuel cell technologies. *Water Res.* 39 (2005), pp. 4673–4682.

Pesola, J., Vaarala, O., Heitto, A & Hakalehto, E.: Use of portable enrichment unit in rapid characterization of infantile intestinal enterobacterial microbiota. *Microb. Ecol. Health Dis.* 21 (2009), pp. 203–210.

Petersson, A. & Wellinger, A.: Biogas upgrading technologies – developments and innovations. IEA Bioenergy, Task 37 — Energy from biogas and landfill gas, 2009, http://www.iea-biogas.net/_download/publi-task37/upgrading_rz_low_final.pdf (accessed July 2012).

Ping, L., Wang, Y., Kun, L. & Lei, T.: Construction of a microbial system for efficient degradation of cellulose. 2008. *International Workshop on Education, Technology and Training, and 2008 International Workshop on Geoscience and remote sensing*, Shanghai, China, 2008.

Pitkänen, T., Bräcker, J., Miettinen, I., Heitto, A., Pesola, J. & Hakalehto, E.: Enhanced enrichment and detection of thermotolerant Campylobacter species from water using the Portable Microbe Enrichment Unit (PMEU) and realtime PCR. *Can. J. Microbiol.* 55 (2009), pp. 849–858.

Porter, J.B.: *Bacterial chemistry and physiology*. Wiley, New York, 1948.

Quan, X., Shi, H., Liu, H., Lv, P. & Qian, Y.: Enhancement of 2,4-dichlorophenol degradation in conventional activated sludge systems bioaugmented with mixed special culture. *Water Res.* 38 (2004), pp. 245–253.

Raoult, D. & Forterre, P.: Redefining viruses: lessons from Mimivirus. *Nature Rev. Microbiol.* 6 (2008), pp. 315–319.

Rättö, M., Vikman, M. & Siika-aho, M.: Yhdyskuntajätteiden hyödyntäminen, biojalostamossa (Utilization of municipal waste in biorefinery). *VTT Tiedotteita Research Notes* 2494. Espoo, Finland, 2009, http://www.vtt.fi/inf/pdf/tiedotteet/2009/T2494.pdf (accessed July 2012).

Ren, Z., Ward, T.E. & Regan, J.M.: Electricity production from cellulose in a microbial fuel cell using defined binary culture. *Environ. Sci. Technol.* 41 (2007), pp. 4781–4786.

Repuske, R & Clayton, M.A.: Control of *Escherichia coli* growth by CO_2. *J. Bacteriology* 135 (1968), pp. 1162–1164.

Rosenberg, M.L., Langseth, E., Krivokapic, A., Gupta, N.S. & Tilset, M.: Investigation of ligand steric effects on a highly cis-selective Rh(i) cyclopropanation catalyst. Dedicated to Prof. Didier Astruc on the occasion of his 65th birthday. CCDC reference numbers 800306& 800307. *New J. Chem.* 35 (2011), pp. 2306–2313.

Shkilnyy, A., Dubois, J., Sabra, G., Sharp, J., Gagnon, S., Proulx, P. & Vermette, P.: Bioreactor controlled by PI algorithm and operated with a perfusion chamber to support endothelial cell survival and proliferation. *Biotechnol. Bioeng.* 109:5 (2012), pp. 1305–1313.

Singh, A. & Mishra, P.: Microbial pentose utilization — Current applications in biotechnology. *Progress Ind. Microbiol.* 33 (1995). Elsevier Science BV Amsterdam, The Netherlands.

Tanisho, S., Suzuki, Y. & Wako, N.: Fermentative hydrogen evolution by *Enterobacter aerogenes* strain E 82005. *Int. J. Hydrogen Energy* 12 (1987), pp. 623–627.

Thorin, E., Daianova, L., Lindmark, J., Nordlander, E., Song, H., Jääskeläinen, A., Malo, L., den Boer, E., den Boer, J., Szpadt, R., Belous, O., Kaus, T. & Käger, M.: State of the art in the waste to energy area – Technology and systems. REMOWE Report No. O4.1.1, 2011, http://www.remowe.eu (accessed July 2012).

Tischer, W. & Wedekind, F.: Immobilized enzymes: methods and applications topics. *Current Chemistry* 200 (1999), pp. 95–126.

Välimäki, P.: Öljy-yhtiöt opettelevat yhteiskuntavastuuta. *Kuluttaja* 8 (2005), http://www.kuluttaja.fi/Page/b6639a3e-0625-4f56-a491-0b1fae08dd48.aspx (accessed July 2012).

Van Etten, J.L.: Giant viruses. *Am. Sci.* 99 (2011), p. 304.

Venter, J.C.: Sea of dreams. *Economist* 371:8373 (2004), pp. 81–82.

Veriö, T.: Öljyvahinkojen torjunta 1. Yleinen osa ja maaöljyvahinkojen torjunta. Suomen palontorjuntaliitto. Mäntän kirjapaino Oy, Mänttä, Finland, 1990.

Vlysidis, A., Binns, M., Webb, C. & Theodoropoulos, C.: A techno-economic analysis of biodiesel biorefineries: assessment of integrated designs for the co-production of fuels and chemicals. *Energy* 36 (2011), pp. 4671–4683.

Weiland, P.: Biogas production: current state and perspectives. *Appl. Microbiol. Biotechnol.* 85 (2010), pp. 849–860.

Xu, L. & Tschirner, U.: Improved ethanol production from various carbohydrates through anaerobic thermophilic co-culture. *Bioresour. Technol.* 102 (2011), pp. 10,065–10,071.

Zyskin, A., Avetisov, A., Kuchaev, V., Shapatina, E. & Christiansen, L.: Simulation of the kinetics of complex heterogeneous catalytic reactions under diffusion limitations. *Kinet. Catal.* 48 (2007), pp. 337–344.

CHAPTER 14

Concluding remarks and perspectives on the future of energy systems using biomass

Erik Dahlquist, Elias Hakalehto & Semida Silveira

Biomass has gone from being more or less the only energy resource used for heating and cooking on a normally quite inefficient way to be a very valuable energy and material resource used for all kind of applications.

As a consequence of this, it has become very important to develop the most efficient energy system for all kind of societies. Some countries like Brazil have high production capacity of sugar cane for ethanol and trees for pulp and paper while others as China are driving from burning straw in the fields to starting to use the straw as an energy resource.

The system solutions include conversion technologies going towards biorefineries, new markets for biomass like vehicle fuels and to new business models where many different markets compete for the biomass. With biomass, we have many more actors than for oil industry with a relatively small number of very big companies. The biomass resources also differ a lot from regions to regions depending on the needs as well as the climatic conditions. In northern countries, the heating demand is very high, while in southern countries this demand is negligible.

To promote the development of biomass resources good policy incentives are needed. The oil industry has in reality been subsidized in many ways by many governments during the last century, by not having to pay for the negative effects of air pollution leading to acidification, emissions of neglect able substances, effect of oil spillage from ships and oil production, effect on global warming etc. Instead, the governments have handled these types of problems using tax money. In many countries like Iran and Nigeria, the gas-price has been strongly subsidized and thereby the consumption has become very high. The good is that the population gets a share of the oil exports, but the negative effect is the vast overuse of energy that is in no way sustainable. Here the aim to be fair has become a problem. It can be compared to Norway where the price of gas for private people is among the highest in the world and the income from oil export instead is invested by governmental companies all over the world to get benefits for the country long term, but not for individuals short term. This shows very different possible strategies and that there are good and bad effects of all strategies, but long term the Norwegian is probably much better than in countries strongly subsidizing the private use. We also can see that in both Iran and Nigeria the subsidies have been decreased with riots in Nigeria as a consequence 2011–2012. In Iran the price has increased more than five times 2005–2012, and the goal now is to get a price in the same range as in many other countries, with the aim to get the oil and gas resources to last longer. This also will favor use of other resources like biomass, sun and wind in reality.

In Sweden, the government succeeded in getting an agreement between most political parties to introduce a carbon tax in 1992. This gave an additional tax of approximately 1.6 €cent/kWh for carbon, which is giving the full cost on coal, and roughly half of it on natural gas with only 50% carbon and the rest hydrogen of the heating value. This gave a dramatic increase of the use of biomass and created many incentives to use biomass for replacement of fossil fuels. From 1992 to 2012, the use of biomass has increased from 55 to almost 140 TWh/y in Sweden, as a direct effect of the carbon tax and green electricity certificates, another control mechanism.

This shows how important policy decisions are for driving the business and the system for how to select where to invest. Wind power and solar power also have been favored and not only biomass. A difficulty still is to get similar conditions in many countries.

We have both national and international regulatory set of rules and incentives. In Germany, we have a feed in tariff to promote electricity production using solar power, wind and biomass. Here the incentives are highest for sun, followed by wind but significantly less for biomass, as biomass is considered economic competitive already as it is. Feed-in tariffs also is used in most other EU countries, where Sweden with its high carbon tax primarily want to get rid of fossil fuels while the other also want to promote the other new technologies significantly. Instead, Sweden is using subsidies for the erection of new wind and solar plants with respect to the investment cost.

Another tool is the use of green certificates where a certain amount of the electricity price is to be used for building new production capacity using renewables, where the power company should increase its share of renewables successively. This mechanism is common in most EU countries but was also introduced in India spring 2011. In Sweden, the effect has been that many local power companies have invested in biomass-fired CHP-plants.

In any process solution, the impact on the environment and the climate has to be evaluated carefully. It is of crucial importance to strive for processes that correspond to the circulation of matter in nature as much as possible. This approach is essential in lowering the environmental burden of the waste treatment plants. Microbes are in fact the organisms carrying out the degradation of organic materials in the ecosystems. By exploiting their vast metabolic capabilities in producing biogas, or liquid biofuels, human economies could be better adapted into nature's own system. Microbes could also be used for modifying the waste masses into direction where their further treatment is taking place in a more eco-friendly fashion.

In the US, there has been a program to promote bio fuels for vehicles, where the price per liter of ethanol has been fixed at a level economically interesting for many farmers. Instead of having to abandon their farms, they have been able to earn their living on producing ethanol from mostly corn for vehicles, and the US has become a major producer of ethanol due to this. Here the major goal with the program has been to get more independent of imported oil.

In Brazil there is a system called Pro-Infa, which promotes all kind of small-scale power production, but is not promoting biomass very much. Instead, Brazil created a system for promoting ethanol cars already during the 1970s and now no economic subsidies are needed anymore, as the production is very competitive. Ethanol should not cost more than 70% of gas was one rule to promote the use. With an efficient technology, the production cost today is in the same range or cheaper for ethanol than many qualities of gas and diesel and thus is keeping a significant share of the business volume. Most cars sold today in Brazil are flexi-fuel cars using primarily 100% ethanol, while gas cars have 22–24% ethanol in all fuels. This is showing that it is possible to have a system for transportation built up on biomass fuels also in big countries like Brazil.

In the chapter on global biomass resources we have seen that biomass can fulfill most of the energy resources needed as well as for replacement of fossil fuels for production of plastics and similar. What we have to do still is to use all material and resources in an efficient system, where the same fibers for instance are used many times for different purposes before they eventually are combusted, instead of combusting stem wood directly. What is considered being waste should instead be seen as a valuable resource. One example of this is the biorefinery in Finland where waste is used after sorting to produce a whole set of different valuable products. In China, the example with a huge biorefinery using at first cereals, but long-term straw and other organic agricultural wastes, is very important to demonstrate that this type of more large-scale factories is not only a fantasy but also true reality. Good examples always are very important to give investors encouragement to risk their money. Bad examples on the other hand have a tendency to really threaten the same investors for long time periods, and thus delay the introduction. It also will determine what technology that will be the most successful on a large scale in the future, as good technology will give a driving force to really promote investments. Still, success or failure very often is not only due to the performance of technical solutions but also due to economic conditions where cost of raw material, tax regulations, investment incentives etc. all is important. For large-scale investments, also the set of rules have to be long term and that investors believe governments will stay to certain set of rules. Here we sometimes have a very big obstacle, as investors are skeptical to what politicians say if they have proven unreliable before.

We now have been talking about the necessity of long-term rules and regulations giving incentives for investments in biomass production and conversion. From an economic point normally large scale plants are more economic, but as the risk with these are higher it may as well be a feasible strategy to develop more small scale solutions, that can be afforded also by less capitalized investors like villages or even single persons. Even if these investments may have longer pay-back time it may still be more feasible as you may eventually have to pay taxes, distribution costs etc. if you produce for yourself. Then instead it should be technical demands on the equipment to avoid emissions of pollutants. Even if biomass is an excellent energy resource, we have to be aware of that if it is combusted in a bad way it will give very strong negative effects on nature, although perhaps not on the global warming.

There are some basic principles regarding the evaluation of the environmental aspects of various treatment methods. For instance, the combustion processes produce remarkable amounts of ashes, which might contain heavy metals and other harmful pollutants. Therefore, any such process plant should include also a plan for the treatment of all fractions of the outlets, not only the volatile emissions.

The utmost potential of the microorganisms could be exploited for the management of the xenobiotic compounds within the dry waste or wastewaters (Slater and Bull 1982). Microbiological pretreatments could be effectively used for lowering the environmental and health hazards related to later process steps of treating the mixed wastes or sludges.

In a Dutch review on the climatological effects of the energy production, various methods for lowering the carbon emissions are summarized (Okken *et al.*, 1989). One potential method presented there is the storage of carbon dioxide in the oceans (de Baar and Stoll, 1989). For example, in this case most of the bacteria in the seas are free-living and utilizing dissolved organic carbon (DOC) rather than carbon in the particles as a food source. Therefore, the route of sinking water masses in the sub polar ocean regions is suggested for the biological carbon fixation. Other carbon dioxide capture and storage methods have been overviewed e.g. in (Benson and Surles, 2006).

The above-mentioned example is showing clearly that the global environment cannot be divided into separate segments but needs to be considered as an entity. For example, the environmental load to the Baltic Sea is the heaviest from Poland, corresponding to 34% of the total phosphorus and 27% of the total carbon. The same figures for the Russian impact are 19% and 14%, and for the Finnish organic pollution 10% and 11% (Poutanen, 2010). Carbon is often not estimated in these models directly, but it contributes to the oxygen demand in the waters. Moreover, the fate of the carbon residues in the water ecosystems is related also to the proportion of anaerobic niches to the aerobic ones. In the Baltic Sea 1/7 of the sea bottom is completely deprived of oxygen. In fact, it could be much better to treat and recycle the wastes in a sustainable way including the microbiological and biotechnological solutions, than by just discharging the organic loads into the water and maritime ecosystems, or to the atmosphere.

Therefore, nothing is only good or bad, but depends on how it is used!

This fact that nothing is absolute can also be illustrated by the following example of ethanol production from wheat in the region close to Västerås city. The data on input to energy from production of cereals comes from the farmer Gustaf Forsberg, who is operating a 450 ha farm with mostly cereals; 350 ha are used for cereals in the calculation below:

- Harrowing twice in autumn with cultivator, 6 ha/h. 350 ha/6 = 60 h × 2 = 120 h
- Harrowing 1–2 times with smaller fingers, 10 ha/h. 350/10 = 35 h × 1.5 = 50 h
- Sowing 3–4 ha/h including filling seed + fertilizer. 350/3.5 = 100 h
- Spraying 1–2 h/y (herbicides-outsourced. 24 m width. Very fast. Can neglect)
- Total hours 270 h × 40 L/h = 10 800 L × 10 kWh/L = 108 MWh
- Harvesting 2.5 ha/h including transport machines, emptying etc. 350/2.5 = 140 h × 32 L/h = 4480 L × 10 = 45 MWh
- Drying from 20% to 13% moisture = 13 L/tonne × 10 = 130 kWh/tonne
- Production of wheat 5 tonne/ha × 350 ha = 1750 tonne cereals per year
- Energy input for drying: 1750 tonne × 0.130 kWh/tonne = 230 MWh
- Addition of fertilizers/nutrients: 150 kg N/ha + 30 kg P-K/ha.

Energy input for the production of the fertilizers comes from data from a neighboring farm. 72 ha oats consume N-fertilizer corresponding to 7 tonne oil for 300 tonne oats. This corresponds to $350 \times 7/72 = 34$ tonne $\times$ 10 MWh/tonne $= 340$ MWh for the 350 ha.

The total energy consumption for the 1750 tonne cereals would then be $108 + 45 + 230 + 340 = 723$ MWh. This gives 0.41 MWh/tonne which corresponds to $0.41/(5.4\,\text{MWh/tonne}) = 7.7\%$ of the heating value of the cereals.

If we produce ethanol we would get out approximately 250 liters per tonne cereals, which corresponds to $0.25 \times 7\,\text{MWh/L} = 1.75$ MWh ethanol/tonne cereals.

The output as ethanol compared to input of primarily fossil fuels for the production then is $1.75/0.41 = 4.3$ times. To this we have to add also evaporation of water to achieve pure ethanol, but on the other hand we get a lot of other products out as well as 250 kg protein fodder, and also heat and electricity corresponding to approximately $0.5 \times 5.4 \times 0.35 = 0.95$ MWh.

If we also make use of the straw, we could use approximately 3 tonne DS/ha $\times 5.4/5 = 3.2$ MWh per tonne cereals. The total then would be 1.75 MWh ethanol, 250 kg protein fodder $(0.25 \times 5.4 = 1.35\,\text{MWh}) + 0.95 + 3.2 = 4.15$ MWh heat + power. The total output then would be 7.25 MWh per tonne cereal grain. $7.25/0.41 = 17.7$ times the fossil fuel input.

As seen the efficiency depends very much on what is included! If we should evaporate from 5% ethanol to 99% in one step this would mean an energy consumption of 20 liter water per liter ethanol or $250 \times 20 \times 2260\,\text{kJ/kg} = 45{,}200$ kJ/L ethanol or 11,300,000 kJ/tonne cereal. This corresponds to 3.16 MWh per tonne cereal. In reality, we can regain most of this energy as heat for e.g. district heating and thus the real energy usage will depend on the economic optimum. Capital *versus* energy!

This calculation was done to show that depending on what constraints we set we can achieve very different results. If our purpose is to show that producing ethanol is bad, we just select the conditions giving a low output, like only including the primary energy input but not using the energy efficiently. If on the other hand you want to show that ethanol production is good you show all the possibilities to reuse input energy and emphasize that protein is not destroyed but used as fodder or even food for humans. It is sometimes very disturbing that facts can be used for so different purposes and thereby affects the total energy system dramatically in the future, depending on the part of the facts that wins!

In some countries, we have a very positive attitude to bioenergy use, like in Sweden, where also the farmer organization has a very strong influence. In other countries or regions, we may have a very different situation depending on other strong lobbying groups with other interests!

We therefore will have to both give facts to politicians and good arguments for the positive aspects of biomass use for energy purposes. Only facts are not enough. Good examples are also very important and these have to be presented in a convincing way. Then both regulatory frameworks and interest from investors can be achieved, and thereby system development can take place.

REFERENCES

Benson, S.M. & Surles, T.: Carbon dioxide capture and storage: an overview with emphasis on capture and storage in deep geological formations. *Proceedings of the IEEE* 94, 2006, pp. 1795–1805.

de Baar, H.J.W. & Stoll, M.H.C. Storage of carbon dioxide in the oceans. In P.A. Okken, R.J. Swart & S. Zwerver (eds): *Climate and energy — The feasibility of controlling CO_2 emissions.* Kluwer Academic Publishers, Dordrecht, The Netherlands, 1989, pp. 143–177.

Okken, P.A., Swart, R.J. & Zwerver, S. (eds): *Climate and energy — The feasibility of controlling CO_2 emissions.* Kluwer Academic Publishers, Dordrecht, The Netherlands, 1989.

Poutanen, E.-L.: Itämeren Ekosysteemi Uhattuna, In Finnish (The Vulnerable Ecosystem of the Baltic Sea, English summary). In: O. Pitkänen & P. Toivanen (eds): Kansalaisen Itämeri (The Citizen's Baltic Sea). The Foreign Ministry of Finland and Edita Publishing Co, Helsinki, Finland, 2010.

Slater, J.H. & Bull, A.T.: Environmental microbiology: biodegradation. In: J.R. Quayle & A.T. Bull (eds): New dimensions in microbiology: mixed substrates, mixed cultures and microbial communities. *Phil. Trans. R. Soc. Lond.* B 297 (1982), pp. 575–597.

Subject index

Sustainable Energy Developments

Series Editor: Jochen Bundschuh

ISSN: 2164-0645

Publisher: CRC/Balkema, Taylor & Francis Group

1. Global Cooling – Strategies for Climate Protection
   Hans-Josef Fell
   2012
   ISBN: 978-0-415-62077-2 (Hbk)
   ISBN: 978-0-415-62853-2 (Pb)

2. Renewable Energy Applications for Freshwater Production
   Editors: Jochen Bundschuh & Jan Hoinkis
   2012
   ISBN: 978-0-415-62089-5 (Hbk)

3. Biomass as Energy Source: Resources, Systems and Applications
   Editor: Erik Dahlquist
   2013
   ISBN: 978-0-415-62087-1 (Hbk)